Developing Structured Procedural and Methodological Engineering Designs

Yohannes Yebabe Tesfay

Developing Structured Procedural and Methodological Engineering Designs

Applied Industrial Engineering Tools

 Springer

Yohannes Yebabe Tesfay ⓘ
Fremont, CA, USA

ISBN 978-3-030-68404-4   ISBN 978-3-030-68402-0   (eBook)
https://doi.org/10.1007/978-3-030-68402-0

This Springer imprint is published by the registered company Springer Nature Switzerland AG
The registered company address is: Gewerbestrasse 11, 6330 Cham, Switzerland

While writing and editing this book, many regions in Tigray, my father's birthplace, Axum, and my mother's home, Raya, are at war. The war on Tigray was initiated by Isaias Aforki, the Eritrean government leader, Abiy Ahmed Ali, the Ethiopian government leader, and the 2019 Nobel Peace Prize Laureate, the Amhara Special force, the Amhara Militia, with the assistance of the Dubai modern Drone Jets. It was the harshest time in my life, and I wondered why the world was silent just looking at such a genocide targeting the Tegarus. With all my desire, I, therefore, dedicate this book to the memory of those who lost their lives at war; priests and believers who massacred at Axum St. Mary Church, the elders, mothers, and babies who displaced to Sudan due to the conflict; and the women who raped by the military forces.

Foreword

I am over the moon to write this foreword for the first edition of *Developing Structured Procedural and Methodological Engineering Designs*, not only because I know Yohannes Yebabe Tesfay as a senior engineer-scientist at Soraa Inc., USA, but also the book has essential contributions in quality and reliability engineering.

To bring the current complex industrial challenges of manufacturers and service providers, I accept as true deeply in the educative significance of explanatory discussion in developing procedural and methodological engineering designs are essential for all industrial engineers, manufacturing engineers, quality and reliability engineers, and engineering researchers and students.

I also believe that engineers and researchers at each stage of their professional career can enrich and strengthen their ability of process modeling to produce a high-quality product with acceptable cost of production is the key performance indicator of successful and competent companies in the industrialized zones of the world. However, in real-world observation, cost-efficient production of a high-quality product is challenging to achieve. Thus, most successful companies apply continuous improvement programs to achieve their goals.

This book gives the framework of continuous improvement models such as Lean Engineering (LE), Six Sigma Methodologies (SSM), and Total Quality Management (TQM). The models accentuate each personnel's involvement and effective engineering teamwork. That can help measure, schematize, and systematize processes layouts to reduce cycle times, variations, and defects.

The book discusses essential components and brings a proper understanding of quality functions for a New Product Introduction (NPI). Especially in the current trends of highly innovative technologies in the Automotive Industry (AI), the Medical Device Industry (MDI), and Aerospace Industry (AI), and the models of quality functions for a New Product Introduction (NPI) are very critical and essential. These quality function models of NPI in the Automotive Industry (AI) and the Aerospace Industry (AI) help for the execution of effective and efficient Advanced Product Quality Planning (APQP) and Production Part Approval Process (PPAP). For those Medical Device companies who engaged in the New Product Introduction (NPI) procedure, necessity supports their determinations with the formation and conservation of Medical Device quality systems that endorse high-quality products and processes through supporting acquiescence compliance with the Federal Drug Administration (FDA).

Furthermore, the book helps engineers, operation analysis, and researchers to know and implement the models of the Eight Disciplines (8D) of Problem-Solving, Hyper-Hybrid Coordination, and the Kraljic-Tesfay Portfolio Matrix of industrial buying.

In conclusion, I highly encourage manufacturing companies, service providers like the airline industry, engineering researchers, and students to use the models discussed in this book.

North Carolina State University   Hervey Jamey West
(NCSU), Raleigh-Durham, NC, USA

Foreword

When I first interviewed Yohannes in May 2019, our new facility in Ohio was ramping production. It was in dire need of a strong Quality Engineer to manage the Quality Control process in the manufacturing line. As Vice President of Quality, Reliability, and Manufacturing at a high-tech startup in the LED industry, I urgently needed a Quality Engineer with an eye for problem identification and the technical prowess to implement changes in our manufacturing process. During my first conversation with Yohannes, I was struck by his ability to articulate complex problem-solving concepts and his deep knowledge of the statistical underpinnings of Six Sigma and Lean Manufacturing. Moreover, I felt that Yohannes' non-judgmental demeanor and passion for Quality Management Systems would make him an excellent mentor and an asset to our team. In his final test, I took Yohannes onto the "Staged" Manufacturing area and asked him to survey our manufacturing process and describe any opportunities for improvement. I was blown away by Yohannes's perceptiveness and insight; where other quality engineers fumbled through inadequate responses, Yohannes identified areas for improvement at a level of detail that truly impressed me. At that moment, I knew he was the engineer our organization needed to facilitate our quality improvement targets and recruited him to our team.

Yohannes quickly fell into his role as Senior Quality Engineer at Soraa, Inc. Aside from his knack for statistical analysis and his command of Six Sigma and Lean Manufacturing, Yohannes exhibited an air of kindness, humility, and friendliness that captivated his coworkers. Yohannes was the ideal colleague: a team player with unmatched statistical expertise and a genuine desire to help and uplift those around him. Yohannes quickly became an indispensable member of our Quality Department, implementing high-level quality improvement processes while helping assist the teams he trained to grow. When I became the VP of Manufacturing Operations at Tactus Technologies, Inc., in October 2020, I was presented with a new challenge that would require an elite team of engineers and managers to transfer and implement state-of-the art technology in Contract Manufacturers in the USA and Asia and meet very demanding customer specifications. Among the candidates for Quality Management, Yohannes stood out once again. I knew his technical capabilities and amicable disposition will bring incredible value to our team. As a senior executive with over 35 years of experience working in Quality, Reliability, and Manufacturing Operations, it is seldom that an employee makes such an impact that you want to take them from one organization to another. I feel fortunate to have Yohannes on my staff, knowing that he supplies the team with accurate data, always with a smile. I am excited that Yohannes has decided to expand his reach and write this textbook. I have witnessed Yohannes' ability to articulate complex topics first-hand and know he will provide students with a straightforward approach to understanding the intricacies of Quality Control Management.

I look forward to the next generation of Quality Engineers. In Yohannes's footsteps, they will solve process problems using statistical tools and data and deliver high-quality products with the best costs to consumers.

Tactus Technology Inc., Fremont, CA,   Nahid Afshar
USA

Foreword

Any company dealing with technology churns out ever-increasing strings of data, and data analysis is Yohannes' expertise. Yohannes combines his vast knowledge of statistical analysis and experience working in high-tech manufacturing companies in writing this book. That includes knowledge of standards and methodologies in today's modern industry dealing with evaluating Quality. With that, he can provide multiple approaches to analyze and improve a diverse array of problems, be it improving the manufacturing efficiency of a finished product or designing appropriate tests for a product or equipment early in the design stages.

I have the privilege of drawing from Yohannes' knowledge and industry experience in my professional career and greatly benefited from it. Although Manufacturing Quality may be the primary focus of the topics presented, I personally feel the chapters he contributes will be highly valuable to anyone interested in concurrent top engineering quality tools with an adequate SPC approach to data analysis.

IEEE (Silicon Valley Chapter), Bay
Area, CA, USA

Alan Swe

Book Summary

This book is organized to help various industrial companies to develop procedural and methodological engineering tools to meet multiple standards. The book contains procedures of Quality Engineering Practices (QEP), SPC, the American Society of Quality Engineers (ASQE) Acceptance Sampling, Robust Design, Quality Function Deployment (QFD), Design for X (DFX), Design for Six-Sigma (DFSS), Quality by Design, Critical Quality Attributes (CQAs), Plan-Do-Check-Act (PDCA), DMAIC of Six-Sigma Methodologies, Eight Disciplines of Problem-Solving (8D), and Corrective and Preventive Action (CAPA). The book's last three chapters introduce some new methodological models in Operations Research (OR) and their engineering applications.

Preface

Introduction

This book is about using modern quality and reliability engineering tools and how to implement them. The book is responsible for a wide-ranging coverage of the subject matters from statistical process control to state-of-the-art concepts and quality functions applications. The book aims to give the reader or the practitioner a sound understanding of the quantity principles and function tools to apply the tools in various industrial manufacturing and service provider sectors. Even though statistical methods are emphasized all through, the book has a robust engineering and management framework. This book's readers need encompassing knowledge of process modeling, and engineering tools are prerequisites for using this book.

Overview of the Subject Matter

This book addresses the current challenges and procedural and methodological mitigation mechanisms in industrial quality and reliability engineering. The book is organized to help various industrial companies develop procedural and methodological engineering tools to meet multiple standards. The book contains procedures of Quality Engineering Practices (QEP), Statistical Process Control (SPC), American Society of Quality Engineers (ASQE) Acceptance Sampling, Robust Design (RD), Engineering of Continuous Improvement Techniques (ECIT), Quality Design for New Product Introduction (NPI), Quality Function Deployment (QFD), Design for X (DFX), Design for Six-Sigma (DFSS), Quality by Design, Critical Quality Attributes (CQAs), Plan-Do-Check-Act (PDCA), DMAIC of Six-Sigma Methodologies, Eight Disciplines (8D) of Problem-Solving, and Corrective and Preventive Action (CAPA). Chapters also introduce some new methodological models in Operations Research (OR) and their applications in engineering. The book aimed to assist various industrial engineers, manufacturing engineers, quality and reliability engineers, and researchers to solve their company's most important problems.

Briefing the Topics of the Book

The key physiognomies of this book are:

It shoulders that the reader's objective is to accomplish an appropriate balance of quality parameter of the product, the optimal cost of production, and customer satisfaction. The book provides a high level of organizing procedural and innovative engineering tools. That aims to meet either the International Organization for Standardization (ISO), International Automotive Task Force (IATF), or National Aerospace Standards (NAS).

It focuses on the critical systems of developing quality functions, which high reliability and capability to produce products or services that could be obtained by considering production costs.

It presents different techniques suitable for Quality Management Systems (QMS) with an adequate procedural background to comprehend their applicability domain and think through variations to outfit practical and industrial production constraints.

It helps to understand the proper implementation of Continuous Improvement Engineering (CIE) practices via the Six Sigma methodologies. Cost efficiency and process effectiveness are the integral parts of the modern industrial production engineering challenges, and Six Sigma helps realize them. Thus, the proper implementations of Six Sigma are equally as the key and technically arduous as any other aspect of product development. This idea is commonly dependable with productivity analysis's current intelligence. Some leading organizations in the USA, Europe, and some countries in Asia are moving toward it.

It gives quality function parameters and ways of analysis techniques together in a coherent production framework, as corresponding methodologies for accomplishing the product's adequate quality at an acceptable production cost. Quality and product improvement are often made on a Project-By-Project (PBP) basis. They encompass teams led by engineering personnel with specialized knowledge of statistical methods and experience in applying them. So, the book assists how to use SPC in the realization of quality and product improvement.

It includes a review of ANSI/ASQ Z1.4 and Z1.9 standards, an explanation of single, double, multiple, sequential, and continuous sampling plans, and a review of Dodge-Romig sampling tables.

It gives the development steps of the Eight Disciplines (8D) of Problem-Solving approach typically employed by quality engineers or other professionals and is most used by the automotive industry but has also been successfully applied in healthcare, retail, finance, government, and manufacturing.

It provides a rigorous understanding of the Bullwhip Effect (BE) concerning the distribution network configuration is responsible for the number, the location, and the network tasks of suppliers, production services, delivery centers, warehouses, cross-docks customers.

It provides the details of the hyper-hybrid coordination to control intra-organizational coordination on the Bullwhip effects.

It includes the significant step in the process of transforming our customer's needs (design inputs) into an epic design concept using tools like Quality Function Deployment (QFD), Robust Design, Design for X (DFX), and Design for Six-Sigma (DFSS), and Quality by Design.

It illustrates the Tesfay process of the impact of experience with the firm's opportunism and Bounded Rationally (BR) behavior that are analyzed as follows. According to the Tesfay-process, the simplest form of the dynamic aspect of the opportunism and Bounded Rationally (BR) behavior in the inter-organizational coordination is modeled. The formula is discussed.

It gives a comprehensive application of the Kraljic-Tesfay Matrix in the industrial buying process which is presented in this book. Later, the book showed a complete application of Tesfay Coordination (TC) in the airline industry.

It presents the applications of Tesfay Dynamic Regression Model (TDRM) in the airline industry, which upgrades the airline capacity's technical aspect.

Why This Book?

In today's complex and dynamic manufacturing environment, as per the requirements of the International Organization for Standardization (ISO), International Automotive Task Force (IATF), and National Aerospace Standards (NAS), quality parameters are the fundamental Key Performance Indicators (KPIs) of the company. Thus, in a competitive industry, meeting the quality standards is the primary challenge of each company.

The objective of adequate product quality at an acceptable production cost is both a managerial and technical challenge. Meeting the goal necessitates an understanding of both the technical subjects and their background framework in product development.

It is widely approved; Quality Assurance (QA) should be an integral part of product development. Quality Assurance (QA) integrates First Article Inspection (FAI) for the approval, or the rejection of incoming materials, Process Quality Control (PQC), and Outgoing Quality Control (OQC). Thus, a quality engineer involves the vast vital engineering and management tasks of the manufacturing company. In turn, the quality assurance (QA) department's function and performance determine the company's profitability, reputability, and sustainability.

The question is how manufacturing or service provider companies implement effective and efficient Quality Assurance (QA) systems. However, apprehensions of the implementation of productive Quality Assurance (QA) systems, in practice, are repeatedly only partial due to several factors. Quality Assurance (QA) systems require careful choices and combinations of the technical requirements fit to the company, products, and processes. However, only a few key people are acquainted with the full range of the quality functions' techniques.

Those best configured to shape the company and its processes are not often accustomed to the technical matters. In this regard, this book fills significant gaps in the development of Quality Assurance (QA) systems. The author has extensive experience in quality and reliability engineering. The author used to work in collaboration with various companies in the USA, EU, and Asia. Furthermore, the author has substantial knowledge of most manufacturing and service sector companies' current challenges. The author intended to share his experience with those companies and researchers in quality and reliability engineering. The book benefits the companies to bring cost-effective quality production.

Description of the Target Audience

The book provides the procedural and methodological engineering designs. The models' can be used by manufacturing, industrial, quality, and reliability engineers, engineering researchers, and students, with a comprehensible understanding of the state and practice of the manufacturing and service provider company's technical quality functions.

Sections of the Book

This book has five parts.

Part I is Chap. 1 of the book, which introduces the engineering definition of quality.
Part II consists of Chaps. 2, 3, 4, 5, and 6. These chapters talk about acceptance sampling and statistical process control (SPC).
Part III consists of Chaps. 7 and 8. These chapters deal with engineering problem-solving tools and continuous improvement techniques.
Part IV consists of Chap. 9. The chapter presents the notion of the New Product Introduction (NPI).
Part V consists of Chaps. 10, 11, and 12. These chapters introduce innovative engineering quantitative models and their applications in the airline industry.

Description of the Chapters in the Book

Chapter 1 deals with quality engineering essential terminologies. Quality engineering involves management, development, operation, maintenance, and related departments to produce a high-quality standard product or service. Quality engineering deals with the models, principles, and practice of product or service quality assurance (QA) and quality control. Quality engineering emphasizes making sure those goods and services are designed, developed, and made to meet or exceed consumers' requirements. This chapter introduces quality tools such as quality systems, auditing, product and process control, design, quality methods and tools, and SPC.

Chapters 2 and 3 focus on Acceptance Sampling. The American Society of Quality (ASQ) engineers define acceptance sampling as a statistical measure system used in quality control. Acceptance sampling helps determine the quality of a batch of products by selecting a specified number for testing, which sample's quality will be viewed as the quality level for the entire group of products. Acceptance sampling solves these problems by testing a representative sample of the product for defects. Based on the results—how many of the predetermined number of samples pass or fail the testing—the company decides whether to accept or reject the entire lot. The inferential statistic generally measures the statistical reliability of a sample.

Chapters 4 and 5 deal with Statistical Process Control (SPC). The core objective of SPC is assuring the quality of product or service via maximizing productivity (effectiveness and efficiency), reduce (scrap, cost, rework, inspections, and warranty claims), improve (operational efficiency, traceability, process control, and continuous improvement), advance (analysis, analytics, modeling, reporting, and documentation report). The selection of proper variables in the study is one of the essential concepts of SPC. The most relevant SPC variable charts cover the I-MR, the X-bar and R, and the X-bar and S-chart. For attribute charts, the p-chart, np-chart, c-chart, and u-chart. Furthermore, Chap. 6 gives details about process capability analysis.

Chapter 6 deals with process capability analysis. The primary concern of process capability analysis is whether the part that we want to produce can meet the engineering specifications or not. Capability analysis applies to understand the process competence for engineering decision-making and process control. Process capability indices measure how much the normal variation the process performs compared to the part specification limits and permits various processes related to how well the manufacturer controls it. This chapter introduces Cp, Cpk, Pp, and Ppk methodologies of process capability analysis.

In Chap. 7, we will cover the topics of the Eight Disciplines of Problem-Solving (8D). The Eight Disciplines of Problem-Solving (8D) is a problem-solving model premeditated to find the Root Cause (RC) of a problem, formulate a short-term fix, and implement a long-term Corrective and Preventive Action (CAPA) solution to prevent recurring problems. An 8D model is a key tool for continuous process improvement. Therefore, this document is prepared to properly implement the 8D problem-solving model for company process operations that includes suppliers and internal operations.

In Chap. 8, we will discuss models of Continuous Improvement: Total Quality Management (TQM), Six-Sigma DMAIC process, Lean Manufacturing (LM) methodologies, Kaizen, and the Plan-Do-Check-Act (PDCA) cycle, the Theory of Constraints (ToC). To truly be successful with a continuous improvement program, an organization must be intentional about using a process to achieve improvements and sustain them over time. Total in TQM implies that all individuals in an organization are responsible for continuous improvement. This requires that an organization engages all employees in the pursuit of improvement. The word Quality reminds people that the ultimate goal of continuous improvement is to meet the customer's sustainable needs.

Chapter 9 emphasizes the different phases and models of analysis to a New Product Introduction (NPI). The first is to identify all different customers who will be sources of design inputs, and then effectively capture customers' wants/needs and expectations to guide our design. The next major step in the process of transforming our customer's needs (design inputs) into an epic design concept using tools like Robust Design, Quality Function Deployment (QFD), Design for X (DFX), Design for Six-Sigma (DFSS), and Quality by Design. Finally, Quality Engineer must utilize the design review process periodically to confirm that our design efforts are still on track to deliver an epic product. When we have completed these three steps, we have done a lot of the heavy lifting required for developing a new product. The Design Outputs should also identify the product features that are essential for the device's proper functioning. In some industries, these are Critical Quality Attributes (CQAs), or Critical to Quality (CTQ), or Critical Material Attributes (CMAs).

In Chap. 10, we will analyze the potential cause of the Bullwhip Effect (BE). To study the BE, we generated experimentally simulated data from the Beer Distribution Game (BDG). The game represents a simple supply chain consisting of a factory, distributor, wholesaler, and retailer. The study applied bootstrap estimation techniques on ANOVA models, Signal Processing, and Recessive Autoregression (RAR) estimation methods. The result confirmed that the BE has different cost implications for the business partners in the production supply chain. The result emphasizes that BE attacks all the business partners in the supply chain one way or another. The study discovers that BE can be caused by intra-organizational and inter-organizational coordination of the business partners' production supply chain. The effect of the intra-organizational coordination on the BE showed the drawback of the existing theory of organizational coordination. The analysis showed that to make the supply chain more efficient and effective, hyper-hybrid coordination is introduced.

Chapter 11 is an extension of Chap. 10, which introduces further study hypotheses on the BE. This study extends the stochastic analysis by applying various panel data regression models and structural equations of Seemingly Unrelated Regression (SUR) models on the experimental data from the Beer Distribution Game (BDG). A consistent result with Chap. 7, the Recursive Autoregressive-SUR model estimation result confirmed that the BE can be caused by intra-organizational coordination and inter-organizational coordination of the supply's business collaborates chain. To control the effect of intra-organizational coordination on the Bullwhip effects, the author outlines a new coordination type known as the Tesfay coordination. As a final point, the author has shown some important Tesfay coordination applications in the airline industry.

Chapter 12 introduces the Tesfay Dynamic Regression Model (TDRM) analysis of the Passenger Load Factor (LF) of the airline industry, a complex metric in the airline industry. This study modeled the LF for Europe's North Atlantic (NA) and Mid Atlantic (MA) flights in the Association of European Airlines. The model fit aimed to forecast the LF of flights within these geographical regions and evaluate the airline's demand and capacity management. The results show that in the airline industry, the LF has both periodic and serial correlations. Therefore, to control the periodic correlation structure, the author modified the existing model by introducing dynamic time effects, the T-panel data regression model.

Moreover, to eradicate serial correlation, the author applied the Prais–Winsten methodology to fit the model. The study finds that AEA airlines have greater demand and capacity management for both NA and MA flights. In conclusion, this study will find an effective and efficient Tesfay dynamic regression model fit, which empowers engineers to forecast AEA airlines' load factor.

What Are the Impacts of This Book?

The book is organized to assist engineers and production managers in mitigating the current engineering and production challenges.

The book benefits the practitioners on implementing adequate Statistical Process Control (SPC), which takes account of the capability to ensure a stable process and regulate if variations occur due to variables other than a random variation. When transferable cause variation occurs, the SPC facilitates identifying the source's root cause so that engineers will get a nice room to fix the problem. Furthermore, it helps engineers about how to formulate the acceptance sampling for in-process and outgoing inspections.

The book contains powerful engineering models of New Product Introduction (NPI), Continuous Improvement (CI), and the Eight Disciplines (8D) of Problem-Solving techniques.

The book contains innovative models in engineering, supply chain analysis, and operations management. The book introduces hyper-hybrid coordination for process effectiveness and production efficiency. The book also introduces the Kraljic-Tesfay Portfolio Matrix of industrial buying.

Fremont, CA, USA   Yohannes Yebabe Tesfay

Acknowledgments

First and foremost, I want to thank God, who gave me space to live and time to think. I want to show my gratitude to Soraa Inc. Engineers and Laboratory Experts, Hitachi Powdered Metals (HPMA) Inc. Engineers and Laboratory Experts, Hitachi University, Cree Inc., Tactus Technologies, Texas A&M University, North Carolina State University, and the Editorial Board for their innovative notions, critical valuation, positive commentaries, constructive suggestions, and on-time responses on the development of this book. I want to thank my mother, Birhan Endalew Berhe. I would also like to thank my aunt Shashitu Endalew Berhe and Lidetu Amare, who gave me real adoration.

I have many reasons to give my special thanks to my father, Yebabe Tesfay Gebremariam. Although I was good at mathematics, I was so sloppy in focusing on school when I was a first-grade student. My father heard about my school activity from my teacher and was very annoyed by what he heard. My father was from Axum, which is located in Tigray and one of the world's historic cities. My father told me that the kingdom of Axum lasted from the first to the eighth centuries AD. It was positioned at the crossroads of three continents, Africa, Arabia, and the Mediterranean. The kingdom of Axum was the most powerful nation between the Roman Empire and Persia, and it was the first state to adopt Christianity around 325 AD formally. It also briefly granted asylum to some of the early followers of Mohammed in the seventh century. Ruins of palaces, buildings, and tombs cover a wide area in the Tigray Plateau, the most impressive being the Axum, Ethiopia obelisks at the stelae field of present-day Axum. The main towering obelisk of Axum is about 1700 years old. It is made of a single granite and weighs 160 tones.

Obelisk of Axum (In Tigrigna: ሓወልቲ አክሱም) is made from a single graphite weighing 160 tones and has a height of 24 m

My father then gave me one of the most essential pieces of advice that changed me for one and the last time. He told me that the Axumites built many obelisks from single stones, and they erected them in the center of the city. He continued, imagine how the Axumites are extraordinary geniuses. They did such magnificent obelisks and pitched them perpendicular to the surface before 1700 years ago, and still, the obelisks stand. He said that the Axumites had significant knowledge about earth science, material science, physics, and astronomy because they learn how nature functions. Remember, most of the obelisks of Axum withstand earthquakes, ocean waves, thermal effects, and other factors and still stand there. He added that with the recent technology in the world, I do not think the world's concurrent engineers can rebuild the Axum obelisks.

Finally, he told me that you are one of these great thinkers, so if you study hard and learn how nature works, you will hopefully discover new ideas to change the world. I was deeply fascinated and got highly motivated. I remember that after he finished his advice, I pondered that I am responsible for something. In my heart, I have said that I must follow the footsteps of my ancestors.

Today I have discovered more than ten equations, models, and applied theories in mathematics, statistics, engineering, supply chain analysis, operations research, physics, and economics. It was my father who directed the academic turning point of my career.

Editorial Advisory Board

Contents

About the Author

Yohannes Yebabe Tesfay is a senior engineer-scientist of new product introduction (NPI), reliability and quality functions. He has innovated many equations, models, and applied theories which include Tesfay coordination. He is a Lean Six Sigma Black Belt, IATF, and ISO professional. He is experienced with lean manufacturing engineering practices via process and design development, job-shadowing, and modeling. He is an expert in the process flow, plant layout of vertically integrated manufacturing operations, and leading projects utilizing his strong analytical background, emphasizing the design of new product development. He can interpret intricate drawings (GD&T) and technical standards in many engineering sectors. In his most recent role as a new product development engineer scientist at HPMA; he is accountable for the supervision of leading research, junior industrial, process, manufacturing engineers in developing the design and implementing improvements to eliminate operational wastes within the process and identify potential gaps associated with quality control via applying SPC, engineering DoE, APQP, FMEA, MSA, and PPAP standard engineering methodologies of ISO, IATF, and AS.

Part I
The Engineering Definition of Quality

Chapter 1
Quality in the Context of Engineering

1.1 Introduction

In the usual context, quality means the fitness of products or services for use. However, such traditional definitions are very narrow to explain "what quality means?" Peter Drucker (1985) states that quality is not what the supplier or trader puts in on the product or service. Quality is the collection of parameters of the usability, durability, and reliability of the product that the customer is willing to pay to acquire.

In production engineering, quality is inversely proportional to variability. Quality engineers need to make sure that the manufacturer makes the goods according to specifications: engineers emphases the quality of the product and the production process and reduces waste. Quality engineers plan, design, and monitor the quality of operations. They work in different industries and play a vibrant role in correcting, reducing, and fixing defects.

Quality engineers aim to ensure that the manufacturer makes the products or services according to the product or service specifications. The vibrant roles of quality engineers design and monitor the quality of product or service and production processes. They work in a multiplicity of industries and play a vital role in correcting or fixing defects. Furthermore, engineers emphasize the quality of the product and the reduction of waste.

The Quality Management System (QMS) should be one of the critical policies of the company. Therefore, the company's quality goals should also be defined and clear, realizable, and quantifiable. The techniques discussed in this book form start of the basic methodology used by engineers and other technical professionals to achieve these goals and extended to advanced quality topics.

1.2   Some Interesting Definitions of Quality

- The International Organization for Standardization in ISO 9000 defines quality as the degree to which a collection of essential characteristics fulfills requirements. Precisely, quality is the standard outlines the requirement needs.
- The American Society for Quality (ASQ) describes quality as a pursuit of optimal solutions to confirm successes and fulfillment to be accountable for the product or service's requirements and expectations.
- Quality from the customer perspective is the qualification, fitness, or appropriateness to meet customer expectations and increase customer satisfaction.
- From the process perspective, quality is the degree to conform to the process standards, design, and specifications.
- From the product perspective, quality is the degree of excellence that optimizes the whole system at a reasonable price.
- From the cost point of view: the best combination of cost and feature defines quality.
- Philip Crosby (1979) defines quality as conformance to standard requirements. According to Crosby, the formal requirements may not characterize customer expectations.
- Walton, Mary and W. Edwards Deming (1988), quality accompanying management emphasizes the effective production of the marketplace's quality. According to Deming (1988), through improved processes, firms can do cost-effective and efficient production. Deming suggested that it enhances the quality and can accomplish by improving the management system and engineering.
- Noriaki Kano (1984) states that several professionals explain quality from the two-dimensional model. (1) Must-be quality describes fitness to use and (2) appealing quality to describe what a customer wants. However, according to Kano, the product's quality is all the requirements to meet or even exceed customer expectations.
- According to the project management triangle, quality is the core parameter constrained by cost, time frame, and project scope. Thus, quality should be analyzed together with the project's capacity to achieve the project's overall success.

1.2.1   Definition of Quality Based on Quality Attribute Dimensions

The quality of the product or service can be defined and assessed in quite a lot of ways. Some people perceive quality as excellence, adding value, conformance to customer requirements, the essential characteristics to the customer, the product, or service without error. Others see quality as conveying faultless in the system in fundamental respects to provide the best solution.

Fig. 1.1 The eight dimensions of quality

As we have seen above, several ways can define quality. A lot of people have an intangible (conceptual) understanding of the definition and description of quality. Some people relate quality with some desirable characteristics that the product or service possesses.

Even though this intangible (conceptual) understanding is unquestionably an excellent initial point, we should have a more precise, systematic, and useful definition. Hence, if we continue this way, we will have an endless synonym for quality, which is not helpful for management and engineering. Therefore, we can give an adequate definition of quality based on its characteristic dimensions. It is essential to distinguish the different dimensions of quality.

As shown in Fig. 1.1, Garvin (1987) delivers an outstanding discussion of eight components or dimensions of quality.

We review Garvin (1987) essential points regarding these dimensions of quality as follows:

1. *Performance*

   Performance is about "can the product do the planned job?" Customers typically assess the product to determine if it can do specific job functions. That characterizes how good the product is when it performs the job functions. For instance, we may perhaps assess a software package's performance by its outputs or execution speed.

2. *Reliability*

Reliability is about "how frequently does the product is out of order (fail)?" Multifaceted products, such as many applications, automobiles, airplanes, or machines, will typically need restoration, repair, or cyclic maintenance over their service life span. For instance, a given car may need irregular and unexpected repair. Nonetheless, if the vehicle needs such repair, we can call that the car is unreliable. The reliability dimension of quality affects numerous industries. Hence, customers do not want unreliable products or services.

3. *Durability*

Durability is about "how long the given product will last?" Durability is about the real service lifetime of the product. Obviously, customers want products that perform the job function fittingly over an extended period. In the automobile industry and primary appliance industries, durability is one of the customers' most essential quality parameters.

4. *Serviceability*

Serviceability is about "how easy it is to repair or restore the product?" Numerous industries are directly influenced by how fast the repair (or maintenance) activities can be accomplished with minimum cost. Most customers are unwilling to buy the product if the maintenance is expensive or the duration of time to fix it.

5. *Aesthetics*

Aesthetics is about the visual appearance of the product. These can be style, tactile, shape, packaging alternatives, color, external characteristics, and other sensual features. So, it is known that the visual appearance of the product affects the customer's viewing platform on the product. Aesthetic-characteristics may be the result of religion, psychology, opinion, or personal bias. For instance, beverage manufacturers depend on their packaging style's visual appearance to distinguish their product from other soft-drink producer competitors.

6. *Features*

The feature is about all the primary and other functionalities of the product. Beyond the actual performance, high-quality products have added extra functionalities to assist the customer whenever needed. The feature is considered as one of the competitive advantages of the producer over its competitors. For instance, the feature of Samsung mobile may have several applications than other mobiles.

7. *Perceived Quality*

Perceived Quality is about the company's reputation in its products. In several cases, customers depend on the previous reputation and the company's status regarding its quality. In general, reputation influences the community's perception of the company's brand. The apparent quality of company loyalty and business are meticulously interrelated. For instance, a delta airline's service may operate preferably by the customers to other airlines due to the punctuality of scheduled flights, quality of service, and baggage handling.

8. *Conformance to Standard Requirements*

   Conformance to Standard Requirements is about the matching of products made with the original design. A high-quality product is the one that accurately and precisely meets all the requirements placed on it. Conformance is a day-to-day quality language in the manufacturing environment. For instance, we may question ourselves that "how well does a screw fits the threaded hole?" Factory-made screws do not precisely meet the customer's requirements and is a significant quality problem. A car engine comprises numerous parts if each part should conform to the machine's original design to be assembled and function correctly.

Therefore, a product's quality is defined in its Performance, Reliability, Durability, Serviceability, Aesthetics, Feature, Perceived Quality, and Conformance to Standards Requirements.

1.2.2   The Three Crucial Elements of Quality

In Sect. 1.2.1, we have seen that quality can be defined based on the eight quality dimensions (performance, durability, reliability, serviceability, features, aesthetics, conformance, and perceived quality). Furthermore, quality can be defined based on the three elements (quality of design, quality of conformance, and reliability). The three essential factors that affect the product or service to satisfy customer expectations are:

1. *Quality of Design*: the product or service needs to have a well-defined design.
2. *Quality of Conformance*: the ability to match the product or service with its design specifications.
3. *Quality of Reliability*: The product's capability or service to perform the job without any trouble over an adequate period.

1.3   Quality Engineering Terminology

Every product owns numerous specification parameters that conjointly define what the consumer thinks about its quality. We call these specification parameters Critical-to-Quality (CTQ) characteristics, or simply Quality Characteristics (QC). Critical-to-Quality (CTQ) characteristics are either physical, sensory, or time orientation:

Physical: length, height, area, weight, current, voltage, viscosity
Sensor: taste, smell, appearance, color
Time orientation: durability (how long it last?), reliability (how well it functions?), serviceability (is maintenance possible with reasonable cost and time?)

Directly or indirectly, different categories of Critical-to-Quality (CTQ) characteristics can interrelate to the dimensions of quality discoursed in Sect. 1.2.1. Quality engineering is the set of operational and managerial activities that the company practices to confirm that its quality characteristics are nominal or required. The quality characteristics of the products also need to verify that the variability around nominal levels is minimum.

Most organizations find it challenging and expensive to deliver products with the same quality characteristics, products that are alike from unit to unit at levels that tie with the customer requirements and expectations. The primary reason for this is process variability, and most companies cannot control process variability.

As a rule of thumb, there is a certain degree of variability in every product. Consequently, products from the equivalent process may not be identical. For instance, the blade edges' width on an aircraft turbine engine impeller is not the same even on the identical impeller. The blade width will also vary among impellers. If the variation in blade width is little (too small), it may not significantly affect the customer.

Nevertheless, if the variation is considerably large, the product will create a significant customer problem. The cause of variation could be materials, operation, equipment, personnel, or other known or unknown factors. Identifying and resolving the cause of variations are the critical job functions of the quality engineer.

Variability can only be defined and described in terms of statistical tools. Therefore, statistical approaches play a central ground role in quality engineering and continuous improvement efforts. While applying statistical methods and tools to quality engineering problems makes it distinctive to categorize data on quality characteristics either as attributes (features or discrete characteristics) or variables data.

Variables data are typically continuous quantities, such as height, length, current, voltage, temperature, or viscosity. On the other hand, attribute data generally are discrete data. Habitually they are taking a form of counts.

Attribute data can be:

- The color type of mobile phone
- The number of machines running at packaging operation
- The number of processes for producing sprocket
- The number of errors that an operator makes
- The number of loan applications
- The number of calls an IT department received per day

Categorizing the data is very important to select the type of statistical tool we will apply to analyze them. Before the analysis, we should define the statistical tool for quality engineering applications to deal with both data categories.

Attribute data can be the color type of mobile phone, the number of machines running at packaging operation, the number of processes for producing sprocket, the number of errors that an operator makes, the number of loan applications, the number of calls that an IT department received per day.

Categorizing the data is very important to select the type of statistical tool you will apply to analyze them. Before the analysis, we should define the statistical tool for quality engineering applications to deal with both data categories.

Quality characteristics are habitually assessed based on the specifications of the product. For example, for the product, the quality specification requirements are anticipated based on all the essential components and the subassemblies that make the product, besides the desired standards for the quality characteristics in the end product.

1.3.1   Specification Limits and Tolerance

A measured quantity corresponds to the anticipated value for that quality characteristic termed as the nominal value, simply, the target value for that characteristic. Target values are typically bounded (limited) by a range of adequately near to the targeted value. They do not impact the functionality or performance of the product. Suppose the quality characteristic is in that range. In that case, the quality engineer is obliged to accept the past as a good part:

Absolute perfection and zero deviation in the process is unattainable.
As we focus more on quality, the quantity of production will decline.
Production of the less tolerable process is costly and time-consuming.

The smallest and the largest acceptable (tolerable) values for the quality characteristic are, respectively, called the lower specification limit (LSL) and the upper specification limit (USL). The given quality characteristics do not need to have both the lower specification limit (LSL) and the upper specification limit (USL). Sometimes the quality characteristics may have only one of the specification limits. For instance, the compressive strength of a fundamental part used in a car bumper probably has a nominal (target) value with a lower specification limit (LSL) with no upper specification limit (USL).

Design engineers produced the product design structure and configuration using engineering and scientific principles, which habitually outcomes in the designer specifying the nominal (target) values for the critical and noncritical design parameters. The specification of quality characteristics limits is determined and decided by the design engineer through the wide-ranging procedure. Thus, the lower specification limit (LSL) and the product's efficient engineering design results.

Tolerance is the range between the upper specification limit (USL) and the lower specification limit (LSL). Since variability in the process is expected, we must accept and set allowable measurements for the differences. In simple terms, tolerances are these acceptable allowances. To illustrate the concept, consider the drawing in Fig. 1.2.

From the drawing, we see that the nominal (target) diameter of the part is 5.45, with the lower specification limit (LSL) of 5.447 and the upper specification limit (LSL) of 5.453. The tolerance, in this case, is 0.006. Furthermore, the upper specification limit (LSL) of the surface flatness is 0.003 with a tolerance of 0.003.

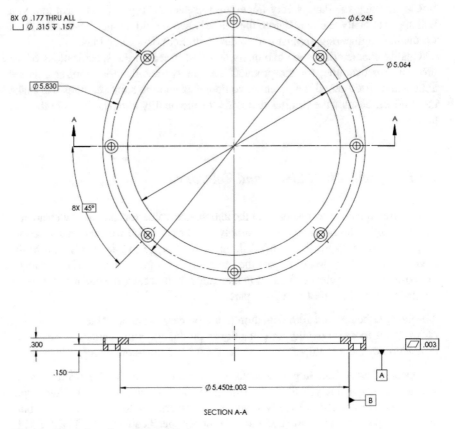

Fig. 1.2 Illustration of specification limits

1.4   Over-the-Wall Model: The Old Design

As technology becomes advanced, complexity leads to specialization in a more specific area. For instance, a large company may have:

1. Marketing specialist (who analyze customer demand)
2. Research and Development expert (develop methods, models, and tools of technology to satisfy future marketing demand)
3. Design engineer (who work the design of the product)
4. Manufacturing engineer (who change the design into production)
5. Sales professional (who sell the product)

Thus, marketing, R&D, product design, manufacturing, and delivery will be the specialist's cumulative effect.

In the Over-the-Wall engineering method to product design, each team associate performs his or her responsibilities and then passes the succeeding team associate.

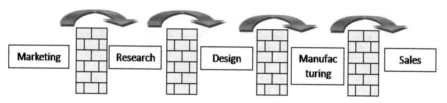

Fig. 1.3 Over-the-Wall approach

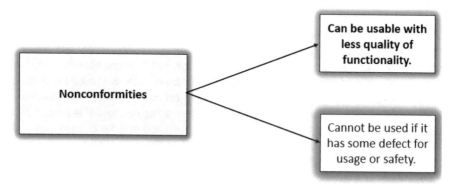

Fig. 1.4 Nonconformities does not necessarily imply unfit for use

As shown in Fig. 1.3, the term Over-the-Wall method came about for the cause that each person's work is given over the theoretical or actual compartment wall to the succeeding person.

Complications in the product quality typically are higher when the Over-the-Wall method to design is applied. In the Over-the-Wall method, specifications are so frequently set without considering the intrinsic variability in the materials, assembly, processes, or other operation components. That results in the products have many variabilities to lead the products to have a nonconformance issue. Nonconforming products are products with at least one defect or unable to meet at least one specification requirement. The explicitly such kind of specific failure (not necessarily a defect) is called nonconformity.

A nonconforming part or product is not essentially unfit for usage. For instance, a soap may have a concentration of constituents (ingredients) lower than the lower specification limit. Nonetheless, it can still perform the function acceptably. The nonconforming part or product is considered defective (malfunctioning) if it has at least one defect. Thus, defects are the real reasons for nonconformities, which significantly affect the product's proper functionality or safety. As shown in Fig. 1.4, we can see that a nonconformance part can be usable, however, with less functionality or cannot be used if it has a defect or is risky for safety.

The Over-the-Wall design approach and its processes have been the theme of great attention for nearly half a century. Design for manufacturability has developed as an essential component of overcoming the intrinsic complications with the

Over-the-Wall method to design. AutoCAD, Solidworks, and CAM systems auto-mated the drawings designs and process more efferently and effectively. Conse-quently, manufacturing activity challenges are resolved by much via the introduction of automation.

1.5   Concurrent (Simultaneous) Engineering Model

The concurrent (simultaneous) engineering model is also called integrated product development (IPD). The Simultaneous engineering model is an engineering approach where team members from different specialized departments work together early in the project initiation. In this model, there are no walls that block the various fields of experts. Everyone contributes to the project and works together until the project is successfully executed. The overall idea overdue this model is to establish an effective and efficient team that can identify, recognize, and resolve any issues at each stage of the project. A simple representation of the simultaneous engineering model is given in Fig. 1.5.

The introduction of automation changes the educational curriculum of engineers. Furthermore, most of the engineering designs start developed by the team than working individually. Experts in manufacturing, reliability, validation, quality engi-neering, and other professional disciplines work together to create its design. The team involvement from each domain brings the product's efficient design if they

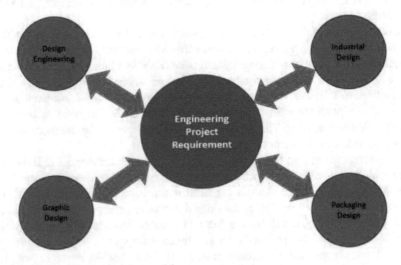

Fig. 1.5 Simultaneous engineering model

work at the initial phases of the product's design process. Besides, the operative use of the quality enhancement methodology, at all stages of the method used in skill, technology, knowledge commercialization, product comprehension, product design, progress, manufacturing, delivery, and customer provision, plays a vital role in quality enhancement.

1.6 Process Planning

At first, to meet the quality objectives, the higher management must have a good process plan. Planning is thinking, analyzing, and conceptualizing the goal and activities essential to accomplish the anticipated goal.

The process planning (design) is the key to choosing the control chart's process variables. The process design is a set of techniques to analyze the whole project and breaking down the project plan into controllable sub-plans. Process design characteristically customizes numerous tools and methods such as flowcharts, simulations, software, time management tools, and scale models to analyze the project.

Use the process to define the steps needed to tackle each project and remember to hold all our ideas and sketches throughout the process. Workflow activity determines the equipment needs and implementation of quality requirements for the given process.

Process design necessitates a comprehensive insight into the entire organization and must not have a narrow-minded viewpoint. That is, the process design must contemplate the suitability of the process to the general organizational objective. The process design must address and deliver customer value with persistent engrossment of the management at different phases. Therefore, to accomplish such process design, an operative and effective process strategy is compulsory.

Operative process strategy should include acquisition of raw materials (effective and efficient procurement), technology investment, personnel, supplier involvement (tire 1, when necessary tire 2, tier 3, etc.), customer involvement, etc. The process design has endured change, improvement, or modification. New methods like the make-to-order plan of Flexible Manufacturing Systems (FMS) have been established.

FMS is a premeditated manufacturing technique to quickly acclimate to necessary and adequate production changes both by quantity and by type. FMS can enhance efficiency and thus drop a company's production cost. FMS allows customers to modify or customize the product. That delivers productive (efficient and effective) product design per the requirements of the customer. Study the process flow of design development in Fig. 1.6.

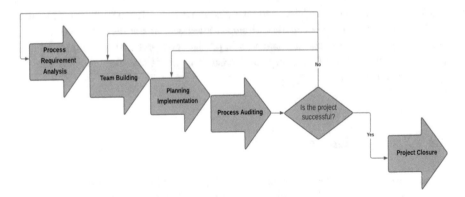

Fig. 1.6 Process design development involves the following steps

1.6.1   Process Requirement Analysis (PRA)

Process requirement analysis (PCA) is a set of activities that help comprehend, including analyzing risk for producing a product. Requirements analysis is the method of defining, describing, analyzing, documenting, validating, and managing customer needs, which can be a new product or a modified one. Requirements analysis encompasses all the techniques and activities that need to comprehend and identify different shareholders' desires. Requirement documentation should be measurable, actionable, traceable, verifiable, and must reflect customer needs.

Data collection is a necessary action in process requirement analysis. The process requirement development involves various phases such as demand forecasting, raw material requirement and acquisition, plant layouts and different stages of production operations, available technology for production, documentation structures and systems, risk, and stakeholders.

The following are some of the techniques of process requirements analysis:

Flowcharts technique: Flowcharts are used to visualize the sequential flow interrelated activities and actions. Flowcharts can be cross-functional, linear, or Top-down.

Business process modeling notation (BPMN): This method is well known in process improvement. BPMN is like generating process flowcharts, even though BPMN has its elements and symbols. BPMN has its unique characteristics and symbols and is applied to create graphs that simplify and comprehend the process input and output.

Unified Modeling Language (UML): These techniques comprise integrated diagrams that are created to specify, construct, document, and visualize the objects of the process.

Data Flow Diagram (DFD): DFD is a flow of information and is used to visually characterize processes and systems compound and difficult to define in text. DFD defines several objects and their relations with the aid of standardized symbolizations.

And other techniques like Gantt Charts (GC), Workflow Technique (WT), Gap Analysis (GA), Role Activity Diagrams (RAD), Integrated Definition for Function Modeling (IDEF), etc.

1.6.2   Team Building

After the process requirements are completed, a cross-functional team should be assembled and established to execute the project task based on experience, knowledge, and skill levels. The cross-functional team (CFT) needs to take the necessary training to accomplish each task. After the required training is completed, the cross-functional team must execute the tasks effectively and efficiently (per the project's budget under a time constraint).

1.6.3   Planning Implementation

Based on the requirements analysis for producing the product, the cross-functional team will create documentation of policies, procedures, work instructions, modules, etc. Then, after documentation review and then the approval process, implementation of the plan follows.

1.6.4   Process Auditing

A process audit is an inspection of outcomes to regulate whether the activities are done based on the planned resources and the project's scheduled (allowed) time frame. An organized audit will be carried out to confirm whether the process implementation is in line with the plan to deliver the product to the customer.

1.6.5   Project Closure

When the project is completed and the product is tested. If the product testing confirms the fulfillment of all the customer requirements, it is considered a successful project. Now, it is time to disclose the project. In the project closure stage, the cross-functional team will finalize reports and celebrate the project.

1.7 Total Quality Control (TQC)

In 1951, Feigenbaum was familiar with Total Quality Control (TQM), which was inclined by the 1950s viewpoint of Japan's quality management system. Through the 1950s, several Japanese companies used the term "Total Quality Control" to define, describe, and pronounce their operational efforts.

Total Quality Control (TQM) uses a three-step method for improving quality. These are Quality Leadership, Quality technology, and Organizational commitment.

1.7.1 Quality Leadership (QL)

Quality Leadership is a precondition and requirement for executing a quality management system. This step emphasizes "how the structure of organizational leadership and direct the organization and what manner they accomplish within the organization are vigorous elements to the realization and comprehension of the effective and efficient quality management process."

Quality Leadership needs to fulfill the following criteria:

- Genuine enthusiasm and interest
- Integrity with all the organizational system
- Great communication skills
- Loyalty to work, employee, and the mission of the organization
- Determination and accountability to take risk
- Managerial competence and managerial quality
- Enabling authorization on employees
- Charisma, well-spoken, approachable, and friendly

1.7.2 Quality Technology (QT)

Quality Technology refers to the technology infrastructures that the organization provides. Quality Technology represents the organization's technological-competency parameter to measure, analyze, and interpret quality-related parameters. These are:

- Datacenter facilities
- Managed services
- SPC and statistical methods
- Engineering designs and methods
- Technical tools and systems

1.7.3   Organizational Commitment (OC)

Organizational commitment integrates psychological preparedness, awareness, requirement understanding, and employees' professional ability to meet its mission and goals. Many studies identified that organizational commitment is critical to productivity and employee performance. It is one of the vital fundamentals in attaining the organization's goals. Committed employees contribute weighty enhancement in various dimensions of the organization's mission. Analyzing what factors inspire and motivate employees to get substantial commitment is significant to improve organizational performance. The factors that affect organizational commitment are:

- Job satisfaction
- Task employee's self-efficacy
- Employee engagement
- Age and tenure in the organization
- Stress and depression related to specific job functions and roles
- Employee's job control and lack of awareness for the responsibility and power concerning their role at work
- Lack of proper decision-making
- Job insecurity of the employee
- The trust of the employee in the organization
- The motivation of the leader
- A good leader can make his subordinates more committed toward the goal of the company
- Competitive salary
- Employee personal and professional behavior
- Employee's position and roles in the organization
- Employee's career advancement
- Performance assessment programs of the organization and acknowledgment
- Positive experience in team involvement of the employee

1.8   Quality Assurance (QA)

Quality Assurance (QA) is the collection of activities that ensures and certifies the quality requirements for products or services. Quality Assurance (QA) emphasizes improving the processes to deliver high-quality products or services to the customer. In receipt of the customer feedback, Quality Assurance (QA) appropriately upheld the customer quality subject matters to be adequately determined and addressed.

Documentation of the quality system is an essential component of Quality Assurance (QA). Based on the ISO 9001:2015, quality management system documentation involves four fundamental elements: (1) policy, (2) procedures, (3) work instructions and specifications, and (4) records.

1.8.1   Quality Policy

In quality management system (QMS), a quality policy is the primary strategic document created by the top-quality management to direct its overall quality objectives and standards.

Quality Policy generally deals with "what quality means the organization" and "why?" The purpose of the quality policy is to safeguard improvement over and done with self-evaluation and action planning. The given organization's quality policy is published in public for employees, customers, suppliers, government, and regulatory agencies.

For instance, a policy is used to guide and influence decisions in the ISO 9001 system (found in the ISO 9001:2015 Procedure). The purpose of quality policy documentation is stated as one of the objectives of the organization. At a minimum, ISO 9001 requires a quality manual, which defines and describes the scope of the given organization's quality procedures and processes.

The following documentation can be embodied in the quality policy:

- *Customer Needs*: to indicate that the products or services will meet customer expectations or needs. Example: affordable price, durable, safe, etc.
- *Customer Preferences*: to describe whether the product is designed to meet (or exceed) the customer needs.
- *Service and Experience*: to customer experience (CE) principle, which is the customer's insight, perception, or awareness of its brand on a specific product.
- *Listening*: to describe the company's commitment to listen to the customer and improve customer expectations.
- *Compliance*: to describe the company's ability to meet standards and regulations.
- *Health and Safety*: to describe the company's concern toward the production of safe products.
- *Defects*: to describe the company's commitment to efficient and effective production (zero defects).
- *Accuracy*: to describe the company's level of honesty and accountability on the product or service. For instance, the company's policy of product description.
- *Testing*: to describe the company's ability to test the product or service before it moves to the customer.
- *Waste*: to describe the company's commitment to zero defects.
- *Improvement*: to describe the company's effort for continuous improvement.
- *Industry Specific*: to describe the company's effort to have an impact on some specific industry.
- *People*: to describe the company's effort to serve the community.
- *Environment*: to describe the company's commitment to protecting the environment.
- *Sourcing*: to describe the company's effort to source the materials with responsibility.

1.8.2   Procedures

Procedures are a systematic method for engineering or process documentation. These procedures can be text-based or graphic or a process map to communicate the information. Procedures help make the job flow-full and modest as per the end user's capability to set up or improve their specific engineering requirements documentation. Procedures focus on the methods, techniques, models, and personnel that will implement the quality policy. The quality assurance policy and accompanying procedures will encompass personnel. The process's management will be through the prevailing company's structure.

After the company's top managers identify the quality policy, the next step will be procedure documentations. The procedures bring a more comprehensive and describe the activities of the company's operation functions in various departments, processes, or specific task functions. Thus, in most high-technology innovative companies, procedures are confidential information of the company.

Suppose the company needs to produce or advance an engineering procedure. In that case, the documentation requirement needs to be analyzed based on introducing a new process or the improvement of the existing process. The development of a new procedure is called procedure origination. Moreover, the person who writes the document is called the procedure originator. After the procedure is developed and revised, the originator or any other certified and skillful person can upgrade the document.

When generating and developing the procedure, confirm the requirements against the quality standard organizations (like ISO, IATF, AS, etc.) to ensure all requirements are addressed.

Definition of standard criteria should be all-inclusive (operation, personnel, process, environment, customer, and supplier). Hence, the involvement of related factors in the definition refines the goals and the ways to achieve them. The goals can be a competitive advantage and position of the company in the market.

1.8.3   Work Instructions

Work instructions (also called work controllers or guides) are the standard operating procedures (SOPs) user manuals for the given process job function. Work instructions can be subsections of the procedure or can also be the procedure itself. Under any circumstance, the purpose and scope of work instructions should clearly and undoubtedly clarify how the specific work function (task) will be done.

Typically, work instructions are applied for product process steps, how the department accomplishes its tasks, how a tool or a machine functions, and how to perform calibration or maintenance. Therefore, specific work instructions must be very comprehensive and must clearly show all the steps about "how" to realize the specific job task.

Fig. 1.7 Plan-Do-Check-Act (PDCA)

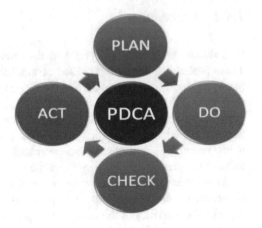

For instance, the work procedure can be established for assembling the ending cover of a product, with step-by-step directives and instructions on all activities, including the details of torque specification requirements for holding and fasten screws. Each work instruction (SOP) is specific to the product, process, department, company, or industry. Supplementary documentation like the user's manual for engineering, technical support, manufacturing notes, etc., details the work instructions.

Procedures define and describe the whole process, whereas the work instruction describes (SOPs) how to achieve the procedure. Process definitions and descriptions include specifications about how the inputs (contributors) changed into outputs (results) and the comment essential to confirm reliable and consistent outcomes.

Typically, the PDCA (Plan, Do, Check, Act) process model is applied to get the appropriate and relevant information about the process. As seen in Fig. 1.7, The Plan-Do-Check-Act (PDCA) model is a quality management system for planning, operation, and continuous improvement tools.

Questions or requests that are essential to be responded in a procedure comprising:

Q1: Who are the suppliers (about sourcing the inputs)?
Q2: Who are the customers (about where are the markets of the product)?
Q3: Who does the actions and takes responsibility?
Q4: What are the effectiveness measures or criteria (assuring doing the right)?
Q5: What metrics must be controlled?
Q6: How do we interpret results (graphs, charts, or reports)?
Q7: What regulations or standards should we apply?
Q8: Supplementary related questions.

1.8.4 Records

Records are systems of documenting quality policies, procedures, and work instructions (SOPs). The documentation comprises forms applied to generate records, surveys, checklists, or other relevant documents used to produce the product or service. Records can be formed from a data registry, information sharing, internal and external auditing, and process improvement enterprises. These are essential and critical for production control and decision-making tools.

Records are used in traceability analysis (to track precise units or specific batches of the production). Traceability analysis is a process of identification and determination of how the product was produced precisely.

Records are habitually vibrant in providing information or data to analyze customer requests, complaints, root cause analysis, and corrective actions. Generation, development, control, and maintenance of documentation are central tasks and functions of quality assurance. Document control confirms and safeguards specific requirements. Work instructions (or SOPs) developed for functioning personnel provide up-to-date design, engineering, and management changes.

Quality improvement encompasses the set of activities, which enable to confirm that the product meets customer requirements. Furthermore, it enhanced on a continuous improvement basis (for instance, the Plan-Do-Check-Act (PDCA) process). Typically, variability is the primary cause of quality problems. Therefore, statistical techniques, including statistical process control (SPC) and designed experiments (DoE), are the indispensable tools of modern quality engineering applications for process control and improvement.

Process improvement is made based on a project that comprises teams. Suppose the team leader has substantial knowledge of statistical models and statistical packages. In that case, he/she can monitor the process effectively. Hence the use of statistics for project management becomes extensively advantageous.

The continuous improvement projects should be carefully chosen based on their significance on the company's bottom line. The tasks of consideration should be cost and time effective and improve the company's profit. So, each project should have clear objectives to address its needs and be qualified to achieve the goals.

1.9 Quality Control (QC)

Quality control (QC) is a process (or set of procedures) planned to confirm and validate that a product or service obeys a well-defined quality criterion in the manufacturing process. Quality control (QC) aims to fulfill the requirements of the customer.

To implement operative quality control (QC) systems, the company must first adopt specific quality standards the product or service must have. Then the degree of quality control (QC) activities must be determined via appropriate acceptance

Table 1.1 The difference between quality assurance (QA) and quality control (QC)

Quality Assurance (QA)	Quality Control (QC)
QA is the procedure able to grant that quality requirement to be accomplished.	QC is the process that fulfills the quality request.
QA object is to stop the defect.	QC object is for recognizing, identifying, and then eliminating the defects.
QA is the method of managing quality.	QC is a technique for confirming and verifying quality.
QA does not include executing quality programs.	QC actively involves executing the quality program.
The entire team members are accountable for QA.	The quality inspection team is accountable for QC.
QA is all about verification.	QC is for validation and confirmation.
QA means the development of planning for doing the process.	QC means activities for performing the planned quality process.
QA uses Statistical Process Control (SPC).	QC uses Statistical Quality Control (SQC).
QA makes certain the process is doing the right.	QC makes certain the outcomes of what has been as expected.
QA outlines standards, tools, and methods to meet customer requirements.	QC guarantees that the standard quality tools and methods are followed during the production of the product.
QA is the procedure to generate the deliverables.	QC is the procedure to verify the deliverables.
QA is accountable for full product development.	QC is accountable for each process of product development.

sampling designs. Real on-time data acquisition performance of the company reflects its ability to achieve effective quality control programs. Suppose the collected data predicts that the process has something going bad. In that case, effective corrective action should be performed to fix the process.

Suppose several unit failures (or occasions) occur in the process. In that case, a plan must be developed to advance and improve the production process.

Quality control (QC) is like, but then again not the same as quality assurance (QA). Quality assurance (QA) denotes the approval that the product or service has fulfilled specified and precise requirements. However, quality control (QC) represents the real quality inspection of the product during manufacturing elements. See the detailed difference between quality assurance (QA) and quality control (QC) in Table 1.1.

1.10   Total Quality Management (TQM)

In this section, we will assess a brief outline of the fundamental components of total quality management. We emphasize to discuss the essential quality systems, methods, and standards. Total quality management links quality, productivity, and regulatory factors to achieve the goals of the company. All the characteristics and aspects of quality planning, quality assurance, quality control, and quality improvement are the central concerns of total quality management.

The phenomenon is widespread. Quality is the key factor whether the consumer is an individual, an industrial organization, a retail store, a bank, a financial institution, a military, etc.

Basically, quality is one of the utmost essential consumer decision factors in the selection among the competing products and services. Therefore, considerate quality programs and improving qualities are the prime factors to business success and enhanced competitiveness. There is a considerable return on investment (ROI) from the enhanced quality and successfully applying for quality programs as an integral part of a complete business strategy.

Total quality management (TQM) is the overall strategy for implementing and managing quality assurance and improvement activities. Based on the focal point of Deming and Juran's philosophies, total quality management (TQM) began in the early 1980s. Total quality management is a systematic tool to integrate the company's culture, operation, and leadership. It goes all-out to deliver high-quality products and services to customers based on their needs and requirements. The culture necessitates quality in all the company's processes to avoid defects and reduce waste from each operation.

Total quality management is a management model assists in participating in all company departments (finance, marketing, engineering, design, and operation, customer, suppliers, and relevant experts) to emphasize fulfilling the customer requirements. As a result, total quality management aims to meet organizational goals.

Total quality management (TQM) model viewpoints a company as an organization that possesses various activities and processes. The model strongly emphasizes that companies can continuously improve these activities or processes by integrating new technologies, knowledge, skills, and workers' experiences from various departments in the company. The simple goal of total quality management (TQM) is "do the right things, right the first time, then every time."

In simple terms, total quality management (TQM) is a technique by which employees and management should participate in the company's continuous improvement efforts to achieve high-quality products. However, the total quality management approach evolved into wide-ranging concepts, involving experiencing management and work environment (culture), customer focus (considers customer requirements in all the stages, from planning to final delivery), supplier quality control, and improvement programs, and integrates and aligns the quality systems of the process with the business goals of the company. Typically, companies that implement total quality management (TQM) approach to quality control and improvement need quality councils. The quality council is a high-level cross-functional team that deals with planned quality initiatives. The council focuses on production and business activities and addresses specific quality improvement issues.

Total quality management (TQM) is found substantially flexible and has more comprehensive applications. Although the model is initially functional to manufacturing operations, total quality management (TQM) becomes a generic tool in various industries.

Total quality management (TQM) is the basis for activities, which comprise:

- Commitment (all personnel of the organization)
- Emphasis on processes and continuous improvement plans
- Fulfilling customer requirements and exceeding customer expectations
- An efficient methodology for product development cycle times
- Cost efficiency techniques
- Just in time (JIT) production
- Demand flow manufacturing
- Process improvement core functional teams
- Contour management ownership
- Employee empowerment and involvement
- Acknowledgment for a successful employee
- Challenging measured goals and benchmarking
- Effective strategic planning

That shows that total quality management (TQM) necessary be adept in entire operations by all personnel in the company. The principle of total quality management is given in Fig. 1.8.

Total quality management (TQM) has modest success for assortments of "the ins and outs." However, often insufficient effort is devoted to applying the TQM tools of variability eradiation and reduction programs.

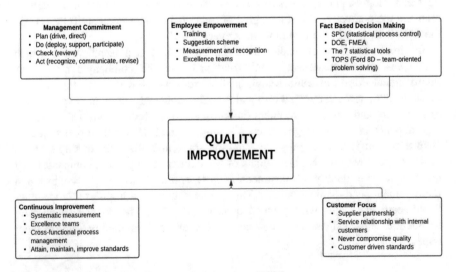

Fig. 1.8 The principle of quality management system (TQM)

Several companies implement total quality management (TQM) systems as training pieces for their personnel. For instance, human resources departments control total quality management training, which is ineffective in many cases. Sometimes the coaches are external people, so they often had no exact real idea about its operations. Thus, the training becomes more theoretical than being applied in the quality improvement process. Therefore, a wide range of process gaps are needed to be analyzed and to realize the advantage of total quality management (TQM) systems.

Some general reasons for the lack of conspicuous success of total quality management (TQM) systems include:

- Lack of top-down, high-level management commitment and involvement
- Ineffective use of statistical methods and inadequate perception of variability reduction as the main objective
- General as opposed to specific business-results-oriented goals
- Too much emphasis on widespread training as opposed to focused technical education

An additional cause for the unpredictable accomplishment of total quality management (TQM) is that several managers and executives have regarded it as just another "program" to improve quality. All through the 1950s and 1960s, programs such as Zero-Defects and value engineering thrived. Nonetheless, they had a slight influence on product quality and productivity enhancement.

During the heyday of total quality management in the 1980s, an additional popular method program was quality. The product or service is a free inventiveness—the administration function on categorizing the cost of quality. As the cost of quality increases, the managers are unwilling to invest more.

Quality is unrestricted to practitioners who frequently had no awareness about advancing the tools and improving several kinds of multifaceted industrial processes. Indeed, suppose this inventiveness's management leaders had no statistical methodology and analysis. In that case, this leads to unsuccessful comprehension and implementation of quality improvement programs. When such obstacles surround the total quality management (TQM), this creates unproductive quality improvement.

Based on TQM's principle, errors or faults originate from personnel, but not least to faulty systems or processes. Therefore, there are three main mechanisms of prevention and elimination of faults:

1. Mistake-proofing methods like poka-yoke to prevent mistakes or defects.
2. Inspection process at the early stages before the next operation of the value-added process.
3. When mistakes occur, we should stop production and immediately fix the problem.

1.11   Statistical Process Control (SPC)

In the 1920s, Walter Stewart industrialized the notion of process control via the applications of statistical charts. Using these statistical models, Walter Shewart substantially improved the process quality, cost, and output. Statistical process control (SPC) is a scientific data-driven analytical decision-making technique applied in the manufacturing or service industry to optimize the process output.

Total quality management (TQM)'s primary goal is to recognize, identify, control, and eliminate variation from process standards and improve productivity (efficiency and effectiveness). Statistical process control (SPC) is influential in analyzing the sources of process variations. Statistical process control is a collection of systematic decision-making techniques. SPC allows seeing the process of real-time data that helps to know when a process is operating based on standard work instruction and specification. Variation exists in any process. Determining when the variation is acceptable or whether the process needs an immediate correction is a piece of essential information about quality control.

In the current complex manufacturing environment, it becomes progressively superficial that necessary decisions be made based on facts. Thus, data must be collected, evaluated, and then should be analyzed to bring the right decision at the right time. Therefore, manufacturing performance has benefited from statistical process control (SPC), which significantly refined the actual process's decision-making.

Control charts and graphs are essential inputs for a continuous quality control process. Control charts can monitor processes to display how the process is being carried out. Furthermore, the control charts can show how the capabilities to process are affected (alerted) by internal or external changes. More importantly, these control charts can help identify the effect of special cause variations on the process performance in manufacturing operations. For instance, the control charts have assisted in determining whether special cause variations are existing in the ongoing process or not. The presence of special cause variations in the process implies that immediate action is required to fix them. Fixing the special cause variations can eradicate that cause (if it has a negative effect).

On the other hand, if it is beneficial, we will include it in the process manual, work instruction, or procedure. Suppose there are no special cause variations in the process. In that case, the statistical process control (SPC) tools help calculate the process capability. The information forms the capability analysis to determine whether the operating process is right in the current form or needs further improvement.

Inferential statistics is a technique that brings a way to generalize the population characteristics (parameter estimation) from an effective and efficient sample. Statistics is an assortment of useful techniques for the decision-making process or making inferences about the population based on data analysis.

Therefore, statistical methods play a vibrant role in quality control and improvement and provide essential predictive information about the process. Statistical models provide a framework to sample and analyze the data and the interpretation from the analysis. Therefore, statistical methods play a vibrant role in quality control and improvement and provide essential predictive information about the process. Statistics is also a powerful tool for research and development management, engineers, manufacturing, procurement, and other business functional components.

1.12   Quality Systems and Standards

The International Standards Organization (founded in 1946 in Geneva, Switzerland), known as ISO, has developed a series of quality systems standards. ISO 9000 is the major family of quality management systems (QMS), which guides and helps organizations certify they meet customer's requirements and regulatory requirements. ISO 9000 is a generic standard, broadly applicable to any organization. It is frequently used to validate a supplier's ability to control its processes.

1.12.1   ISO 9001

ISO 9000 defines and describes the vital notions and principles of quality management, which are globally applied to the following major areas:

- Organizations in quest of sustained reputability via the operation of a quality management system (QMS).
- Customers in quest of self-assurance in an organization's capability to deliver products or services compliant reliably and consistently with the requirements.
- Organizations in pursuit of self-assurance in their operational supply chain for their products or services meet the requirements.
- Organizations and conservative parties in a quest to advance and enhance communication via a common sympathetic of the quality management system.
- Organizations are accomplishing conformity evaluation assessments against the ISO requirements (for instance, ISO 9001).
- Consultants and training service providers for training and evaluation of personnel to meet customer requirements or related standards.
- ISO 9000:2015 delivers and specifies the vocabularies (definitions) used in all quality management system (QMS) standards established by ISO/TC 176.

1.12.2   Examples of Standards

Countless organizations have compulsorily required their suppliers to be certified the ISO 9000. As the industry becomes more specific, the requirement becomes more specialized. Illustrations of these industry-specific quality management system (QMS) standards are:

- Aerospace industry-AS 9100
- Automotive Industry-International Automotive Task Force (IATF) 16949 and QS 9000
- Telecommunications industry-TL 9000

Much focuses of the ISO 9000 (and of the industry-specific standards) are on official and recognized documentation of the quality management system. Organizations typically must make all-embracing efforts to convey their documentation to align the requirements of the standards.

There is a distant too abundant effort dedicated to paperwork documentation not virtually adequate to improve processes and products via reducing variability. Besides, countless third-party registrars, inspectors, and advisors in the process improvement area are not adequately educated or skilled enough to apply the technical tools required for quality improvement. That is the current big challenge across many industries. Every so often, they are unaware of what establishes contemporary engineering and statistical analysis. So, this book intended to assist quality professionals in fulfilling their requirements of quality management systems.

1.12.3   The Seven Quality Management Principles

ISO 9000 emphasizes the essentials of quality management systems (QMSs) and the seven quality management principles that trigger the family of quality standards. See the seven quality management principles in Fig. 1.9.

The ISO 9000 family comprehends the following well-known standards:

- ISO 9000 (latest edition 2015): Quality Management System (QMS) fundamentals and vocabulary (definitions), which specialized universal standardization agency that consists of more than 160 countries.
- ISO 9001 (latest edition 2015): Quality Management System (QMS)-Requirements.
- ISO 9004 (latest edition 2018): Quality Management System (QMS)-Quality of an Organization Leadership to accomplish sustainable success via continuous improvement.

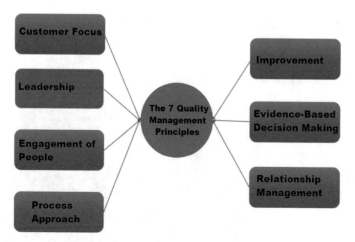

Fig. 1.9 The seven quality management principles

- ISO 19011 (latest edition 2018): Quality Management System (QMS)-Guidelines for Inspection and Auditing Management Systems.
- ISO 9001 deals with the necessities and all the requirements that organizations plan to meet the standards. The first ISO 9001 standards were issued in 1987, and the contemporary document was released in 2015. The ISO 9001 standard has eight clauses:

1. Scope
2. Normative References
3. Definitions
4. Quality Management Systems
5. Management Responsibility
6. Resource Management
7. Product (or Service) Realization
8. Measurement, Analysis, and Improvement

To become qualified under the ISO standard, the company must choose a registrar and prepare for a certification audit. There is no single self-determining authority that licenses, regulates, monitors, or qualifies registrars. Preparing for the ISO certification, the audit involves many activities, including all the quality management systems against the standard. The audit is usually followed by forming a team to confirm and endorse that all components of the vital clause are developed and implemented. Then the certification audit takes place. If the company is certified, then periodic surveillance audits by the registrar continue, usually on an annual (or perhaps 6-month) schedule.

1.13   ISO 9001-Quality Management Systems of Requirements (Source: Adapted from the ISO 9001:2008 Standard, International Standards Organization, Geneva, Switzerland)

1. Quality Management System

1.1. **General Requirements**
The organization shall establish, document, implement, and maintain a quality management system and continually improve its effectiveness following the requirements of the international standard.
1.2. **Documentation Requirements**
Quality management system documentation will include a quality policy and quality objectives; a quality manual; documented procedures; documents to ensure effective planning, operation, and control of processes; and records required by the international standard.

2. Management System

2.1. **Management Commitment**
a. Communication of meeting customer, statutory, and regulatory requirements
b. Establishing a quality policy
c. Establishing quality objectives
d. Conducting management reviews
e. Ensuring that resources are available
2.2. Top management shall ensure that customer requirements are determined and are met with the aim of enhancing customer satisfaction.
2.3 **Management shall establish a quality policy.**
2.4 **Management shall ensure that quality objectives** shall be established. Management shall ensure that planning occurs for the quality management system.
2.5 **Management shall ensure that responsibilities** and authorities are defined and communicated.
2.6 **Management shall review** the quality management system at regular intervals.

3. Resource Management

3.1 **The organization shall determine** and provide needed resources.
3.2 **Workers will be provided necessary education:** training, skills, and experience.
3.3 The organization shall determine, provide, and maintain the infrastructure needed to achieve conformity to product requirements.
3.4 **The organization shall determine and manage** the work environment needed to achieve conformity to product requirements.

4. Product or Service Realization

4.1 **The organization shall plan** and develop processes needed for product or service realization.
4.2 **The organization shall determine** requirements as specified by customers.
4.3 **The organization shall plan and control** the design and development for its products or services.
4.4 **The organization shall ensure** that purchased material or product conforms to specified purchase requirements.
4.5 **The organization shall plan and carry out production** and service under controlled conditions.
4.6 **The organization shall determine the monitoring** and measurements to be undertaken and the monitoring and measuring devices needed to provide evidence of conformity of products or services to determined requirements.

5. Measurement, Analysis, and Improvement

5.1 **The organization shall plan** and implement the monitoring, measurement, analysis, and improvement process for continual improvement and conformity to requirements.
5.2 **The organization shall monitor information** relating to customer perceptions.
5.3 **The organization shall ensure** that product that does not conform to requirements is identified and controlled to prevent its unintended use or delivery.
5.4 **The organization shall determine, collect, and analyze data** to demonstrate the suitability and effectiveness of the quality management system, including
a. Customer satisfaction
b. Conformance data
c. Trend data
d. Supplier data
5.5 **The organization shall continually improve** the effectiveness of the quality management system.

1.13.1 ISO 9001-Quality Management Systems of Requirements (Source: Adapted from the ISO 9001:2015 Standard, International Standards Organization, Geneva, Switzerland)

ASQ/ANSI/ISO 9001:2015

1.14 Quality Management Systems: Requirements

1.14.1 Scope

This International Standard specifies requirements for a quality management system when an organization:

(a) Needs to demonstrate its ability to consistently provide products and services that meet customer and applicable statutory and regulatory requirements.
(b) Aims to enhance customer satisfaction through the effective application of the system, including processes for improvement of the system and the assurance of conformity to customer and applicable statutory and regulatory requirements.

All the requirements of this International Standard are generic and are intended to be applicable to any organization, regardless of its type or size, or the products and services it provides.

NOTE 1:In this International Standard, the terms "product" or "service" only apply to products and services intended for, or required by, a customer.
NOTE 2: Statutory and regulatory requirements can be expressed as legal requirements.

1.14.2 Normative References

The following documents, in whole or in part, are normatively referenced in this document and are indispensable for its application. For dated references, only the edition cited applies. For undated references, the latest edition of the referenced document (including any amendments) applies.

ISO 9000:2015, Quality management systems—Fundamentals and vocabulary

1.14.3   Terms and Definitions

For the purposes of this document, the terms and definitions given in ISO 9000:2015 apply.

1.14.4   Context of the Organization

1.14.4.1   Understanding the Organization and Its Context

The organization shall determine external and internal issues that are relevant to its purpose and its strategic direction and that affect its ability to achieve the intended result(s) of its quality management system.

The organization shall monitor and review information about these external and internal issues.

NOTE 1: Issues can include positive and negative factors or conditions for consideration.

NOTE 2: Understanding the external context can be facilitated by considering issues arising from the legal, technological, competitive, market, cultural, social, and economic environments, whether international, national, regional, or local.

NOTE 3: Understanding the internal context can be facilitated by considering issues related to values, culture, knowledge, and performance of the organization.

1.14.4.2   Understanding the Needs and Expectations of Interested Parties

Due to their effect or potential effect on the organization's ability to consistently provide products and services that meet customer and applicable statutory and regulatory requirements, the organization shall determine:

(a) The interested parties that are relevant to the quality management system.
(b) The requirements of these interested parties that are relevant to the quality management system.

The organization shall monitor and review information about these interested parties and their relevant requirements.

1.14.4.3   Determining the Scope of the Quality Management System

The organization shall determine the boundaries and applicability of the quality management system to establish its scope.

When determining this scope, the organization shall consider:

(a) The external and internal issues referred to in Sect. 1.14.4.1.
(b) The requirements of relevant interested parties referred to in Sect. 1.14.4.2.
(c) The products and services of the organization.

The organization shall apply all the requirements of this International Standard if they are applicable within the determined scope of its quality management system. The scope of the organization's quality management system shall be available and be maintained as documented information. The scope shall state the types of products and services covered, and provide justification for any requirement of this International Standard that the organization determines is not applicable to the scope of its quality management system.

Conformity to this International Standard may only be claimed if the requirements determined as not being applicable do not affect the organization's ability or responsibility to ensure the conformity of its products and services and the enhancement of customer satisfaction.

1.14.4.4 Quality Management System and Its Processes

1. The organization shall establish, implement, maintain, and continually improve a quality management system, including the processes needed and their interactions, in accordance with the requirements of this International Standard.

   The organization shall determine the processes needed for the quality management system and their application throughout the organization, and shall:

   (a) Determine the inputs required and the outputs expected from these processes.
   (b) Determine the sequence and interaction of these processes.
   (c) Determine and apply the criteria and methods (including monitoring, measurements, and related performance indicators) needed to ensure the effective operation and control of these processes.
   (d) Determine the resources needed for these processes and ensure their availability.
   (e) Assign the responsibilities and authorities for these processes.
   (f) Address the risks and opportunities as determined in accordance with the requirements of Sect. 1.14.6.1.
   (g) Evaluate these processes and implement any changes needed to ensure that these processes achieve their intended results.
   (h) Improve the processes and the quality management system.

2. To the extent necessary, the organization shall:

   (a) Maintain documented information to support the operation of its processes.
   (b) Retain documented information to have confidence that the processes are being carried out as planned.

1.14.5 Leadership

1.14.5.1 Leadership and Commitment

General

Top management shall demonstrate leadership and commitment with respect to the quality management system by:

(a) Taking accountability for the effectiveness of the quality management system.
(b) Ensuring that the quality policy and quality objectives are established for the quality management system and are compatible with the context and strategic direction of the organization.
(c) Ensuring the integration of the quality management system requirements into the organization's business processes.
(d) Promoting the use of the process approach and risk-based thinking.
(e) Ensuring that the resources needed for the quality management system are available.
(f) Communicating the importance of effective quality management and of conforming to the quality management system requirements.
(g) Ensuring that the quality management system achieves its intended results.
(h) Engaging, directing, and supporting persons to contribute to the effectiveness of the quality management system.
(i) Promoting improvement.
(j) Supporting other relevant management roles to demonstrate their leadership as it applies to their areas of responsibility.

NOTE: Reference to "business" in this International Standard can be interpreted broadly to mean those activities that are core to the purposes of the organization's existence, whether the organization is public, private, for profit, or not for profit.

Customer Focus

Top management shall demonstrate leadership and commitment with respect to customer focus by ensuring that:

(a) Customer and applicable statutory and regulatory requirements are determined, understood, and consistently met.
(b) The risks and opportunities that can affect the conformity of products and services and the ability to enhance customer satisfaction are determined and addressed.
(c) The focus on enhancing customer satisfaction is maintained.

1.14.6   Policy

1.14.6.1   Establishing the Quality Policy

Top management shall establish, implement, and maintain a quality policy that:

(a) Is appropriate to the purpose and context of the organization and supports its strategic direction.
(b) Provides a framework for setting quality objectives.
(c) Includes a commitment to satisfy applicable requirements.
(d) Includes a commitment to continual improvement of the quality management system.

1.14.6.2   Communicating the Quality Policy

The quality policy shall:

(a) Be available and be maintained as documented information.
(b) Be communicated, understood, and applied within the organization.
(c) Be available to relevant interested parties, as appropriate.

1.14.6.3   Organizational Roles, Responsibilities, and Authorities

Top management shall ensure that the responsibilities and authorities for relevant roles are assigned, communicated, and understood within the organization.

Top management shall assign the responsibility and authority for:

(a) Ensuring that the quality management system conforms to the requirements of this International Standard.
(b) Ensuring that the processes are delivering their intended outputs.
(c) Reporting on the performance of the quality management system and on opportunities for improvement (see Sect. 1.14.11.1), in particular to top management.
(d) Ensuring the promotion of customer focus throughout the organization.
(e) Ensuring that the integrity of the quality management system is maintained when changes to the quality management system are planned and implemented.

1.14.7   *Planning*

1.14.7.1   Actions to Address Risks and Opportunities

1. When planning for the quality management system, the organization shall consider the issues referred to in Sect. 1.14.4.1 and the requirements referred to in Sect. 1.14.4.2 and determine the risks and opportunities that need to be addressed to:

   (a) Give assurance that the quality management system can achieve its intended result(s).
   (b) Enhance desirable effects.
   (c) Prevent, or reduce, undesired effects.
   (d) Achieve improvement.

2. The organization shall plan:

   (a) Actions to address these risks and opportunities.
   (b) How to:

      • Integrate and implement the actions into its quality management system processes (see Sect. 1.14.4.4).
      • Evaluate the effectiveness of these actions.

   Actions taken to address risks and opportunities shall be proportionate to the potential impact on the conformity of products and services.

NOTE 1: Options to address risks can include avoiding risk, taking risk in order to pursue an opportunity, eliminating the risk source, changing the likelihood or consequences, sharing the risk, or retaining risk by informed decision.

NOTE 2: Opportunities can lead to the adoption of new practices, launching new products, opening new markets, addressing new customers, building partnerships, and using new technology and other desirable and viable possibilities to address the organization's or its customers' needs.

1.14.7.2   Quality Objectives and Planning to Achieve Them

1. The organization shall establish quality objectives at relevant functions, levels, and processes needed for the quality management system.
   The quality objectives shall:

   (a) Be consistent with the quality policy.
   (b) Be measurable.
   (c) Consider applicable requirements.
   (d) Be relevant to the conformity of products and services and to the enhancement of customer satisfaction.
   (e) Be monitored.

(f) Be communicated.
(g) Be updated as appropriate.

 The organization shall maintain documented information on the quality objectives.

2. When planning how to achieve its quality objectives, the organization shall determine:

(a) What will be done.
(b) WHAT resources will be required.
(c) Who will be responsible.
(d) When it will be completed.
(e) How the results will be evaluated.

1.14.7.3   Planning of Changes

When the organization determines the need for changes to the quality management system, the changes shall be carried out in a planned manner (see Sect. 1.14.4.4).

 The organization shall consider:

(a) The purpose of the changes and their potential consequences.
(b) The integrity of the quality management system.
(c) The availability of resources.
(d) The allocation or reallocation of responsibilities and authorities.

1.14.8   Support

1.14.8.1   Resources

General

The organization shall determine and provide the resources needed for the establishment, implementation, maintenance, and continual improvement of the quality management system.

 The organization shall consider:

(a) The capabilities of, and constraints on, existing internal resources.
(b) What needs to be obtained from external providers.

People

The organization shall determine and provide the persons necessary for the effective implementation of its quality management system and for the operation and control of its processes.

Infrastructure

The organization shall determine, provide, and maintain the infrastructure necessary for the operation of its processes and to achieve conformity of products and services.
   NOTE: Infrastructure can include:

(a) Buildings and associated utilities.
(b) Equipment, including hardware and software.
(c) Transportation resources.
(d) Information and communication technology.

Environment for the Operation of Processes

The organization shall determine, provide, and maintain the environment necessary for the operation of its processes and to achieve conformity of products and services.
   NOTE: A suitable environment can be a combination of human and physical factors, such as:

(a) Social (e.g., nondiscriminatory, calm, nonconfrontational).
(b) Psychological (e.g., stress-reducing, burnout prevention, emotionally protective).
(c) Physical (e.g., temperature, heat, humidity, light, airflow, hygiene, noise).

   These factors can differ substantially depending on the products and services provided.

Monitoring and Measuring Resources

General

The organization shall determine and provide the resources needed to ensure valid and reliable results when monitoring or measuring is used to verify the conformity of products and services to requirements.
   The organization shall ensure that the resources provided:

(a) Are suitable for the specific type of monitoring and measurement activities being undertaken.
(b) Are maintained to ensure their continuing fitness for their purpose.

   The organization shall retain appropriate documented information as evidence of fitness for purpose of the monitoring and measurement resources.

Measurement Traceability

When measurement traceability is a requirement, or is considered by the organization to be an essential part of providing confidence in the validity of measurement results, measuring equipment shall be:

(a) Calibrated or verified, or both, at specified intervals, or prior to use, against measurement standards traceable to international or national measurement standards; when no such standards exist, the basis used for calibration or verification shall be retained as documented information.
(b) Identified in order to determine their status.
(c) Safeguarded from adjustments, damage, or deterioration that would invalidate the calibration status and subsequent measurement results.

The organization shall determine if the validity of previous measurement results has been adversely affected when measuring equipment is found to be unfit for its intended purpose, and shall take appropriate action as necessary.

Organizational Knowledge

The organization shall determine the knowledge necessary for the operation of its processes and to achieve conformity of products and services. This knowledge shall be maintained and be made available to the extent necessary. When addressing changing needs and trends, the organization shall consider its current knowledge and determine how to acquire or access any necessary additional knowledge and required updates.

NOTE 1: Organizational knowledge is knowledge specific to the organization; it is generally gained by experience. It is information that is used and shared to achieve the organization's objectives.
NOTE 2: Organizational knowledge can be based on:

(a) Internal sources (e.g., intellectual property; knowledge gained from experience; lessons learned from failures and successful projects; capturing and sharing undocumented knowledge and experience; the results of improvements in processes, products, and services).
(b) External sources (e.g., standards; academia; conferences; gathering knowledge from customers or external providers).

1.14.8.2 Competence

The organization shall:

(a) Determine the necessary competence of person(s) doing work under its control that affects the performance and effectiveness of the quality management system.
(b) Ensure that these persons are competent based on appropriate education, training, or experience.
(c) Where applicable, take actions to acquire the necessary competence, and evaluate the effectiveness of the actions taken.
(d) Retain appropriate documented information as evidence of competence.

NOTE: Applicable actions can include, for example, the provision of training to, the mentoring of, or the reassignment of currently employed persons; or the hiring or contracting of competent persons.

1.14.8.3 Awareness

The organization shall ensure that persons doing work under the organization's control are aware of:

(a) The quality policy.
(b) Relevant quality objectives.
(c) Their contribution to the effectiveness of the quality management system, including the benefits of improved performance.
(d) The implications of not conforming with the quality management system requirements.

1.14.8.4 Communication

The organization shall determine the internal and external communications relevant to the quality management system, including:

(a) On what it will communicate.
(b) When to communicate.
(c) With whom to communicate.
(d) How to communicate.
(e) Who communicates.

1.14.8.5   Documented Information

General

The organization's quality management system shall include:

(a) Documented information required by this International Standard.
(b) Documented information determined by the organization as being necessary for the effectiveness of the quality management system.

   NOTE: The extent of documented information for a quality management system can differ from one organization to another due to:

- The size of organization and its type of activities, processes, products, and services.
- The complexity of processes and their interactions.
- The competence of persons.

Creating and Updating

When creating and updating documented information, the organization shall ensure appropriate:

(a) Identification and description (e.g., a title, date, author, or reference number).
(b) Format (e.g., language, software version, graphics) and media (e.g., paper, electronic).
(c) Review and approval for suitability and adequacy.

Control of Documented Information

1. Documented information required by the quality management system and by this International Standard shall be controlled to ensure:

   (a) It is available and suitable for use, where and when it is needed.
   (b) It is adequately protected (e.g., from loss of confidentiality, improper use, or loss of integrity).

2. For the control of documented information, the organization shall address the following activities, as applicable:

   (a) Distribution, access, retrieval, and use.
   (b) Storage and preservation, including preservation of legibility.
   (c) Control of changes (e.g., version control).
   (d) Retention and disposition.

Documented information of external origin determined by the organization to be necessary for the planning and operation of the quality management system shall be identified as appropriate and be controlled. Documented information retained as evidence of conformity shall be protected from unintended alterations.

NOTE: Access can imply a decision regarding the permission to view the documented information only, or the permission and authority to view and change the documented information.

1.14.9   Operation

1.14.9.1   Operational Planning and Control

The organization shall plan, implement, and control the processes (see Sect. 1.14.4.4) needed to meet the requirements for the provision of products and services, and to implement the actions determined in Clause 6, by:

1. Determining the requirements for the products and services.
2. Establishing criteria for:

   (a) The processes.
   (b) The acceptance of products and services.

3. Determining the resources needed to achieve conformity to the product and service requirements.
4. Implementing control of the processes in accordance with the criteria.
5. Determining, maintaining, and retaining documented information to the extent necessary:

   (a) To have confidence that the processes have been carried out as planned.
   (b) To demonstrate the conformity of products and services to their requirements.

The output of this planning shall be suitable for the organization's operations. The organization shall control planned changes and review the consequences of unintended changes, taking action to mitigate any adverse effects, as necessary. The organization shall ensure that outsourced processes are controlled (see Sect. 1.14.9.4).

1.14.9.2   Requirements for Products and Services

Customer Communication

Communication with customers shall include:

(a) Providing information relating to products and services.
(b) Handling inquiries, contracts, or orders, including changes.

(c) Obtaining customer feedback relating to products and services, including customer complaints.
(d) Handling or controlling customer property.
(e) Establishing specific requirements for contingency actions, when relevant.

Determining the Requirements for Products and Services

When determining the requirements for the products and services to be offered to customers, the organization shall ensure that:

1. The requirements for the products and services are defined, including:

   (a) Any applicable statutory and regulatory requirements.
   (b) Those considered necessary by the organization.

2. The organization can meet the claims for the products and services it offers.

Review of the Requirements for Products and Services

1. The organization shall ensure that it could meet the requirements for products and services to be offered to customers. The organization shall conduct a review before committing to supply products and services to a customer, to include:

   (a) Requirements specified by the customer, including the requirements for delivery and postdelivery activities.
   (b) Requirements not stated by the customer, but necessary for the specified or intended use, when known.
   (c) Requirements specified by the organization.
   (d) Statutory and regulatory requirements applicable to the products and services.
   (e) Contract or order requirements differing from those previously expressed.

   The organization shall ensure that contract or order requirements differing from those previously defined are resolved. The customer's requirements shall be confirmed by the organization before acceptance, when the customer does not provide a documented statement of their requirements.

   NOTE: In some situations, such as internet sales, a formal review is impractical for each order. Instead, the review can cover relevant product information, such as catalogs.

2. The organization shall retain documented information, as applicable:

   (a) On the results of the review.
   (b) On any new requirements for the products and services.

Changes to Requirements for Products and Services

The organization shall ensure that relevant documented information is amended, and that relevant persons are made aware of the changed requirements, when the requirements for products and services are changed.

1.14.9.3   Design and Development of Products and Services

General

The organization shall establish, implement, and maintain a design and development process that is appropriate to ensure the subsequent provision of products and services.

Design and Development Planning

In determining the stages and controls for design and development, the organization shall consider:

(a) The nature, duration, and complexity of the design and development activities.
(b) The required process stages, including applicable design and development reviews.
(c) The required design and development verification and validation activities.
(d) The responsibilities and authorities involved in the design and development process.
(e) The internal and external resource needs for the design and development of products and services.
(f) The need to control interfaces between persons involved in the design and development process.
(g) The need for involvement of customers and users in the design and development process.
(h) The requirements for subsequent provision of products and services.
(i) The level of control expected for the design and development process by customers and other relevant interested parties.
(j) The documented information needed to demonstrate that design and development requirements have been met.

Design and Development Inputs

The organization shall determine the requirements essential for the specific types of products and services to be designed and developed. The organization shall consider:

(a) Functional and performance requirements.
(b) Information derived from previous similar design and development activities.
(c) Statutory and regulatory requirements.
(d) Standards or codes of practice that the organization has committed to implement.
(e) Potential consequences of failure due to the nature of the products and services.
(f) Inputs shall be adequate for design and development purposes, complete and unambiguous.

Conflicting design and development inputs shall be resolved.

The organization shall retain documented information on design and development inputs.

Design and Development Controls

The organization shall apply controls to the design and development process to ensure that:

(a) The results to be achieved are defined.
(b) Reviews are conducted to evaluate the ability of the results of design and development to meet requirements.
(c) Verification activities are conducted to ensure that the design and development outputs meet the input requirements.
(d) Validation activities are conducted to ensure that the resulting products and services meet the requirements for the specified application or intended use.
(e) Any necessary actions are taken on problems determined during the reviews, or verification and validation activities.
(f) Documented information of these activities is retained.

NOTE: Design and development reviews, verification, and validation have distinct purposes. They can be conducted separately or in any combination, as is suitable for the products and services of the organization.

Design and Development Outputs

The organization shall ensure that design and development outputs:

(a) Meet the input requirements.
(b) Are adequate for the subsequent processes for the provision of products and services.
(c) Include or reference monitoring and measuring requirements, as appropriate, and acceptance criteria.
(d) Specify the characteristics of the products and services that are essential for their intended purpose and their safe and proper provision.

The organization shall retain documented information on design and development outputs.

Design and Development Changes

The organization shall identify, review, and control changes made during, or after, the design and development of products and services, to the extent necessary to ensure that there is no adverse impact on conformity to requirements.

The organization shall retain documented information on:

(a) Design and development changes.
(b) The results of reviews.
(c) The authorization of the changes.
(d) The actions taken to prevent adverse impacts.

1.14.9.4   Control of Externally Provided Processes, Products, and Services

General

The organization shall ensure that externally provided processes, products, and services conform to requirements. The organization shall determine the controls to be applied to externally provided processes, products, and services when:

(a) Products and services from external providers are intended for incorporation into the organization's own products and services.
(b) Products and services are provided directly to the customer(s) by external providers on behalf of the organization.
(c) A process, or part of a process, is provided by an external provider because of a decision by the organization.

The organization shall determine and apply criteria for the evaluation, selection, monitoring of performance, and reevaluation of external providers, based on their ability to provide processes or products and services in accordance with requirements. The organization shall retain documented information of these activities and any necessary actions arising from the evaluations.

Type and Extent of Control

The organization shall ensure that externally provided processes, products, and services do not adversely affect the organization's ability to consistently deliver conforming products and services to its customers.

The organization shall:

1. Ensure that externally provided processes remain within the control of its quality management system.
2. Define both the controls that it intends to apply to an external provider and those it intends to apply to the resulting output.
3. Take into consideration:

   (a) The potential impact of the externally provided processes, products, and services on the organization's ability to consistently meet customer and applicable statutory and regulatory requirements.
   (b) The effectiveness of the controls applied by the external provider.

4. Determine the verification, or other activities, necessary to ensure that the externally provided processes, products, and services meet requirements.

Information for External Providers

The organization shall ensure the adequacy of requirements prior to their communication to the external provider.

The organization shall communicate to external providers its requirements for:

1. The processes, products, and services to be provided.
2. The approval of:

   (a) Products and services.
   (b) Methods, processes, and equipment.
   (c) The release of products and services.

3. Competence, including any required qualification of persons.
4. The external providers' interactions with the organization.
5. Control and monitoring of the external providers' performance to be applied by the organization.
6. Verification or validation activities that the organization, or its customer, intends to perform at the external providers' premises.

1.14.9.5   Production and Service Provision

Control of Production and Service Provision

The organization shall implement production and service provision under controlled conditions.

Controlled conditions shall include, as applicable:

1. The availability of documented information that defines:

   (a) The characteristics of the products to be produced, the services to be provided, or the activities to be performed.
   (b) The results to be achieved.

2. The availability and use of suitable monitoring and measuring resources.
3. The implementation of monitoring and measurement activities at appropriate stages to verify that criteria for control of processes or outputs, and acceptance criteria for products and services, have been met.
4. The use of suitable infrastructure and environment for the operation of processes.
5. The appointment of competent persons, including any required qualification.
6. The validation, and periodic revalidation, of the ability to achieve planned results of the processes for production and service provision, where the resulting output cannot be verified by subsequent monitoring or measurement.
7. The implementation of actions to prevent human error.
8. The implementation of release, delivery, and postdelivery activities.

Identification and Traceability

The organization shall use suitable means to identify outputs when it is necessary to ensure the conformity of products and services. The organization shall identify the status of outputs with respect to monitoring and measurement requirements throughout production and service provision. The organization shall control the unique identification of the outputs when traceability is a requirement and shall retain the documented information necessary to enable traceability.

Property Belonging to Customers or External Providers

The organization shall exercise care with property belonging to customers or external providers while it is under the organization's control or being used by the organization.

   The organization shall identify, verify, protect, and safeguard customers' or external providers' property provided for use or incorporation into the products and services. When the property of a customer or external provider is lost, damaged, or otherwise found to be unsuitable for use, the organization shall report this to the customer or external provider and retain documented information on what has occurred.

   NOTE: A customer's or external provider's property can include materials, components, tools and equipment, premises, intellectual property, and personal data.

Preservation

The organization shall preserve the outputs during production and service provision, to the extent necessary to ensure conformity to requirements.

NOTE: Preservation can include identification, handling, contamination control, packaging, storage, transmission or transportation, and protection.

Postdelivery Activities

The organization shall meet requirements for postdelivery activities associated with the products and services. In determining the extent of postdelivery activities that are required, the organization shall consider:

(a) Statutory and regulatory requirements.
(b) The potential undesired consequences associated with its products and services.
(c) The nature use and intended lifetime of its products and services.
(d) Customer requirements.
(e) Customer feedback.

NOTE: Postdelivery activities can include actions under warranty provisions, contractual obligations such as maintenance services, and supplementary services such as recycling or final disposal.

Control of Changes

The organization shall review and control changes for production or service provision, to the extent necessary to ensure continuing conformity with requirements. The organization shall retain documented information describing the results of the review of changes, the person(s) authorizing the change, and any necessary actions arising from the review.

1.14.9.6 Release of Products and Services

The organization shall implement planned arrangements, at appropriate stages, to verify that the product and service requirements have been met.

The release of products and services to the customer shall not proceed until the planned arrangements have been satisfactorily completed, unless otherwise approved by a relevant authority and, as applicable, by the customer.

The organization shall retain documented information on the release of products and services. The documented information shall include:

(a) Evidence of conformity with the acceptance criteria.
(b) Traceability to the person(s) authorizing the release.

Control of Nonconforming Outputs

1. The organization shall ensure that outputs that do not conform to its requirements are identified and controlled to prevent their unintended use or delivery. The organization shall take appropriate action based on the nature of the nonconformity and its effect on the conformity of products and services. This shall also apply to nonconforming products and services detected after delivery of products, during or after the provision of services.
   The organization shall deal with nonconforming outputs in one or more of the following ways:

   (a) Correction.
   (b) Segregation, containment, return, or suspension of provision of products and services.
   (c) Informing the customer.
   (d) Obtaining authorization for acceptance under concession.
       Conformity to the requirements shall be verified when nonconforming outputs are corrected.

2. The organization shall retain documented information that:

   (a) Describes the nonconformity.
   (b) Describes the actions taken.
   (c) Describes any concessions obtained.
   (d) Identifies the authority deciding the action in respect of the nonconformity.

1.14.10   Performance Evaluation

1.14.10.1   Monitoring, Measurement, Analysis, and Evaluation

General

The organization shall determine:

(a) What needs to be monitored and measured.
(b) The methods for monitoring, measurement, analysis, and evaluation needed to ensure valid results.
(c) When the monitoring and measuring shall be performed.
(d) When the results from monitoring and measurement shall be analyzed and evaluated.

The organization shall evaluate the performance and the effectiveness of the quality management system. The organization shall retain appropriate documented information as evidence of the results.

Customer Satisfaction

The organization shall monitor customers' perceptions of the degree to which their needs and expectations have been fulfilled. The organization shall determine the methods for obtaining, monitoring, and reviewing this information.

   NOTE: Examples of monitoring customer perceptions can include customer surveys, customer feedback on delivered products and services, meetings with customers, market-share analysis, compliments, warranty claims, and dealer reports.

Analysis and Evaluation

The organization shall analyze and evaluate appropriate data and information arising from monitoring and measurement.

   The results of the analysis shall be used to evaluate:

(a) Conformity of products and services.
(b) The degree of customer satisfaction.
(c) The performance and effectiveness of the quality management system.
(d) If planning has been implemented effectively.
(e) The effectiveness of actions taken to address risks and opportunities.
(f) The performance of external providers.
(g) The need for improvements to the quality management system.

   NOTE: Methods to analyze data can include statistical techniques.

1.14.10.2   Internal Audit

1. The organization shall conduct internal audits at planned intervals to provide information on whether the quality management system:

   (a) Conforms to:

      • The organization's own requirements for its quality management system.
      • The requirements of this International Standard.

   (b) Is effectively implemented and maintained.

2. The organization shall:

   (a) Plan, establish, implement, and maintain an audit program(s) including the frequency, methods, responsibilities, planning requirements, and reporting, which shall take into consideration the importance of the processes concerned, changes affecting the organization, and the results of previous audits.
   (b) Define the audit criteria and scope for each audit.

(c) Select auditors and conduct audits to ensure objectivity and the impartiality of the audit process.
(d) Ensure that the results of the audits are reported to relevant management.
(e) Take appropriate correction and corrective actions without undue delay.
(f) Retain documented information as evidence of the implementation of the audit program and the audit results.

NOTE: See ISO 19011 for guidance.

1.14.10.3   Management Review

General

Top management shall review the organization's quality management system, at planned intervals, to ensure its continuing suitability, adequacy, effectiveness, and alignment with the strategic direction of the organization.

Management Review Inputs

The management review shall be planned and carried out taking into consideration:

1. The status of actions from previous management reviews.
2. Changes in external and internal issues that are relevant to the quality management system.
3. Information on the performance and effectiveness of the quality management system, including trends in:

    (a) Customer satisfaction and feedback from relevant interested parties.
    (b) The extent to which quality objectives have been met.
    (c) Process performance and conformity of products and services.
    (d) Nonconformities and corrective actions.
    (e) Monitoring and measurement results.
    (f) Audit results.
    (g) The performance of external providers.

4. The adequacy of resources.
5. The effectiveness of actions taken to address risks and opportunities (see Sect. 1.14.6.1).
6. Opportunities for improvement.

Management Review Outputs

The outputs of the management review shall include decisions and actions related to:

1. Opportunities for improvement.
2. Any need for changes to the quality management system.
3. Resource needs.

   The organization shall retain documented information as evidence of the results of management reviews.

1.14.11   Improvement

1.14.11.1   General

The organization shall determine and select opportunities for improvement and implement any necessary actions to meet customer requirements and enhance customer satisfaction.
   These shall include:

(a) Improving products and services to meet requirements as well as to address future needs and expectations.
(b) Correcting, preventing, or reducing undesired effects.
(c) Improving the performance and effectiveness of the quality management system.

   NOTE: Examples of improvement can include correction, corrective action, continual improvement, breakthrough change, innovation, and reorganization.

1.14.11.2   Nonconformity and Corrective Action

1. When a nonconformity occurs, including any arising from complaints, the organization shall:

   (a) React to the nonconformity and, as applicable:

   • Take action to control and correct it.
   • Deal with the consequences.

   (b) Evaluate the need for action to eliminate the cause(s) of the nonconformity, in order that it does not recur or occur elsewhere, by:

   • Reviewing and analyzing the nonconformity.
   • Determining the causes of the nonconformity.
   • Determining if similar nonconformities exist or could potentially occur.

(c) Implement any action needed.

(d) Review the effectiveness of any corrective action taken.

(e) Update risks and opportunities determined during planning, if necessary.

(f) Make changes to the quality management system, if necessary.

Corrective actions shall be appropriate to the effects of the nonconformities encountered.

2. The organization shall retain documented information as evidence of:

(a) The nature of the nonconformities and any subsequent actions taken.

(b) The results of any corrective action.

1.14.11.3   Continual Improvement

The organization shall continually improve the suitability, adequacy, and effectiveness of the quality management system. Figure 1.10 shows how the PDCA applies for the continuous improvement scenario.

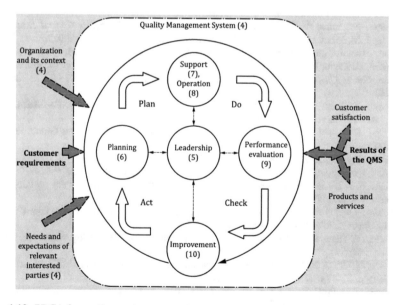

Fig. 1.10 PDCA for continuous improvement process

Bibliography

D. Anton, C. Anton, *ISO 9001 Survival Guide*, 3rd edn. (AEM Consulting Group, Inc, Ashland, 2006), p. 100. ISBN 978-0-9672170-8-6

R.E. Barlow, T.Z. Irony, Foundations of statistical quality control, in *Current Issues in Statistical Inference: Essays in Honor of D. Basu*, ed. by M. Ghosh, P. K. Pathak, (Hayward, CA, Institute of Mathematical Statistics, 1992), pp. 99–112

B. Bergman, Conceptualistic pragmatism: A framework for Bayesian analysis? IIE Trans. **41**, 86–93 (2009)

R. Breu, A. Kuntzmann-Combelles, M. Felderer, New perspectives on software quality. IEEE Softw. **31**, 32–38 (2014)

P. Crosby, *Quality Is Free* (McGraw-Hill, New York, 1979) ISBN 0-07-014512-1

W.E. Deming, On probability as a basis for action. Am. Stat. **29**(4), 146–152 (1975)

W.E. Deming, *Out of the Crisis: Quality, Productivity and Competitive Position*. ISBN 0-521-30553-5 (Cambridge, Cambridge University Press, 1982)

F. Dobb, *ISO 9001:2000 Quality Registration Step-by-Step*, 3rd edn. (Butterworth-Heinemann, Burlington, MA, 2004), p. 292. ISBN 978-0-7506-4949

P. Drucker, *Innovation and Entrepreneurship* (Harper & Row, New York, 1985) ISBN 978-0-06-091360-1

J.R. Evans, Introduction to Statistical Process Control Archived October 29, 2013, at the Wayback Machine, Fundamentals of Statistical Process Control (1994)

D. Garvin, Competing on the eight dimensions of quality. Harv. Bus. Rev. **65**, 101–109 (1987)

A.B. Godfrey, *Juran's Quality Handbook* (McGraw-Hill, New York, 1999) ISBN 0-07-034003-X

E.L. Grant, *Statistical Quality Control* (McGraw-Hill, New York, 1946) ISBN 0071004475

D. Hoyle, *ISO 9000 Quality Systems Handbook*, 5th edn. (Butterworth-Heinemann, Burlington, MA, 2005), p. 686. ISBN 978-0-7506-6785-2

ISO 9001 Auditing Practices Group., committee.iso.org

J.M. Juran, Appendix IV Quality systems terminology, in *Juran's Quality Control Handbook*, ed. by J. M. Juran, (McGraw-Hill, New York, 1988), pp. 2–3. ISBN 0-07-033176-6

N. Kano, Attractive quality and must-be quality. J. Jpn Soc. Quality Control **14**, 39–48 (1984)

N. Majcen, P. Taylor (eds.), *Practical Examples on Traceability, Measurement Uncertainty and Validation in Chemistry*, vol 1 (European Commission, Belgium. ISBN 978-92-79-12021-3, 2010)

A. Monnappa, Pioneers of Project Management: Deming vs Juran vs Crosby. Certified Quality Engineer Certification Preparation - ASQ. asq.org. Accessed 2 Oct 2018

J. Oakland, *Statistical Process Control*. ISBN 0-7506-5766-9 (Taylor & Francis Group, London, 2002)

Process Quality Engineer. automotiveengineeringhq.com

T. Pyzdek, *Quality Engineering Handbook* (M. Dekker, New York, 2003) ISBN 0-8247-4614-7

T. Salacinski, *SPC - Statistical Process Control* (The Warsaw University of Technology Publishing House, Warsaw, 2015) ISBN 978-83-7814-319-2

L. Stebbing, *Quality Assurance: The Route to Efficiency and Competitiveness*, 3rd edn. (Prentice Hall, Upper Saddle River, NJ, 1993), p. 300. ISBN 978-0-13-334559-9

M. Thareja, P. Thareja, The quality brilliance through brilliant people. Quality World **4**(2) (2007) SSRN 1498550

R. Tricker, *ISO 9001:2000 Audit Procedures*, 2nd edn. (Butterworth-Heinemann, Burlington, MA, 2005a), p. 320. ISBN 978-0-7506-6615-2

R. Tricker, *ISO 9001:2000 for Small Businesses* (Butterworth-Heinemann, Burlington, MA, 2005b), p. 480. ISBN 978-0-7506-6617-6

R. Tricker, B. Sherring-Lucas, *ISO 9001:2008 in Brief*, 2nd edn. (Butterworth-Heinemann, Burlington, MA, 2005), p. 192. ISBN 978-0-7506-6616-9

M. Walton, W. Edwards Deming, *The Deming Management Method* (Perigee, New York, 1988), p. 88. ISBN 0-399-55000-3

D.J. Wheeler, *Understanding Variation: The Key to Managing Chaos*, 2nd edn. (SPC Press, Knoxville, TN, 1999) ISBN 0-945320-53-1

D.J. Wheeler, D.S. Chambers, *Understanding Statistical Process Control* (SPC Press, Knoxville, TN, 1992) ISBN 0-945320-13-2

S.A. Wise, D.C. Fair, *Innovative Control Charting: Practical SPC Solutions for Today's Manufacturing Environment* (ASQ Quality Press, Milwaukee, WI, 1998) ISBN 0-87389-385-9

S.L. Zabell, Predicting the unpredictable. Synthese **90**(2), 205 (1992). https://doi.org/10.1007/bf00485351

Part II
Acceptance Sampling and Statistical Process Control (SPC)

Chapter 2
Engineering Designs of Acceptance Sampling

2.1 Introduction

Acceptance sampling in its contemporary industrial form originates in the early 1940s. It was purposeful initially by the US military to test the functionality of bullets during World War II. Harold Dodge industrialized the notion and methodology of acceptance sampling. Dodge was an experienced quality assurance personnel in the Bell Laboratories department.

While the bullet's functionalities had to be tested, the necessity for speed (the time of testing) was extremely crucial. At this time, Dodge had given logical and essential comments. He said that tacking samples might perhaps make verdicts about entire lots (bullets) at random. Harry Romig came up with the idea of determining sample size, the number of acceptable defectives (or defects), and other essential testing criteria of sampling. Thus, he suggested that the sampling plan can be used as a standard.

Acceptance sampling procedures became known throughout World War II and afterward. Industrialized throughout World War II as a swift fix for manufacturing, acceptance sampling should not substitute more systemic acceptance quality control methods. Dodge himself distinguished in 1969 that acceptance sampling is not the identical concept as acceptance quality control. Reliant on a specific sampling plan (procedure), it smears precise lots and is a rapid, short-term test (a spot check). In contrast, acceptance quality control is put on in a long-term quality inspection plan for the entire product line—acceptance quality control purposes as an essential part of manufacturing planning.

There may be a high volume (or number) of items to inspect at a reasonable cost or within a reasonable time frame. Furthermore, the test (for instance, the hardness test) may affect the product's quality. The company cannot test every one of its products. So, adequate sampling is a must practice for manufacturers. Thus, acceptance sampling is a statistical quality-control tool that allows the firm to determine the quality of products in the lot by testing randomly selected samples. If we apply

Y. Y. Tesfay, *Developing Structured Procedural and Methodological Engineering Designs*, https://doi.org/10.1007/978-3-030-68402-0_2

the best practices and implement them correctly, acceptance sampling is a useful quality control tool.

Acceptance sampling resolves these problems by testing a representative sample of the product from the lot. The acceptance sampling process involves determining the size of the lot to be tested, the number of items (products) to be sampled, and the number of defectives (nonconformance or defect) acceptable within the sample lot.

An essential concept in the sampling process is randomization. That process allows the company to quantify (measure) the lot's quality with a specified degree of statistical accuracy without having to inspect every single unit. That is, items from the lot must be selected at random for sampling. The process typically happens at the manufacturing site (the plant or factory). Just beforehand, the items move to the subsequent stage.

Based on how many of the predetermined numbers of samples (pass or fail) the inspecting, the inspector decides whether to accept or reject the entire lot. The sample's statistical trustworthiness is confirmed by performing the t-statistic. The t-statistic is a type of inferential statistic used to determine if there is a significant difference among two groups that share standard features.

In general, sampling is used to reduce cost, time, and resources. Acceptance sampling is habitually used to regulate and determine the disposition of outgoing and incoming material. An operative and effective acceptance sampling plan will let a firm set and continue with having adequate quality assurance standard whereas acquiring materials from outside suppliers and materials ship to the customers. A sufficient sample should be taken to determine the lot's status (i.e., the specification requirements). Then, we should test to verify the condition according to some standards. The result of the inspection determines the acceptance of the lot or not. Habitually, the verdict is either to "reject" or "accept" the lot. If the lot is rejected, it may be reworked, scraped, or returned to the supplier. Conversely, if the lot is accepted, it will move to the next steps.

2.2   Foundations of Acceptance Sampling

Acceptance sampling is a practice whereby the sample is inspected, and a decision to "reject" or "accept" that the entire population (lot) is based on the test results of the sample. There are several definitions given for acceptance sampling. The most important definitions are given in Table 2.1.

The acceptance control chart (ACC) syndicates consideration of control consequences with the elements of acceptance sampling. ACC is a suitable tool for serving to make verdicts concerning process acceptance. The difference between acceptance sampling (AS) and ACC methods is that the ACC emphasizes acceptability and capability. In contrast, acceptance sampling (AS) emphasizes product disposition verdict.

Table 2.1 Essential foundations of Acceptance Sampling

Dodge and Romig	Acceptance sampling is an essential arena of statistical quality control introduced by Dodge and Romig. Initially, acceptance sampling was applied by the U.S. military to test bullets from the lot during World War II. Dodge and Romig reason that if each bullet is tested in advance, no bullets would be left to ship and distribute the soldiers. Conversely, suppose none of the bullets were tested. In that case, failures of the functionality of the bullet might happen in the battlefield, with potentially disastrous results. In conclusion, Dodge and Romig suggested applying the acceptance sampling technique to tackle these significant and risky military challenges.
Definition of Acceptance Sampling	Dodge reasoned that the sample should be selected at random from the lot. Based on the sample's information, an understandable decision must be made concerning the lot's disposition. In general, the verdict is either to "accept" or "reject" the lot. This process is called Acceptance Sampling.
Attributes (i.e., defect counting) and variables (measurable quality parameters)	Acceptance sampling is an effective approach between "no inspection" and "100% inspection." There are two significant classifications of acceptance plans. One is by attributes (for instance, "go" or "no-go"), and the other is by variables (measurable quality parameters). The attribute case is the utmost communal for acceptance sampling.
Objective	The goal of acceptance sampling is to bring scientific evidence to accept or reject the lot.
Scenarios	Acceptance sampling is employed when one or several of the following hold: • Testing is destructive (damages the functionality of the item) • 100% inspection is expensive • 100% inspection takes much time
AQC vs. AS	In 1969, Harold Dodge pointed out that Acceptance Quality Control (AQC) is not the same concept as Acceptance Sampling (AS). The latter is subject to a definite sampling plan, which, when executed, specify the pre-conditions for decision (i.e., acceptance or rejection) of the lot. That is being tested—the implemented in the form of an Acceptance Control Chart (ACC). The ACC's control limits are obtained from the requirement specification limits and the process standard deviation (see Ryan 2000 for details).
Thoughts of Harold Dodge	In 1942, Dodge stated that acceptance quality encompasses the concept of protecting the consumer from getting an unacceptable defective product. That encourages the producer in the use of

(continued)

Table 2.1 (continued)

	process quality control. The varying quantity and severity of acceptance inspections are directly related to the importance of the characteristics inspected, in inverse relation to the quality level's goodness as indicated by the inspections. To repeat the difference in these two methodologies: acceptance sampling plans are one-shot deals, which virtually test short-run effects. Quality control is of the long run variety and is part of a well-designed for lot acceptance.
Comment of Schilling	Schilling (1989) said that each sampling plan has a considerable effect on alone sniper (a gunman). In contrast, the sampling plan system can deliver a bombardment in the encounter for quality improvement.
Control charts of acceptance	According to the ISO (ISO 7870-3:2020), the acceptance control chart is, naturally, used when the process variable is normally distributed. Nevertheless, it can be functional to nonnormally distributed process variables. The ISO 7870-3:2020 document demonstrates various conditions in which the acceptance control chart technique has essential applications. The acceptance sampling details of sample size calculations, process action limits, and different decision criteria.

2.3   Sample Size Calculation

Sampling is a technique used in a statistical study. A predetermined (planned) number of data points (observations, called sample points) are taken from the population. The technique (method or model) was applied to the sample from the population to the analysis.

Precisely, sampling is an application of well-defined techniques, queries, and actions to take a fewer object that reflects the whole population's essential characteristics.

In other words, sampling is a scientific procedure of selecting an efficient and effective small number of objects from the population. The characteristics of the sample taken should enable us to infer the essential parameters of the population.

As shown in Fig. 2.1, the goal of sampling is to get a sample that is best to infer the population characteristics.

Sampling and sample size calculations are essential topics in engineering. That is due to:

- Observations collected on the manufacturing floor are continuously measured to compile process data in the future.
- Counting and testing the population is costly and consumes much time.

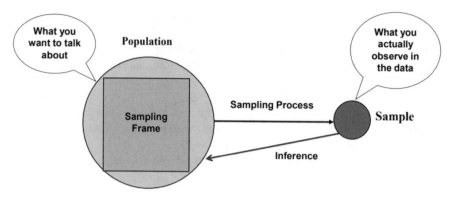

Fig. 2.1 The goal of sampling

- In manufacturing, data is necessary for engineering decisions. Nevertheless, considering the entire population for engineering purposes at an instant is impossible. In that case, the mere solution will be taking an adequate sample for engineering analysis.
- Sampling assists in applying a variety of research designs.
- Quantifying a population parameter could be difficult (impossible), whereas a statistic from the sample is easy to quantify.
- In manufacturing, inspection, material review board, and then scrap are all the functions and consequences of sampling (see the flow charts in the Annex).
- The sampling process encompasses integrated several essential steps. Let us see the steps for sampling for the simple random sampling design as follows.

Step 1: Determine population size (N)

An intermediary point between the population and the sample drawn for the research is called a sampling frame. The first step is defining and describing the population of interest (called the target population). The population is a set of all objects (items, parts, people, etc.).

Step 2: Determine the sampling frame

The sampling frame is a list of objects that we want to draw a sample. Population size (N) denotes the total number of objects that we are emphasized.

Step 3: Setting Sampling Criteria

- Check if an appropriate sampling frame is used or not. If the sampling frame is unsuitable, the result may be sampling bias.
- Check if the device of measurement is calibrated.
- Check if the right population is being sampled.
- Apply restriction criteria.

- Apply the indeterminacy principle. The sample should be taken under the same conditions as all the observations.

Step 4: Determine the margin of error (Effect size, e)

The margin of error or confidence interval denotes the amount of error we let our result wish to allow in the result.

Step 5: Determine the confidence level $(1 - \alpha)$

Confidence level value measures the degree of certainty concerning how to fit the sample characterizes and represents the overall population the margin of error.

Step 6: Determine the power of the test $(1 - \beta)$

A Power Test $(1 - \beta)$ helps adjust the sample size required for the data collection to have sufficient power.

Step 7: Specify the standard of deviation (σ)

The standard deviation designates how much variation we expect among the responses.

If the standard deviation is unknown, we use the proportion value $p = 0.5$ (50%).

Step 8: Set Z-score

The Z-score is a value set based on the confidence level.

Step 9: Sample size formula

- A small population size (N)

$$n = N \left[\frac{\frac{\left(Z_{a/2}+Z_{1-\beta}\right)^2 \sigma^2}{e^2}}{N - 1 + \frac{\left(Z_{a/2}+Z_{1-\beta}\right)^2 \sigma^2}{e^2}} \right]$$

- A large population size

$$n = \frac{\left(Z_{a/2} + Z_{1-\beta}\right)^2 \sigma^2}{e^2}$$

- Sample size using proportion from small population of size N

$$n = N \left[\frac{\frac{\left(Z_{a/2}+Z_{1-\beta}\right)^2 p(1-p)}{e^2}}{N - 1 + \frac{\left(Z_{a/2}+Z_{1-\beta}\right)^2 p(1-p)}{e^2}} \right]$$

Sample size using proportion from large population size

$$n = \frac{\left(Z_{a/2} + Z_{1-\beta}\right)^2 p(1 - p)}{e^2}$$

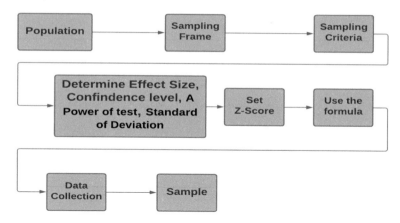

Fig. 2.2 Steps of sampling

Next, once we determine the sample size, we will collect the data to identify the data point to be included in the sample. Then, we will have a sample. Figure 2.2 gives information about the steps of sampling.

However, there are also important notices that we should keep in mind that sampling may cause wrong conclusions. That is when:

- The sampling had a preliminary design.
- The sampling process may have a bias.
- Insufficient experience in the population characteristics.

2.4   Some Examples of Sample Size Calculations

Consider the Drawing in Fig. 2.3.

In the drawing, we have three critical dimensions

- Dimension 1 ($D1$): Target $= 0.25$ mm, LSL $= 0.249$ mm, USL $= 0.25$ mm, Tolerance $= 0.001$.
- Dimension 2 ($D2$): Target $= 0.096$ mm, LSL $= 0.093$ mm, USL $= 0.099$ mm, Tolerance $= 0.006$.
- Dimension 3 ($D3$): Target $= 0.005$ mm, LSL $= 0.005$ mm, USL $= 0.010$ mm, Tolerance $= 0.005$ (Table 2.2).

The question is, how many sample points we need to inspect?

Although the inspection results suggested that all the inspected parts are confirming, there is process variation. So, the variation is the essential factor to decide the sample size for the inspection. In this scenario, since the process is ongoing, we can apply sample size determination for large data.

First, let us calculate the sample standard deviations of these critical dimensions as follows.

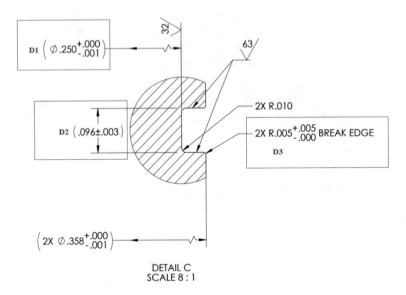

DETAIL C
SCALE 8 : 1

Fig. 2.3 Drawing dimensions and specifications

Table 2.2 Gives the inspection data of these three critical dimensions

Sample point	D1	Status	D2	Status	D3	Status
1	0.250	Accepted	0.096	Accepted	0.005	Accepted
2	0.249	Accepted	0.094	Accepted	0.005	Accepted
3	0.250	Accepted	0.097	Accepted	0.008	Accepted
4	0.249	Accepted	0.093	Accepted	0.006	Accepted
5	0.250	Accepted	0.099	Accepted	0.007	Accepted
6	0.249	Accepted	0.097	Accepted	0.006	Accepted
7	0.249	Accepted	0.094	Accepted	0.008	Accepted
8	0.250	Accepted	0.096	Accepted	0.008	Accepted
9	0.249	Accepted	0.096	Accepted	0.007	Accepted
10	0.250	Accepted	0.098	Accepted	0.010	Accepted
11	0.249	Accepted	0.097	Accepted	0.010	Accepted
12	0.250	Accepted	0.098	Accepted	0.007	Accepted
13	0.249	Accepted	0.096	Accepted	0.005	Accepted
14	0.249	Accepted	0.098	Accepted	0.008	Accepted
15	0.250	Accepted	0.095	Accepted	0.010	Accepted
16	0.249	Accepted	0.095	Accepted	0.006	Accepted
17	0.249	Accepted	0.097	Accepted	0.009	Accepted
18	0.249	Accepted	0.094	Accepted	0.009	Accepted
19	0.248	Accepted	0.095	Accepted	0.006	Accepted
20	0.250	Accepted	0.094	Accepted	0.008	Accepted

	D1	D2	D3
Sample standard deviation	0.0005	0.0017	0.0017

The process variability for $D2$ and $D3$ is more significant than $D1$. However, the tolerance of $D1$ is tighter than $D2$ and $D3$. Therefore, we need to take the sample standard deviation ratios to effect size (tolerance).

	D1	D2	D3
Sample standard deviation	0.0005	0.0017	0.0017
Effect size (tolerance)	0.001	0.006	0.005
Ratio of s to ES	0.495	0.281	0.332

Apply the formula and calculate the sample size at $\alpha = 5\%$.

	D1	D2	D3
Standard deviation	0.0005	0.0017	0.0017
Effect size	0.001	0.006	0.005
Ratio of s to ES	0.495	0.281	0.332
$n = \dfrac{\left(Z_{a/2}\right)^2 \sigma^2}{e^2}$	0.940	0.303	0.423
Sample size	1	0	0

The sample size calculation suggests that the process needs inspection of 1 part is enough. However, the above sample size does not include the power (Type II error). So, we need to adjust the sample size calculation by including power. Let us set $\beta = 10\%$, i.e., Power $= 90\%$.

	D1	D2	D3
Standard deviation	0.0005	0.0017	0.0017
Effect size	0.001	0.006	0.005
Ratio of s to ES	0.495	0.281	0.332
	0.940	0.303	0.423
Sample size	1	1	1
$n = \dfrac{\left(Z_{a/2}+Z_{1-\beta}\right)^2 \sigma^2}{e^2}$	3.178	1.026	1.431
Sample size adjusted to power	3	1	1

After adjusting the power to 90%, the sample size calculation suggested that inspecting three quality assurance parts will be enough. Figure 2.4 also gives the same result.

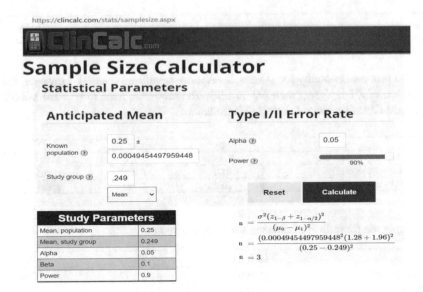

Fig. 2.4 Sample size calculation using ClinCalc.com, which is available online at https://clincalc.com/stats/samplesize.aspx

2.5 Essential Concepts of Sampling

There are three significant categories of sampling concepts. The first sampling concepts include topics such as the operating characteristic (OC) curve, producer and consumer risk (CR), acceptance quality level (AQL), lot tolerance percent defective (LTPD), average outgoing quality (AOQ), and average outgoing quality limit (AOQL).

The second sampling one is various sampling standards. These include assessing ANSI/ASQ Z1.4 and Z1.9 standards, explaining sampling plans (single, double, multiple, sequential, and continuous), and assessing Dodge–Romig sampling tables. We will learn the sampling plan in the following chapter.

The third one is sample integrity, which includes creating and maintaining sample integrity when performing the sampling plan. That also will be discussed in the next chapter.

2.5.1 First Sampling Concept

The first sampling concept is divided into three subsections. There are

1. Operating characteristic (OC) curves
2. Producer and consumer risk (PCR)
3. Terms and definitions

These three subsections are interconnected and are introductory to the concept of acceptance sampling. We will start first with the operating characteristics (OC) curve then, which will lead us to consider risks (producer risk and consumer risk) related to the sampling. Then finally, the related terms of AQL, AOQ, LTPD, and AOQL.

2.5.2 Operating Characteristics Curve (OC Curve)

The operating characteristic (OC) curve describes the planned (proposed) acceptance sampling's discriminatory and inequitable power. The OC curve is a graphic representation of the appropriateness of a specific acceptance sampling plan. The curve displays the probabilities of "accepting" a lot, inspected against the fraction defective. Thus, statistically, the probability tells us how much our sampling is realistic to decide on the lot's acceptance or rejection.

Figure 2.5 is an illustration of a generic OC curve. The Y-axis is the likelihood of acceptance (P_a), and the X-axis is percentage defective, which is dependent on the incoming quality level (p).

The OC curve helps us to (visually) assess and evaluate the sampling plan and comprehend the likelihood of accepting a lot of variable quality levels.

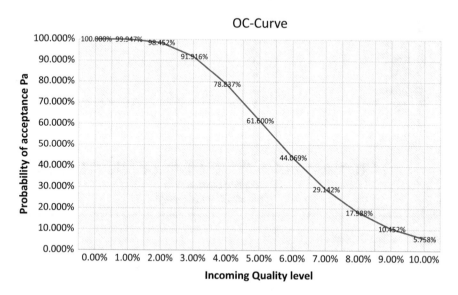

Fig. 2.5 Operating characteristics (OC) curve

2.5.2.1 Operating Characteristics Curve (OC Curve) for a Perfect Sampling Plan

To comprehend how the OC curve functions and its relationship to the risks within acceptance sampling, we must understand the perfect OC curve. Suppose we buy 10,000 resistors from an electronic supplier. Further, assume our acceptance quality level (AQL) is set at 4%. The perfect (ideal) sampling plan is to accept all lots if the lots contain at most 4% of nonconformance (defective) parts (resistors). Otherwise, the lot will be rejected. See the perfect OC curve in Fig. 2.6.

Figure 2.6 shows that the likelihood of acceptance is 100% for all incoming quality levels (IQL) of less than 4% nonconforming. That means 0% for all incoming quality levels (IQL) of more than 3% nonconforming. Consequently, we will accept the incoming lot whose quality level is less than 4% nonconforming. Alternatively, we will reject all lots whose incoming quality level is more than 4%.

Nevertheless, the sampling process does not work that way. Every so often, we accept lots, which have inadequate qualities.

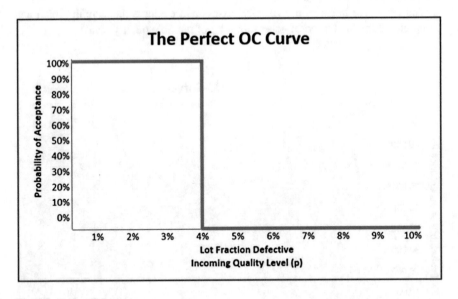

Fig. 2.6 Perfect OC curve

2.5.2.2 Generating an OC Curve

Given various incoming quality levels (p), the OC curves are made by calculating the probability of acceptance (P_a). Most acceptance sampling plans use attribute (discrete) data. Therefore, binominal distribution is the best to model the data and to build the probability of acceptance. Suppose p is the probability of success in each trial, and we have n-successive trials (sample size), then the binomial distribution is given as:

$$\Pr(X = x; p, n) = \binom{n}{x} p^x (1 - p)^{n-x}$$

where $x = 0, 1, 2\ldots, n$ and X is a binomial random variable.

After the inspection, items inspected are categorized into accepted (conformance) or rejected (nonconformance). Here, we assume the probability of an item being defected is fixed and does not change from inspection to inspection. Therefore, the commutative distribution of the binomial distribution, which is less than or equal to the c-number of defectives (nonconformances), is given.

$$\Pr(x \leq C) = \sum_{x=0}^{C} \binom{n}{x} p^x (1 - p)^{n-x}$$

where

d is the number of nonconformances items observed during the inspection
c is the acceptance number obtained from the sampling plan
n is the size of the lot
p is the incoming quality level

Let us see the following example to illustrate the illustration OC curve example using the binomial distribution.

Suppose our sampling plan is to inspect 200 parts. The supplier is accepted if only the nonconformance rate is at a maximum of 1%. That is, the minimum conformance rate is 99%. Since the maximum conformance rate is at most 1%, when we multiply with the lot size (i.e., 200), we have $200 \times 0.01 = 2$.

Acceptance number (c) is the highest number of nonconforming (defective) units allowed for the lot to be accepted.

Suppose we assign the acceptance number of the lot is 5. That means the lot is accepted if it contains five or fewer number of nonconforming items.

Table 2.3 Calculation of cumulative probabilities for higher IQL

Probability of acceptance (P_a) (%)	$\Pr(x \leq 5) = \sum\limits_{x=0}^{1} \binom{100}{x} P_a{}^x (1 - P_a)^{n-x}$ (%)
0.00	100
1.00	98.398
2.00	78.672
3.00	44.323
4.00	18.565
5.00	6.234
6.00	1.772
7.00	0.442
8.00	0.099
9.00	0.020
10.00	0.004

$$
\begin{aligned}
\Pr(x \leq 5) &= \sum_{x=0}^{1} \binom{200}{x} 0.01^x (0.99)^{200-x} \\
&= \binom{200}{0} 0.01^0 (0.99)^{200} + \binom{200}{1} 0.01^1 (0.99)^{199} \\
&\quad + \binom{200}{2} 0.01^2 (0.99)^{198} + \binom{200}{3} 0.01^3 (0.99)^{197} \\
&\quad + \binom{200}{4} 0.01^4 (0.99)^{196} + \binom{200}{5} 0.01^5 (0.99)^{195} \\
&= 0.99398 \\
&= 98.398\%
\end{aligned}
$$

Table 2.3 gives all the cumulative probability for higher incoming quality levels (IQL) for $p = 2\%, 3\%, \ldots, 10\%$.

The OC curve is given in Fig. 2.7.

2.5.2.3   The Poisson Approximation of a Binomial Distribution

As the sample size n is sufficiently large and the probability of success sufficiently small, it is mathematically proved that the Poisson distribution best approximates a binominal distribution. That is

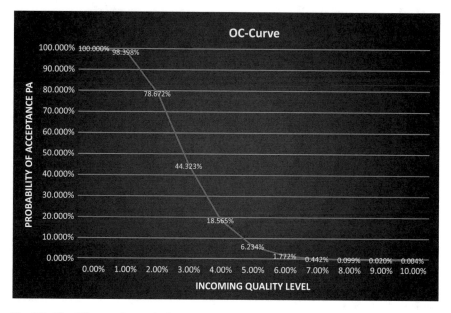

Fig. 2.7   The OC curve for purchasing 200 items from an external supplier

$$\lim_{n \to \infty} [\Pr(X = x; p, n)] \tilde{P}(x; np) = P(x; \lambda)$$

where $np = \lambda$

For instance, if the probability of failure (defect, nonconformance) is at most 10%, we can reasonably approximate the Poisson one's binomial distribution.

The primary benefit of such approximation is that the Poisson distribution is easier for calculation than the binomial distribution. That is often preferred as the calculations are easier, and the Poisson tables are simpler to use.

Using the Poisson distribution, we can use the following equation to compute the probability of acceptance. Suppose λ is the mean of the success, Poisson distribution is given as:

$$P(X = x; \lambda) = \frac{e^{-\lambda} \lambda^x}{x!}$$

where X is a Poisson random variable and $x = 0, 1, 2, \ldots$.
   Where:

- $\lambda = np$ is the mean of the Poisson distribution, which is the product of lot size multiplied by the incoming quality level (IQL)
- X is the number of nonconformances (defectives) revealed in the sample.

We are intended to calculate the likelihood that an accurately "x" number of nonconformances are revealed in the sample consisting of n items. The probability of

Table 2.4 Binomial distribution and the Poisson approximation

Incoming quality (%)	$\lambda = np$	Actual probability calculation using binomial distribution	Probability calculation using Poisson approximation
		Probability of acceptance (%)	Probability of acceptance (%)
0	0	100	100
1	0.6	87.7	87.81
2	1.2	66	66.26
3	1.8	45.7	46.28
4	2.4	30.1	30.84
5	3	19.1	19.91
6	3.6	11.7	12.57
7	4.2	7.1	7.80
8	4.8	4.2	4.77
9	5.4	2.4	2.89
10	6	1.4	1.74

incoming quality level (IQL) is valued as p, which is the ratio of nonconforming items to the lot size.

Now, let us apply the technique of approximating the binomial distribution by the Poisson distribution.

The lot size 60 (i.e., $n = 60$).

The probability of nonconforming items in the lot being inspected is 1%.

The parameter of the Poisson distribution $\lambda = np = 60 \times 0.01 = 0.6$.

Therefore, the Poisson distribution is given as:

$$P(X = x; \lambda) = \frac{e^{-0.6}0.6^x}{x!}$$

where X is a Poisson random variable and $x = 0, 1, 2. \ldots$

Now, we are intended to calculate the commutative probability when $c = 1$.

$$\sum_{x=0}^{1} P(X = x; \lambda) = \frac{e^{-0.6}0.6^0}{0!} + \frac{e^{-0.6}0.6^1}{1!} = 0.8781 = 87.81\%$$

In Table 2.4, we will compare the binomial distribution and the Poisson approximation.

Table 2.4 shows that the binomial distribution is best approximated by the Poisson distribution when p is small. Furthermore, if we try for a large sample size, the approximation will be much better.

2.5.3 Critical Parameters of Acceptance Sampling

Two critical parameters describe the acceptance sampling (AS) plan and their accompanying OC curve. These are:

1. The acceptance number (c) is a planned quantity that tells the maximum allowable (tolerable) quantity of nonconformance items.
2. The sample size (n) is the number of sample points to be inspected.

Furthermore, a rejection number (r) is also considered. The rejection number (r) is the lowest number of nonconformance (defects or defective) items in the given sample that will cause (reason) to the rejection of the lot characterized by the sample. The rejection number (r) should be acceptance number (c) plus one nonconformance item for the single sampling planning methods. That is, $r = c + 1$. However, if the sampling plan is double or multiple, then the rejection number (r) will vary.

The number of nonconformances (x, also symbolized by d) is also considered the acceptance sampling plan parameter. If the actual number of nonconformances (x) is greater than the acceptance number (c), the lot will automatically be rejected. If the actual number of nonconformances (x) is less than the acceptance number (c), the lot will not be rejected.

2.5.3.1 OC Curve with Variable Acceptance Number (c)

When the acceptance number (c) changes, then the geometrical shape (feature) of the OC curve will be changed. So, studying the changes in the geometry of the curve is essential for planning acceptance sampling.

Let us illustrate the concept of varying acceptance number (c) as follows. Suppose we fix the sample size ($n = 100$, which is the lot size) and only change the acceptance number (four varying acceptance numbers, i.e., $c = 0$, $c = 1$, $c = 2$, $c = 3$, $c = 4$, $c = 5$, $c = 6$). Suppose we consider the probabilities of acceptance (P_a) as 1%, 2%, 3%, 4%, 5%, 6%, 7%, 8%, 9%, and 10%. Using binomial distribution we have:

$$\Pr(x \le c) = \sum_{x=0}^{c} \binom{100}{x} p^x (1 - p)^{100-x}$$

Then, we have the following table of acceptance probabilities.

In Table 2.5, we observe that the probability of acceptance increases with the acceptance numbers.

Let us analyze the incoming quality level (IQL) of 10%, where 10 out of 100 units within the lot are nonconforming. The likelihood of sampling 100 companies from the lot and zero nonconformance ($c = 0$ plan) is 0.003%. The result suggests that it is unlikely to find such a lot. However, the likelihood of sampling 100 units from the

Table 2.5 Calculation of acceptance probabilities

Probability of acceptance (%)	Acceptance number						
	$c = 0$	$c = 1$	$c = 2$	$c = 3$	$c = 4$	$c = 5$	$c = 6$
0.00	100%	100%	100%	100%	100%	100%	100%
1.00	36.603%	73.576%	92.063%	98.163%	99.657%	99.947%	99.993%
2.00	13.262%	40.327%	67.669%	85.896%	94.917%	98.452%	99.594%
3.00	4.755%	19.462%	41.978%	64.725%	81.785%	91.916%	96.877%
4.00	1.687%	8.716%	23.214%	42.948%	62.886%	78.837%	89.361%
5.00	0.592%	3.708%	11.826%	25.784%	43.598%	61.600%	76.601%
6.00	0.205%	1.517%	5.661%	14.302%	27.678%	44.069%	60.635%
7.00	0.071%	0.601%	2.579%	7.441%	16.316%	29.142%	44.428%
8.00	0.024%	0.232%	1.127%	3.671%	9.034%	17.988%	30.316%
9.00	0.008%	0.087%	0.476%	1.730%	4.739%	10.452%	19.398%
10.00	0.003%	0.032%	0.194%	0.784%	2.371%	5.758%	11.716%

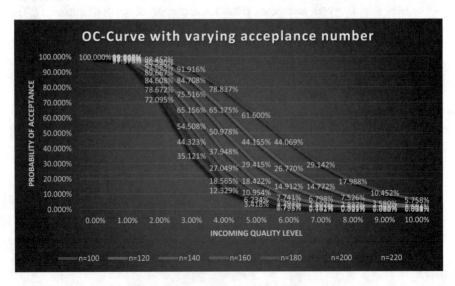

Fig. 2.8 The shape of the OC curves when the acceptance number varies

lot and six nonconformance (the $c = 6$ plan) is 11.716%. The result suggests that the chance of getting such a lot is 1 out of 10, which is likely.

Using Table 2.5, we have the following OC curve for varying acceptance numbers.

Here we learn that the lower acceptance number has a steeper OC curve and are likely to reject incoming lots due to unacceptable quality standard. That means if we are too tight (i.e., a low tolerance) for quality, it is difficult to achieve the required quality standard. Logically, this statement fits with reality.

In Fig. 2.8, we can observe that as the value of the acceptance number increases, the OC curve getting stretch out. As clearly seen, changing the acceptance number will affect accepting the lot. Having the acceptance number of zero ($c = 0$) is a much precipitous OC curve than the other OC curves associated with more significant acceptance numbers (i.e., $c = 1$, $c = 2$, ..., $c = 6$).

2.5.3.2 The OC Curve and Varying Sample Size

Here we learn that the lower acceptance number has a steeper OC curve and are likely to reject incoming lots due to unacceptable quality standard. That means if we are too tight (i.e., a low tolerance) for quality, it is difficult to achieve the required quality standard. Logically, this statement fits with reality.

Let us illustrate the concept of varying lot size (n) as follows. Suppose we fix the acceptance number ($c = 5$) and only change lot size (four varying lot sizes, i.e., $n = 100$, $n = 120$, $n = 140$, $n = 160$, $n = 180$, $n = 200$, and $n = 220$). Suppose we consider the probabilities of acceptance (P_a) as 1%, 2%, 3%, 4%, 5%, 6%, 7%, 8%, 9%, and 10%. Using binomial distribution, i.e.,

$$\Pr(x \leq 5) = \sum_{x=0}^{5} \binom{n}{x} p^x (1 - p)^{n-x}$$

Then, we have the following table of acceptance probabilities.

In Table 2.6, we observe that the probability of acceptance decreases with the lot size.

Let us analyze the incoming quality level (IQL) of 5%, where 5 out of 100 units within the lot are nonconforming with fixed acceptance number $c = 5$. The likelihood of sampling 100 units from the lot and that results in five nonconformance ($c = 5$ plan) is 61.6%. The result suggests that it is very likely to find such a lot.

Table 2.6 Calculation of acceptance probabilities

Probability Acceptance	Lot size						
	n=100	n=120	n=140	n=160	n=180	n=200	n=220
0.00%	100%	100%	100%	100%	100%	100%	100%
1.00%	99.947%	99.862%	99.700%	99.426%	99.004%	98.398%	97.576%
2.00%	98.452%	96.590%	93.683%	89.667%	84.608%	78.672%	72.095%
3.00%	91.916%	84.708%	75.516%	65.156%	54.508%	44.323%	35.121%
4.00%	78.837%	65.175%	50.978%	37.948%	27.049%	18.565%	12.329%
5.00%	61.600%	44.155%	29.415%	18.422%	10.954%	6.234%	3.418%
6.00%	44.069%	26.770%	14.912%	7.741%	3.793%	1.772%	0.795%
7.00%	29.142%	14.772%	6.798%	2.897%	1.161%	0.442%	0.161%
8.00%	17.988%	7.526%	2.837%	0.986%	0.321%	0.099%	0.029%
9.00%	10.452%	3.580%	1.098%	0.310%	0.082%	0.020%	0.005%
10.00%	5.758%	1.604%	0.398%	0.091%	0.019%	0.004%	0.001%

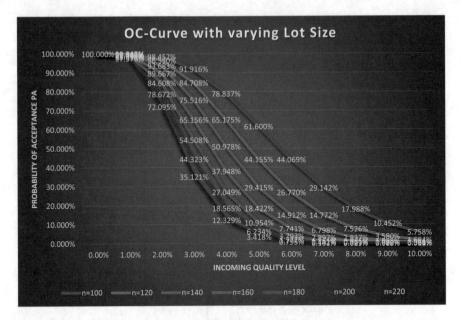

Fig. 2.9 The shape of the OC curves when the lot size (n) varies

However, the likelihood of sampling 220 units from the lot and that results in five nonconformance (the $c = 5$ plan) is 3.418%. The result suggests that the chance of getting such a lot is improbable (approximately 1 out of 30). As the lot size increases, the probability of acceptance decreases.

Using Table 2.6, we have the following OC curve for varying lot sizes (Fig. 2.9).

The acceptance sampling plan for this illustration has an acceptance number of 1. That means if the actual number of nonconformances found during sampling is two or more, we will reject the whole lot.

Let us analyze the probability of acceptance for $n = 100$ and $n = 220$ at the 5% incoming quality level.

- When $n = 100$, from Table 2.4, we observe that the probability of accepting the lot with 5% nonconformances is 74%, approximately 3 out of 4.
- When $n = 220$, from Table 2.4, we observe that the probability of accepting the lot with 5% nonconformances is very low, at 3.418%, which is approximately 1 out of 30.

From the above example, we have learned that sampling more units improves the sampling plan's ability to differentiate between high-quality incoming lots and bad ones.

2.5.4 Risks of Acceptance Sampling

The core objective of sampling is selecting fewer objects from the population to predict the population itself. The sampling design, including the excellent representative sample points, and bias are the challenges of an effective and efficient sample.

2.5.4.1 Sampling risk

Sampling risk is a risk that the sampler collects the sample data and make inferences (decision, or conclusion) about the population is wrong.

Sampling risk is a tool to analyze the risk (quantified in terms of probability). The sampler collects the sample data and makes inferences (decision, conclusion) about the population is wrong. Precisely, sampling risk is the probability that the objects selected by the sampler cannot characterize the population. Sampling risk permanently occurs when the sampler (inspector) does the inspection based on the sample information than the population.

Nevertheless, there may be a circumstance that the inaccurate inferences (extrapolations, implications, or decisions) about a population. The inference from the sample may not be adequate due to the sample may not represent the population. Furthermore, the inspector error or the sampling procedure may not be suitable.

2.5.4.2 Risks Related to Acceptance Sampling

Inferential statistics is a subsection of statistical techniques that infers from a representative sample the population's essential characteristic (parameters).

Definition 1 A parameter is a quantification (measure) of the characteristics of the population.

The following are examples of parameters:

(a) Population mean (μ), standard deviation (σ), population proportion (p).
(b) In regression equation $y_i = \beta_0 + x_{1i}\beta_1 + x_{2i}\beta_2 + \ldots + x_{ki}\beta_k + \varepsilon_i$, where $\varepsilon_i \sim iidN$ $(0, \sigma^2)$, regression coefficients ($\beta_0, \beta_1, \beta_2, \ldots, \beta_k$) and the variance of the random error term σ^2.
(c) In Poisson distribution, i.e., $P(X = x; \lambda) = \frac{e^{-\lambda}\lambda^x}{x!}$, the population mean λ is the parameter.

Definition 2 A statistic is a quantification (measure) description of the characteristics of the sample.

The statistics of the sample are used to estimate the population. In contrast, the parameters are the real values that characterize the population.

The following are examples of statistic

(a) Sample mean ($\bar{x}$), Standard Deviation ($\hat{\sigma}$), Population Proportion ($\hat{p}$)
(b) In regression equation $y_i = \beta_0 + x_{1i}\beta_1 + x_{2i}\beta_2 + \ldots + x_{ki}\beta_k + \varepsilon_i$, where $\varepsilon_i \sim iidN$
    $(0, \sigma^2)$, estimates of regression coefficients $(\hat{\beta}_0, \hat{\beta}_1, \hat{\beta}_2, \ldots, \hat{\beta}_k)$ and the estimate of
    the variance of the random error term $\hat{\sigma}^2$.
(c) In Poisson distribution, i.e., $P(X = x; \lambda) = \frac{e^{-\lambda}\lambda^x}{x!}$, the sample mean $\hat{\lambda}$ is the
    statistic.

Definition 3 Inferential statistics help predict (estimate or infer) the population characteristics (parameters) from the representative sample.

Inferential statistics allow us to make predictions about the population using a small portion of the population. Inferential statistics are classified into two significant subsections. These are:

1. Estimation: Estimation is a variety of statistical techniques that estimate the population parameter from the sample statistic. The estimation can be point estimates and confidence intervals.

   Examples:

   (a) Maximum likelihood estimation (MLE) by maximizing the probability of the sample.
   (b) Generalized method of moments (GMM) estimation by satisfying the moment conditions.
   (c) Least squares estimation (LSE) by minimizing the sum square of Euclidian distances.

2. Hypothesis testing: Hypothesis testing is a statistical technique used to analyze (test) whether the hypothesis presumed about the population from the sample is statistically consistent or not.

   Example: In a regression equation $y_i = \beta_0 + x_{1i}\beta_1 + x_{2i}\beta_2 + \ldots + x_{ki}\beta_k + \varepsilon_i$, where $\varepsilon_i \sim iidN(0, \sigma^2)$. Our hypothesis can:

   > Null hypothesis (H$_0$): The regression coefficients $(\hat{\beta}_0, \hat{\beta}_1, \hat{\beta}_2, \ldots, \hat{\beta}_k)$ are all zero
   > Against
   > The Alternative hypothesis (H$_1$): Some of the regression coefficients $(\hat{\beta}_0, \hat{\beta}_1, \hat{\beta}_2, \ldots, \hat{\beta}_k)$ are not zero.

Like other statistical techniques, acceptance sampling is also inferential statistics. We take representative data points from the population to make the sample and predict the population's essential characteristics using the sample information. Acceptance sampling works by taking data to form the sample and making inferences about the entire population.

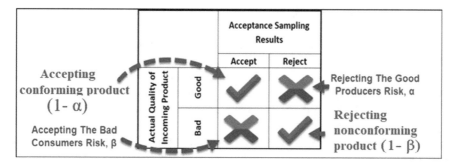

Fig. 2.10 Essential risks of acceptance sampling

Therefore, like regression analysis, ANOVA (or other inferential statistics), there is a probability that the sampling may lead to a wrong conclusion about the population.

As it is seen in Fig. 2.10, producer risk (α) and consumer risk (β) represent the errors in acceptance sampling due to the acceptance sampling. That is used when the product moves from the producer (manufacturer) to the consumer (buyer).

Recall the goal of acceptance sampling, accept conforming (good) product, and reject the nonconforming (bad) product. Therefore, in Fig. 2.10, we observe that two out of the four outcomes (results) are correct (adequately performed). Producer risk (PR) and consumer risk (CR) rise once we make the wrong decisions.

- Producers' risk is the likelihood of rejecting the lot whose correct quality level meets (or exceeds) the required, acceptable quality level (AQL). Typically, the producer risk is represented as α (alpha, the probability of Type I Error). Usually, producer risk (α) is set to be 5% (0.05).
- Consumers' risk is the likelihood of accepting the lot whose correct quality level does not meet the required lot tolerance percent defective (LTPD). Typically, consumer risk is represented as β (the probability of Type II Error) and set to 10% (0.10).
- Minimizing α maximizes the rejection probability (rate) of the lots at the customer. In contrast, maximizing α minimizes the rejection probability (rate) of the lots at the customer. So, a good producer should maximize α to minimize $(1 - \alpha)$.
- Minimizing β maximizes the rejection probability (rate) of the lots at the customer, whereas maximizing β minimizes the lots' rejection probability after the customer receives it. So, strict quality customer maximizes β to minimize $(1 - \beta)$.
- Given any sampling plan, the risks (i.e., the CR and PR) related to that plan can be observed in the OC curve along with their relation to the acceptable quality level (AQL) and lot tolerance percent defective (LTPD) levels.

2.5.4.3　Nonsampling Risk

Any risk that the inspector performs a wrong decision that is not caused due to (related to) sampling risk is called a nonsampling risk. It may happen due to inspectors use unsuitable audit procedures (or mistakenly interpret the audit result). Usually, the nonsampling risk occurs due to the lack of skills, knowledge, and ability in doing the audit. Henceforth, education, appropriate training, and suitable supervision can control such risks.

In the subsequent subsection, we will define essential terms in acceptance sampling.

2.5.5　Essential Terms in Acceptance Sampling

The most common terms in acceptance sampling are acceptance quality limit (AQL), lot tolerance percent defective (LTPD), average outgoing quality (AOQ), and average outgoing quality limit (AOQL). Let us define each term as follows.

2.5.5.1　Acceptance Quality Limit (AQL)

AQL was initially being introduced in the military sector, which denotes the acceptable quality level. Acceptance quality limit (AQL) is the least (most deficient) quality level in the process, which is acceptable for the product from the given lot size (sample).

When we select the acceptance quality limit (AQL) level for the sampling plan, the AQL number gives essential information about the acceptance number (c). The smaller AQL number implies a lower acceptance rate. The sampling plan recommends a high-quality level and increases the probability of rejecting the lot. Equivalently, if we increase the AQL number, the effect will be the opposite. The following are essential points about AQL.

- AQL describes the quantity of the poorest tolerable (acceptable) process quality. That is, the AQL reflects the degree of acceptance of the process output.
- Some people may think that AQL characterizes the required quality level for the production process being sampled. However, that is a big mistake! AQL is about the required quality acceptance level for the production process of being sampled.
- AQL is the connotation of the acceptable level (c) of quality. Acceptance quality limit (AQL) is an applicable term in all the primary acceptance sampling guideline standards, including the American National Standards Institute (ANSI), American Society for Quality (ASQ) Z1.4, International Organization for Standardization (ISO) 2859-1, and The American Society of Mechanical Engineers (ASTME) 2234.

- AQL quantified in terms of the percentage of nonconforming material. For example, if we say an AQL of 5.0. It means the customer allows the supplier at a maximum of 5% (5 out of 100 parts) nonconformance of the given lot.
- For the given sampling plan, setting the AQL must be based on considering: the sampling risk, the quality attribute criticality, and the criticality of nonconformances (i.e., critical, major, minor).
- During the inspection process, the product is inspected based on multiple quality characteristics (parameters). With various failure modes, we can set (specify) unique AQL levels for distinct nonconformance types.

2.5.5.2 Lot Tolerance Percent Defective (LTPD)

Lot tolerance percent defective (LTPD) is another well-known metric in acceptance sampling. LTPD of the sampling plan describes the level of quality normally rejected. Usually, the LTPD is described as the percentage of defectives. The sampling plan rejects the given lot at 90% of the time. The LTPD is the percentage of defective having a 10% (100–90%) likelihood of acceptance employing the operational (accepted) sampling plan. The LTPD has a tight likelihood of acceptance.

LTPD is associated with consumer risk (β) and characteristically tells that the percent defective is related to a 10% likelihood of acceptance. During the inspection, if the lot does not meet (fails) the sampling plan's requirement, with a 90% confidence level, we can say that the lot's quality level is not as good as the LTPD (i.e., the failure rate of the lot is greater than the LTPD). However, suppose the lot meets (passes) the sampling plan, then with a 90% confidence level, in that case, we can say that the lot's quality level is equal to (or better than) the LTPD.

Consider the following sampling plan. Suppose the producer risk sets the AQL at 5% (a 95% probability of acceptance and a 5% probability of rejection). Furthermore, consumer risk (CR) sets the LTPD at 10% (that is, a 10% probability of acceptance and a 90% probability of rejection).

In Fig. 2.11, the OC curve is intended to illustrate the consumer risk (CR) and producer risk (PR) based on the selected sampling plan via considering the AQL and LTPD levels.

Figure 2.11 shows that the AQL related to the selected sampling plan brings a percent defective incoming quality level (IQL) of 2%. The LTPD for the same sampling plan is 8%.

2.5.5.3 Average Outgoing Quality (AOQ)

Average outgoing quality (AOQ) is another essential concept in acceptance sampling. AOQ is the expected value (average or mean) of the quality levels of the shipping (outgoing) product to the customer at a given value of receiving (incoming) product quality level. The AOQ plot shows the association between the product

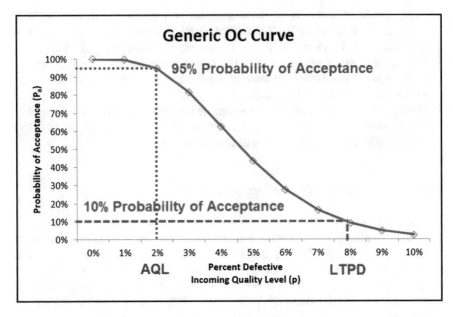

Fig. 2.11 Generic OC curve via considering the AQL and LTPD levels

quality of shipping (outgoing) and receiving (incoming). The simple logic for the AOQ analysis is that if the receiving (incoming) lot is "good," then the shipping (outgoing) quality of the lot will be "good."

First, let us focus on illustrating what do we mean by outgoing quality. If a lot is accepted through sampling, it is converted into an "outgoing" product. Similarly, suppose a lot gets rejected and 100% sorted. In that case, that lot will then be released as an outgoing product as soon as the sorting is completed. The quality of that outgoing product is what we mean when we say outgoing quality. That is quantified in the percentage of nonconforming product. The concept of AOQ assumes that any lot that gets rejected will be 100% sorted, and a conforming unit will replace any nonconformances. If we work off this assumption, we can calculate the average outgoing quality for any sampling plan.

AOQ quantifies the average outgoing quality of the given sampling plan (in percentage nonconforming). Subsequently, lots have been "accepted" or "rejected" and then 100% sorted. Let us consider the example in Fig. 2.12.

In Fig. 2.12, we observe that on the OC curve, there are three points (i.e., A, B, and C). Let us consider the three points as follows on the OC curve for an instance of AOQ. As seen in Fig. 2.12:

- At point A, the incoming lot has satisfactory (good) quality, 95% acceptance (only 5% nonconforming). Subsequently, there is a high likelihood that many lots will be "accepted" and considered an outgoing item (part, product, or material). Once the lot is rejected, then the lot will be sorted and then unrestricted as an

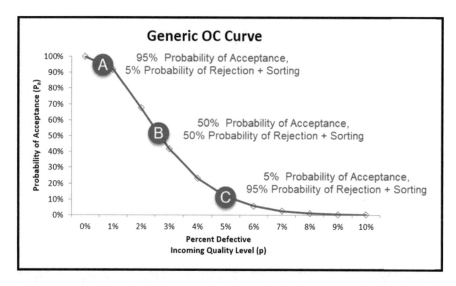

Fig. 2.12 Illustration of average outgoing quality (AOQ)

outgoing item. The high quality of the incoming lot cause to have the outgoing quality is also high.

- At point B, the incoming lot has partial quality. That is, the probability of the lot will be accepted is 50%. When the lot is accepted, it is considered as an outgoing item. The other rejected lot will be sorted and then unrestricted as an outgoing item. Here, the AOQ tends towards being the poorest. That is due to the sampling plan is to accept the lot with half chance of rejection.
- At point C, the incoming lot has the most insufficient quality. That leads the lot to be rejected and then sorted 100%. Therefore, the sorted lot has a high probability of an outgoing quality level.

Suppose P_a is the probability of acceptance, and p is the incoming quality level (IQL). Average outgoing quality (AOQ) is defined as the product of the probability of acceptance (P_a) and incoming quality level (p).

That is

$$AOQ = P_a p$$

To illustrate the above formula, let us consider Table 2.7.

The resulting graph is presented in Fig. 2.13.

Figure 2.13 shows that the higher the incoming quality level (that is, a low percent defective rate), the AOQ also becomes the higher (that is, a low percent defective). The AOQ is maximum when the incoming quality level (IQL) is partial. As we accept lots of partial (marginal) quality, we do sort for the rejected lot. The AOQ recovers as the incoming quality level (IQL) declines since the incoming lots that get rejected will be sorted 100%.

Table 2.7 Calculation of AOQ

Incoming quality level (p)	Probability of acceptance (P_a)	$AOQ = p_a p$
0%	100%	0.00%
1%	92%	0.92%
2%	68%	1.36%
3%	42%	1.26%
4%	23%	0.92%
5%	12%	0.60%
6%	6%	0.36%
7%	3%	0.21%
8%	1%	0.08%
9%	0%	0.00%
10%	0%	0.00%

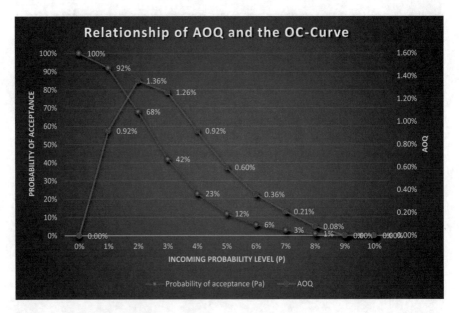

Fig. 2.13 OC curve and AOQ

2.5.5.4 Average Outgoing Quality Limit (AOQL)

The idea of an average outgoing quality limit (AOQL) is the extension of thought of average outgoing quality (AOQ). The average outgoing quality limit (AOQL) signifies the maximum percentage of defective (or the bottommost quality, which will be accepted during the inspection when testing the sample) in the outgoing lot.

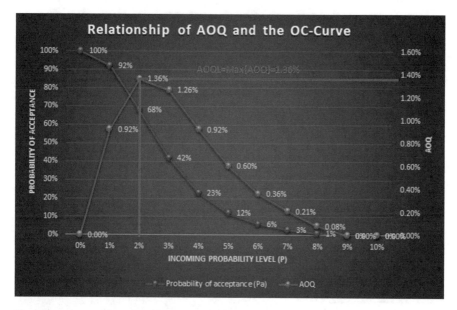

Fig. 2.14 Average outgoing quality limit (AOQL) example

The naturalistic inquiry that arises when looking at an AOQ graph is.

Therefore, the AOQL gives a solution to the question about what is the possible most deficient outgoing quality level related to the sampling plan? For instance, from Table 2.7, the maximum AOQ of 1.36%, obtained at an incoming quality level of 2%. The probability of acceptance of 68%. Graphically, the result is shown in Fig. 2.14.

In the instance above, the maximum of AOQ is 1.36% outgoing quality. That AOQL happens as an incoming quality level of 2% nonconforming.

Annex

Incoming Inspection Flow Chart

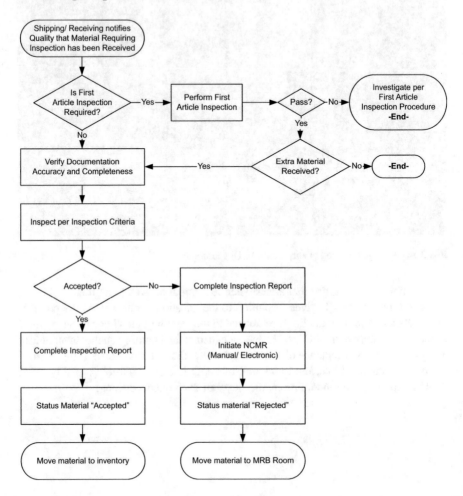

Disposition of Nonconforming Material Flow Chart

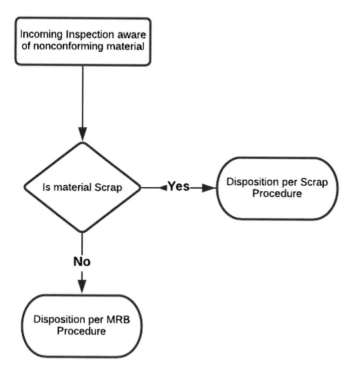

Scrap Ticket Flow Chart

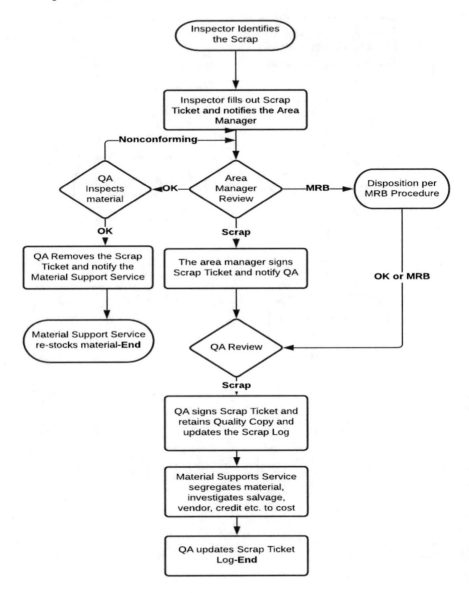

Material Review Board (MRB) Flow Chart

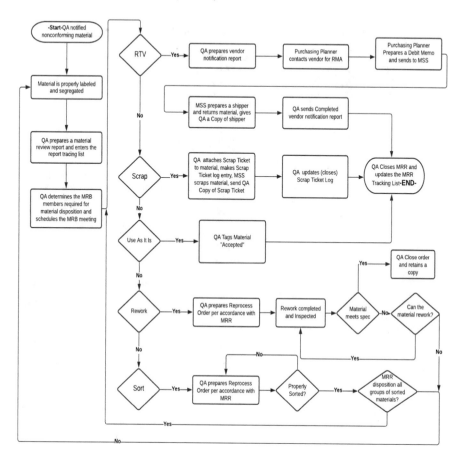

Normal probability Table

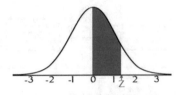

STANDARD NORMAL TABLE (Z)

Entries in the table give the area under the curve between the mean and z standard deviations above the mean. For example, for $z = 1.25$ the area under the curve between the mean (0) and z is 0.3944.

z	0.00	0.01	0.02	0.03	0.04	0.05	0.06	0.07	0.08	0.09
0.0	0.0000	0.0040	0.0080	0.0120	0.0160	0.0190	0.0239	0.0279	0.0319	0.0359
0.1	0.0398	0.0438	0.0478	0.0517	0.0557	0.0596	0.0636	0.0675	0.0714	0.0753
0.2	0.0793	0.0832	0.0871	0.0910	0.0948	0.0987	0.1026	0.1064	0.1103	0.1141
0.3	0.1179	0.1217	0.1255	0.1293	0.1331	0.1368	0.1406	0.1443	0.1480	0.1517
0.4	0.1554	0.1591	0.1628	0.1664	0.1700	0.1736	0.1772	0.1808	0.1844	0.1879
0.5	0.1915	0.1950	0.1985	0.2019	0.2054	0.2088	0.2123	0.2157	0.2190	0.2224
0.6	0.2257	0.2291	0.2324	0.2357	0.2389	0.2422	0.2454	0.2486	0.2517	0.2549
0.7	0.2580	0.2611	0.2642	0.2673	0.2704	0.2734	0.2764	0.2794	0.2823	0.2852
0.8	0.2881	0.2910	0.2939	0.2969	0.2995	0.3023	0.3051	0.3078	0.3106	0.3133
0.9	0.3159	0.3186	0.3212	0.3238	0.3264	0.3289	0.3315	0.3340	0.3365	0.3389
1.0	0.3413	0.3438	0.3461	0.3485	0.3508	0.3513	0.3554	0.3577	0.3529	0.3621
1.1	0.3643	0.3665	0.3686	0.3708	0.3729	0.3749	0.3770	0.3790	0.3810	0.3830
1.2	0.3849	0.3869	0.3888	0.3907	0.3925	0.3944	0.3962	0.3980	0.3997	0.4015
1.3	0.4032	0.4049	0.4066	0.4082	0.4099	0.4115	0.4131	0.4147	0.4162	0.4177
1.4	0.4192	0.4207	0.4222	0.4236	0.4251	0.4265	0.4279	0.4292	0.4306	0.4319
1.5	0.4332	0.4345	0.4357	0.4370	0.4382	0.4394	0.4406	0.4418	0.4429	0.4441
1.6	0.4452	0.4463	0.4474	0.4484	0.4495	0.4505	0.4515	0.4525	0.4535	0.4545
1.7	0.4554	0.4564	0.4573	0.4582	0.4591	0.4599	0.4608	0.4616	0.4625	0.4633
1.8	0.4641	0.4649	0.4656	0.4664	0.4671	0.4678	0.4686	0.4693	0.4699	0.4706
1.9	0.4713	0.4719	0.4726	0.4732	0.4738	0.4744	0.4750	0.4756	0.4761	0.4767
2.0	0.4772	0.4778	0.4783	0.4788	0.4793	0.4798	0.4803	0.4808	0.4812	0.4817
2.1	0.4821	0.4826	0.4830	0.4834	0.4838	0.4842	0.4846	0.4850	0.4854	0.4857
2.2	0.4861	0.4864	0.4868	0.4871	0.4875	0.4878	0.4881	0.4884	0.4887	0.4890
2.3	0.4893	0.4896	0.4898	0.4901	0.4904	0.4906	0.4909	0.4911	0.4913	0.4916
2.4	0.4918	0.4920	0.4922	0.4925	0.4927	0.4929	0.4931	0.4932	0.4934	0.4936
2.5	0.4938	0.4940	0.4941	0.4943	0.4945	0.4946	0.4948	0.4949	0.4951	0.4952
2.6	0.4953	0.4955	0.4956	0.4957	0.4959	0.4960	0.4961	0.4962	0.4963	0.4964
2.7	0.4965	0.4966	0.4967	0.4968	0.4969	0.4970	0.4971	0.4972	0.4973	0.4974
2.8	0.4974	0.4975	0.4976	0.4977	0.4977	0.4978	0.4979	0.4979	0.4980	0.4981
2.9	0.4981	0.4982	0.4982	0.4983	0.4984	0.4984	0.4985	0.4985	0.4986	0.4986
3.0	0.4987	0.4987	0.4987	0.4988	0.4988	0.4989	0.4989	0.4989	0.4990	0.4990
3.1	0.4990	0.4991	0.4991	0.4991	0.4992	0.4992	0.4992	0.4992	0.4993	0.4993
3.2	0.4993	0.4993	0.4994	0.4994	0.4994	0.4994	0.4994	0.4995	0.4995	0.4995
3.3	0.4995	0.4995	0.4995	0.4996	0.4996	0.4996	0.4996	0.4996	0.4996	0.4997
3.4	0.4997	0.4997	0.4997	0.4997	0.4997	0.4997	0.4997	0.4997	0.4997	0.4998

Bibliography

T.W. Anderson, *Introduction to Multivariate Statistical Analysis*, 2nd edn. (Wiley, New York, NY, 1984)

Army Chemical Corps, *Master Sampling Plans for Single, Duplicate, Double and Multiple Sampling, Manual No. 2* (Chemical Corps Engineering Agency, Army Chemical Center, Arsenal, MD, 1953)

A.F. Bissell, How reliable is your capability index? Appl. Stat. **39**, 331–340 (1990)

C.W. Champ, W.H. Woodall, Exact results for Shewhart control charts with supplementary runs rules. Technometrics **29**, 393–399 (1987)

J.-P. Chen, C.G. Ding, A new process capability index for non-normal distributions. Int. J. Qual. Reliab. Mange. **18**(7), 762–770 (2001)

A.J. Duncan, *Quality Control and Industrial Statistics*, 5th edn. (Irwin, Homewood, IL, 1986)

H. Hotelling, Multivariate quality control, in *Techniques of Statistical Analysis*, ed. by C. Eisenhart, M. W. Hastay, W. A. Wallis, (McGraw-Hill, New York, NY, 1947)

R.A. Johnson, D.W. Wichern, *Applied Multivariate Statistical Analysis*, 4th edn. (Prentice Hall, Upper Saddle River, NJ, 1998)

J.M. Juran, Early SQC: A historical supplement. Qual. Prog. **30**(9), 73–81 (1997)

S. Kotz, N.L. Johnson, *Process Capability Indices* (Chapman & Hall, London, 1992)

C.A. Lowry, W.H. Woodall, C.W. Champ, S.E. Rigdon, A multivariate exponentially weighted moving average chart. Technometrics **34**, 46–53 (1992)

J.M. Lucas, M.S. Saccucci, Exponentially weighted moving average control schemes: Properties and enhancements. Technometrics **32**, 1–29 (1990)

D.C. Montgomery, *Introduction to Statistical Quality Control*, 4th edn. (Wiley, New York, NY, 2000)

E.R. Ott, E.G. Schilling, *Process Quality Control*, 2nd edn. (McGraw-Hill, New York, NY, 1990)

W.L. Pearn, Y.T. Tai, F. Hsiao, Y.P. Ao, Approximately unbiased estimator for non-normal process capability index CNpk. J. Test. Eval. **42**(6), 1–10 (2014)

C.P. Quesenberry, The effect of sample size on estimated limits for and X control charts. J. Qual. Technol. **25**(4), 237–247 (1993)

T.P. Ryan, *Statistical Methods for Quality Improvement*, 2nd edn. (Wiley, New York, NY, 2000)

T.P. Ryan, N.C. Schwertman, Optimal limits for attributes control charts. J. Qual. Technol. **29**(1), 86–98 (1997)

E.G. Schilling, *Acceptance Sampling in Quality Control* (Marcel Dekker, New York, NY, 1982)

E.G. Schilling, Elements of process control. Qual. Eng. **2**(2), 121–135 (1989)

N.D. Tracy, J.C. Young, R.L. Mason, Multivariate control charts for individual observations. J. Qual. Technol. **24**(2), 88–95 (1992)

W.H. Woodall, Control charting based on attribute data: Bibliography and REVIEW. J. Qual. Technol. **29**, 172–183 (1997)

W.H. Woodall, B.M. Adams, The statistical design of CUSUM charts. Qual. Eng. **5**(4), 559–570 (1993)

N.F. Zhang, A. Stenback, D.M. Wardrop, Interval estimation of the process capability index. Commun. Stat. Theory Methods **19**(21), 4455–4470 (1990)

Chapter 3
Engineering Sampling Standards and Plans

3.1 Introduction

Total quality management (TQM) is an integrated plan for executing the given company's quality assurance and enhancement of operational goals. In today's manufacturing environment, the company's effort is concentrated on TQM. Hence, TQM enables the company to work as close as possible with the suppliers to ensure high-quality incoming material. As a component of TQM, acceptance sampling encompasses the supplier, manufacturer, and consumer. When the firm implements TQM (total quality management) plan, the role of acceptance sampling is significant.

Chapter 2 defined acceptance sampling as the inspection process (set of procedures) used to accept or reject a lot (the number of items). Incoming or outgoing quality control comprises a set of procedures to establish a well-defined quality inspection process. We have discussed the OC curve and the chapter details the sample size determination and the operating characteristic (OC) curve. The OC predicts the given acceptance sampling plan's power and significance. Furthermore, we have assessed the two fundamental sampling risks and essential terminologies in acceptance sampling. This chapter details about sampling plan. The sampling plan is a comprehensive framework of how the data is taken. The goal of the sampling plans is all about how to get representative data to predict the population.

Stages of quality are well-thought-out in the plan of the acceptance sampling. The acceptance plan should identify the AQL (acceptable quality level), which is the customer's quality expectation. The manufacturer's effort is to meet the AQL before the part is shipped to the customer. The producer risk (α), also called Type I error, is usually set to 5%. The sampling plan is the risk of meeting the lot's acceptable quality level.

Producers want to have a low producer risk. However, the customers have the power to set the acceptance sampling plan. Hence, returning suitable materials to the producer causes bad inventory management (for example, stock-outs and scheduling discrepancy). The customer also wants to have a low customer risk (CR, also called Type II error, β, usually set to 10%). Thus, the sampling plan should address such issues.

Another quality issue is the LTPD (lot tolerance proportion defective), or the poorest-quality level that the consumer can accept. LTPD reflects the bad-quality-level that the customer wants to reject. The section of the sampling plan, i.e., LTPD is the level of quality "rejected." It is usually described as the percent nonconformance (defective) rejection of the 90% $(1 - \beta$, i.e., the percentage of the nonconformance is accepted). Knowing and identifying the high cost of nonconformances (overall operation failure), customers do more vigilant about accepting poor-quality materials from suppliers.

We have already seen the first task of sampling the OC chart, sampling risk, and terminologies (AQL, LTPD, AOQ, and AOQL). In this chapter, we will assess sampling standards, sampling plans, and the Dodge–Romig sampling tables.

3.2   Terms and Definitions

- American National Standards Institute (ANSI) is a nonprofit-making organization accountable for the US firms to be competitive globally by empowering and inspiring conformity valuation systems and brings professional harmonized standards.
- ASQ stands for American Society for Quality. ASQ is a comprehensive data and then analysis-based universal society of quality experts, dedicated to endorsing and progressing quality methods, tools, principles, and practices in manufacturing, service sector industries, and societies.
- AQL stands for acceptance quality limit. AQL specifies the percentage of non-conformance (defective) items (part, product, or material) associated with the lot for outgoing shipment. Based on ISO 2859-1, acceptance quality limit (AQL) is stated as the quality level, which is the poorest tolerable.
- ANSI ASQ Z1.4 is a statistical table cognitive that certifies the sample size (n) inspected makes available a rational depiction of the lot's overall quality. One essential aspect of acceptance quality (AQL) is that we need not know complex formulas and comprehend other detailed technical aspects for inspection and sampling. Nevertheless, completely considerate the ANSI Z1.4 help avoid mistakes and sampling errors when applying AQL intended for inspection.
- ANSI/ASQ Z1.9-2003 (Revision 2018) is a sampling instruction, tables, and procedures for inspection (quality check) for the lot acceptance sampling process via variables for the percentage of nonconforming based on AQL specification. ANSI/ASQ Z1.9-2003 provides reduced and tightened comprehensive plans for normally distributed variables. The estimator of variation is obtained from the calculation of sample standard deviation (s).

- General inspection level (GIL): General inspection levels are three different levels (i.e., G-I, G-II, G-III) that specify the sample sizes (n) of a lot size (N) for nondestructive inspection.
- Special Inspection Levels (SIL): SILs have four different levels (i.e., S-1, S-2, S-3, and S-4) that are prepared correspondingly to the general inspection levels (GIL) in the acceptance quality level (AQL) chart.
- A homogenous population is a population that consists of similar objects. Here the population characteristics (i.e., population parameters like mean, variance, proportion, etc.) equally characterize each object.
- A heterogeneous population is a population that consists of dissimilar, inequivalent, or disparate objects. Alternatively, simply, a population that is not homogenous.
- Sampling procedure: A sampling procedure describes instructions that determine how the sample size is calculated and how the samples are taken. The sampling procedure comprehends the essential information about the assessment and valuation of the inspection process.
- A sampling scheme is a comprehensive characterization and description of how a representative sample will be taken. There are two categories of sampling schemes: probability sampling and non-probability sampling.
- A probability sampling denotes the sampling technique. Each object of the whole population has a prespecified probability to be selected in the sample. Examples of probability sampling are:
  - Simple random sampling (SRS) is a sampling approach that randomly selects n-sample points from a homogenous population. In SRS, each object (sample point) has an equal probability of being selected in the sample. Let us see a population of size N. The probability that each object is included in the sample is determined, and the value is $\frac{1}{N}$.
  - Stratum (plural form strata) is a homogenous subgroup of a heterogeneous population. Stratified random sampling is a sampling approach that randomly selects sample points from each of a heterogeneous population. In stratified random sampling, each object (sample point) from each stratum has an equal probability of being selected in the subsample (i.e., from each stratum). Let us consider the stratum of size N_i. Then the probability that each object included in the subsample is determined, and the value is $\frac{1}{N_i}$. The sample size for stratified random sampling is equated as $\sum_{i=1}^{k} n_i$.
  - Cluster (plural form clusters) a group of similar subpopulation. Clusters are homogenous, but each cluster is internally heterogeneous—cluster sampling a sampling approach by randomly selecting clusters from a heterogeneous population. In cluster sampling, each cluster has an equal probability of being selected as subsections of the population. Then the sample will be each object from each selected cluster.

– Systematic sampling is a sampling approach in which sample points from the population are carefully chosen, affording it to a random preliminary point, with a fixed and periodic interval. This interval is called the sampling interval. The sampling interval is the index that represents the fixed gap between two successive sample points in systematic sampling. The sampling interval is calculated by approximating the next number of the ratio of population size (N) to the sample size (n).

• Non-probability sampling: A non-probability sampling a sampling technique in which the inspector (quality engineer, a product engineer, or any other researcher) elects sample points based on the inspector's subjective interest (judgment). Non-probability sampling is a typical example of a biased and stringent sampling method.
• Sampling error any technical and methodological error that leads to having an unrepresentative sample of the population.

3.3   Sampling Plan

Before we define a sampling plan, we need to know the exact definition of sample points. Sample points are elements of the population that have the same chance of being included in the sample. Thus, the sampling plan is a sequence of methods to snatch the sample points to build the representative sample, enabling the prediction of the population.

The sampling plan provides a framework based on which inspection is conducted. The sampling plan comprises sample size, population parameters, cost of sampling, variability, practicality, accuracy risk, and prediction. Essentially, sampling plans are employed to perform verdicts on the acceptance or the rejection of the lot. Sampling plans are a quality control method for inspection (i.e., it could be in-process, incoming, final, or outgoing).

Sampling standards comprise the assessment of sampling plans. These are

• Single sampling plan (SSP)
• Double sampling plan (DSP)
• Multiple sampling plan (MSP)
• Sequential sampling plan (SqSP)

3.3.1   Single Sampling Plans (SSP)

A single sampling plan (SSP) is the common and easiest sampling plan. In SSP, the whole lot is rejected or accepted based on the inspection of sample points from the sample of size n. The sample is taken from a lot of size N. That is, the decision rule established to reject or accept based on the single sample, which was drawn from the same lot.

The sample size can be calculated or taken from the table. The single sampling plan (SSP) is defined by three essential parameters, which listed as follows:

1. Acceptance number (c)
2. Lot size (N)
3. Sample size (n)

The process is to select a sample of size n using a simple random sampling technique. If the number of nonconformances (defects) surpasses a prespecified acceptance number (c), the entire lot will be rejected. Therefore, the lot will be returned to the producer, sorted at the customer, reworked at the customer, and scrapped at the customer based on the contract's agreement.

The following simple steps will be applied to the exercise:

Step 1: Take n sample points (items, parts, or material) from the lot.
Step 2: Inspector test the sample
Step 3: Record the number of nonconformance items (d)
Step 4: Plan the acceptance number (c)
Step 5: Decision Rule: reject the lot if the number of nonconformance items (d) greater than the acceptance number (c)

3.3.1.1 General Inspection Levels (GIL)

The quality inspector uses the sample size (n) of General Inspection Levels (GIL) to the items' workmanship when visual inspection is performed. Thus, any nonconformances found concerning the workmanship should be considered in the acceptance quality level (AQL). The quality inspector then uses the acceptance quality level (AQL) chart to observe the levels (i.e., G-I, G-II, and G-III) sample size (n) of the lot sizes.

- G-I is called a reduced (also called Level I) inspection. Instead of accepting the incoming parts as they are, the manufacturer performs an inspection on fewer sample points by determining a G-I inspection.
- G-II is called a regular (also called Level II) inspection. The G-II is a default and the utmost commonly used inspection level.
- G-III is called a tightened (also called Level III) inspection. If the vendor (supplier) has quality problems, then the G-III inspection level is suitable. Here, the manufacturer needs more than a few sample points to inspect to assure the incoming parts are conforming.

3.3.1.2 Special Inspection Level (SIL)

Special inspection level (SIL) of the acceptance quality level (AQL) table used by the quality inspector when inspecting the items (concerning the customer's specific requirements). Special inspection levels (SILs) are used when the inspector needs to check only a few samples.

The number of sample points inspected under S-4 is the maximum size. The next is under S-3, and the least is under S-1.

Example of single sampling plans (SSP): Suppose let us consider many sizes of 5000, and we aim to select n sample points from the lot. Let us also use apply the G-II (General Inspection Level II). See the ANSI/ASQ Z1.4 standard in Table 3.1.

Table 3.1 shows that this plan's sample size is determined by the Sample Size Code Letter (SSCL) of L.

3.3.1.3 Acceptance Number

Acceptance number (c) determines the maximum number of tolerable nonconformances (defects) detected in the sample. The acceptance number (c) is consequential from a combination of the sample size (n) and the acceptance quality level (AQL) that the sample plan selects for inspection.

In the above example, suppose we set the quality acceptance level (AQL) as 1 (i.e., 1 out of 100). The ANSI/ASQ Z1.4 standard gives us information to obtain

Table 3.1 The ANSI/ASQ Z1.4 standard

Single sampling plan for normal inspection, ANSI/ASQ standard Z1.4-2003 (R2013)							
Sample size code letters							
Lot size	General inspection levels			Special inspection levels			
	GI	GII	GIII	S1	S2	S3	S4
2 to 8	A	A	B	A	A	A	A
9 to 15	A	B	C	A	A	A	A
16 to 25	B	C	D	A	A	B	B
26 to 50	C	D	E	A	B	B	C
51 to 90	C	E	F	B	B	C	C
91 to 150	D	F	G	B	B	C	D
151 to 280	E	G	H	B	C	D	E
281 to 500	F	H	J	B	C	D	E
501 to 1200	G	J	K	C	C	E	F
1201 to 3200	H	K	L	C	D	E	G
3201 to 10000	J	L	M	C	D	F	G
10001 to 35000	K	M	N	C	D	F	H
35001 to 150000	L	N	P	D	E	G	J
150001 to 500000	M	P	Q	D	E	G	J
500001 and over	N	Q	R	D	E	H	K

Table 3.2 Acceptance quality limit (AQL)

Each cell shows the plan as "Ac Re"; ↓ = use first sampling plan below the arrow; ↑ = use first sampling plan above the arrow.

Sample Size code letter	sample size	AQL (acceptance quality limit) for normal inspection — 0.25	0.4	0.65	1	1.5	2.5	4	6.5	10	15	25	40	65
A	2	↓	↓	↓	↓	↓	↓	↓	↓	↓	0 1	1 2	2 3	3 4
B	3	↓	↓	↓	↓	↓	↓	↓	↓	0 1	1 2	2 3	3 4	5 6
C	5	↓	↓	↓	↓	↓	↓	↓	0 1	1 2	2 3	3 4	5 6	7 8
D	8	↓	↓	↓	↓	↓	↓	0 1	1 2	2 3	3 4	5 6	7 8	10 11
E	13	↓	↓	↓	↓	↓	0 1	1 2	2 3	3 4	5 6	7 8	10 11	14 15
F	20	↓	↓	↓	↓	0 1	1 2	2 3	3 4	5 6	7 8	10 11	14 15	21 22
G	32	↓	↓	↓	0 1	1 2	2 3	3 4	5 6	7 8	10 11	14 15	21 22	↑
H	50	↓	↓	0 1	1 2	2 3	3 4	5 6	7 8	10 11	14 15	21 22	↑	↑
J	80	↓	0 1	1 2	2 3	3 4	5 6	7 8	10 11	14 15	21 22	↑	↑	↑
K	125	0 1	1 2	2 3	3 4	5 6	7 8	10 11	14 15	21 22	↑	↑	↑	↑
L	200	1 2	2 3	3 4	5 6	7 8	10 11	14 15	21 22	↑	↑	↑	↑	↑
M	315	2 3	3 4	5 6	7 8	10 11	14 15	21 22	↑	↑	↑	↑	↑	↑
N	500	3 4	5 6	7 8	10 11	14 15	21 22	↑	↑	↑	↑	↑	↑	↑
P	800	5 6	7 8	10 11	14 15	21 22	↑	↑	↑	↑	↑	↑	↑	↑
Q	1250	7 8	10 11	14 15	21 22	↑	↑	↑	↑	↑	↑	↑	↑	↑
R	2000	10 11	14 15	21 22	↑	↑	↑	↑	↑	↑	↑	↑	↑	↑

the optimal acceptance number (c) via the acceptance quality level (AQL) and the sample size code letter (SSCL). Therefore, AQL is 1, and SSCL is L.

For this example, let us say our AQL is 1.0. We start by finding our AQL across the top and our sample size code letter (L in this case) along the side, and we find their intersection in the matrix.

It is essential to recognize that the ANSI/ASQ Z1.4 Table determines the optimal number of samples that we need to inspect during the inspection.

From Table 3.2, we can observe that for a sample size code letter (SSCL) of L, we should inspect 200 samples (n).

In Table 3.2, we observe that for SSCL of L and sample size of 200, the acceptance number (c) becomes 5, and the rejection number (d) is 6.

The number of nonconformances (d) is the crucial parameter of the sampling plan. It is also called the rejection number, which tells the minimum number of nonconformance that leads us to reject the lot. That is the exact number of non-conformances detected (observed) in a sample. After inspecting the 200 samples, the quality accepts the lot if the rejection number is 5 or less.

3.3.1.4 Limitations of Single Sampling Plan

The single sampling plan (SSP) is the easiest to work with and execute. Nevertheless, the following most essential are limitations (disadvantages) of SSP.

- The SSP may require a large sample size.
- Relay on a single sample size may result in a wrong decision if the sampling is not representative.

If we have limited resources to inspect the sample, we need to find another sampling plan. In the next section, we will assess double and sequential sampling plans, which enable us to optimize sample size (n).

3.3.2 Double Sampling Plan (DSP)

Simple sampling (SSP) is easy to use. The main disadvantage of simple sampling (SSP) is that it needs to consider a larger sample size. Furthermore, just inspecting a sample and giving a decision from the inspection result may not be adequate. We may need to consider another sample to confirm the result we found in the first inspection.

To handle the drawback of SSP, ANSI introduces the double sampling plan (DSP). Though the single sampling plan (SSP) is implemented with a single sample of items, we will consider two different samples in a double sampling plan (DSP).

A double sampling plan (DSP) is useful when the incoming lot may be "good" or "bad" so that the quality inspector can perform a rational decision. That is because the DSP requires a small sample size than using it in the SSP.

The following points characterize the double sampling plan (DSP).

- The DSP uses two different samples.
- The DSP uses a less sample size than the corresponding SSP.
- The DSP is more advanced to work with than the SSP.
- The DSP uses samples whose sizes are less than the sample size in the SSP. But the aggregate of the samples may greater the sample size in SSP.
- Hence, it gives an additional chance, the DSP more convenient and convincing than SSP.
- The DSP encompasses more expenses than SSP.
- The DSP involves additional record-keeping (RK) than the SSP.

We have two distinct sample sizes in the double sampling plan (DSP) approach, two acceptance numbers, and two rejection numbers. Just like the SSP, if the quality level of the lot is "good" or "bad," the manufacturer may "accept" or "reject" the lot based on the first sample. However, we may not rely on the first sample. So, we need an additional sample for inspection.

So, the triples parameter of DSP is (n_i, c_i, d_i) where $i = 1, 2$. The parameter of the first sample in DSP is (n_1, c_1, d_1) and the combined first and second samples is $((n_1 + n_2), c_2, d_2)$.

Where

- n_1 is the sample size of the first sample.
- n_2 is the sample size of the second sample.
- c_1 and c_2 are acceptance numbers in the first and combined samples, respectively.
- d_1 and d_2 are rejection numbers in the first and combined samples, respectively.

3.3.2.1 Decision Rule of the DSP

- If the number of nonconformance in the first sample (n_1) is less than or equal to the acceptance number, c_1, we will accept the lot.
- If the number of nonconformance in the first sample (n_1) is greater than the acceptance number, c_2, we will reject the lot.
- If the number of nonconformance in the first sample (n_1) is in between c_1 and c_2, we will consider the second sample accept n_2. Then, we will combine the samples, n_1 and n_2. If the number of conformances in the combined sample is less than or equal to c_2, then we accept the lot. Else, we will reject the lot.

Recall the previous problem and let us analyze using the double sampling plan (DSP). The lot size is 5000. Here, we also set G-II so that the SSCL is L. Table 3.3 gives the double sampling plan.

From Table 3.3, we see that

- $n_1 = 125$ is the sample size of the first sample
- $n_2 = 125$ is the sample size of the second sample
- $c_1 = 2$ is the acceptance number of the first sample (i.e., $n_1 = 125$)
- $c_2 = 6$ is the acceptance number of the combined samples (i.e., $n_1 + n_2 = 250$)
- $d_1 = 5$ are rejection number for the first sample (i.e., $n_1 = 125$)
- $d_2 = 7$ is the rejection number of the combined samples (i.e., $n_1 + n_2 = 250$)

Using the DSP, we will have the following decisions.

1. If we find a maximum of 2 (i.e., $c_1 = 2$ is acceptance number) nonconformance (defected part, item, or material), then we will accept the lot.
2. If we find a minimum of 5 (i.e., $d_1 = 5$ are rejection number) nonconformance (defected part, item, or material), then we will reject the lot.

Table 3.3 Double sampling plan AQL

Code letter	Sample	Sample size	Total sample	0.25		0.40		0.65		1.0		1.5		2.5		4		6.5		10		15		25		40		65		100	
				Ac	Re	Ac	Re	Ac	Re	Ac	Re	Ac	Re	Ac	Re	Ac	Re	Ac	Re	Ac	Re	Ac	Re	Ac	Re	Ac	Re	Ac	Re	Ac	Re
A																															
B	First	2	2																			0	2	0	3	1	4	2	5	3	7
	Second	2	4																			1	2	3	4	4	5	6	7	8	9
C	First	3	3																	0	2	0	3	1	4	2	5	3	7	5	9
	Second	3	6																	1	2	3	4	4	5	6	7	8	9	12	13
D	First	5	5															0	2	0	3	1	4	2	5	3	7	5	9	7	11
	Second	5	10															1	2	3	4	4	5	6	7	8	9	12	13	18	19
E	First	8	8													0	2	0	3	1	4	2	5	3	7	5	9	7	11	11	16
	Second	8	16													1	2	3	4	4	5	6	7	8	9	12	13	18	19	26	27
F	First	13	13											0	2	0	3	1	4	2	5	3	7	5	9	7	11	11	16		
	Second	13	26											1	2	3	4	4	5	6	7	8	9	12	13	18	19	26	27		
G	First	20	20									0	2	0	3	1	4	2	5	3	7	5	9	7	11	11	16				
	Second	20	40									1	2	3	4	4	5	6	7	8	9	12	13	18	19	26	27				
H	First	32	32							0	2	0	3	1	4	2	5	3	7	5	9	7	11	11	16						
	Second	32	64							1	2	3	4	4	5	6	7	8	9	12	13	18	19	26	27						
J	First	50	50					0	2	0	3	1	4	2	5	3	7	5	9	7	11	11	16								
	Second	50	100					1	2	3	4	4	5	6	7	8	9	12	13	18	19	26	27								
K	First	80	80			0	2	0	3	1	4	2	5	3	7	5	9	7	11	11	16										
	Second	80	160			1	2	3	4	4	5	6	7	8	9	12	13	18	19	26	27										
L	First	125	125	0	2	0	3	1	4	2	5	3	7	5	9	7	11	11	16												
	Second	125	250	1	2	3	4	4	5	6	7	8	9	12	13	18	19	26	27												
M	First	200	200	0	3	1	4	2	5	3	7	5	9	7	11	11	16														
	Second	200	400	3	4	4	5	6	7	8	9	12	13	18	19	26	27														
N	First	315	315	1	4	2	5	3	7	5	9	7	11	11	16																
	Second	315	630	4	5	6	7	8	9	12	13	18	19	26	27																
P	First	500	500	2	5	3	7	5	9	7	11	11	16																		
	Second	500	1000	6	7	8	9	12	13	18	19	26	27																		
Q	First	800	800	3	7	5	9	7	11	11	16																				
	Second	800	1600	8	9	12	13	18	19	26	27																				
R	First	1250	1250	5	9	7	11	11	16																						
	Second	1250	2500	12	13	18	19	26	27																						

If we find nonconformances between 2 and 6, (i.e., between $c_1 = 2$ and $c_2 = 6$), then we will add another sample of size of 125 (i.e., $n_2 = 125$) items. Now, we will combine the two samples (i.e., $n_1 + n_2 = 250$). Then,

1. If we find a maximum of 6 (i.e., $c_2 = 6$ is acceptance number) nonconformance (defected part, item, or material), then we will accept the lot.
2. If we find a minimum of 7 (i.e., $r_2 = 7$ is the rejection number of the combined samples (i.e., $n_1 + n_2 = 250$)) nonconformance (defected part, item, or material), then we will reject the lot.

3.3.3 Multiple Sampling Plan (MSP)

Multiple sampling plan (MSP) is an advancement of the double sampling plan (DSP). In MSP, the quality inspector inspects k-samples. If $k = 2$, then the sampling plan is DSP. So, we can consider the DSP is a special form of the MSP. For sampling efficiency, k ranges from two to seven samples.

The Multiple sampling plan (MSP) is characterized by the following points.

- The MSP uses k-distinct samples.
- The MSP uses a less sample size than the corresponding DSP. However, the aggregate sample size of the k subsamples may greater than the sample size of DSP.
- The MSP uses a similar sampling technique and logic with DSP.
- Hence, it gives more additional chances, the MSP more convenient and convincing than DSP and SSP.
- MSP involves additional record-keeping (RK) than the DSP and SSP.

In the multiple sampling plan (MSP) approach, we have k-distinct sample sizes, $n_1, n_2, n_3, \ldots, n_k$, k-acceptance numbers, $c_1, c_2, c_3, \ldots, c_k$, and k-rejection numbers, $d_1, d_2, d_3, \ldots, d_k$.

Just like the DSP, if the quality level of the lot is "good" or "bad," the manufacturer may "accept" or "reject" the lot based on the first sample. However, we may not rely on the first two samples. So, we need additional samples for inspection.

So, the triples parameter of MSP is (n_i, c_i, d_i) where $i = 1, 2, 3, \ldots, k$. That is, the parameter of the first sample in MSP is (n_1, c_1, d_1), the combined (i.e., first and second) samples is $((n_1 + n_2), c_2, d_2)$, and the combined the k-samples is $((n_1 + n_2 + n_3 + \ldots + n_k), c_k, d_k)$.

Where

- n_1 is the sample size of the first sample.
- n_2 is the sample size of the second sample.
- n_i is the sample size of the ith sample.
- n_k is the sample size of the kth sample.
- c_1 and c_k are acceptance numbers in the first and combined samples, respectively.
- d_1 and d_k are rejection numbers in the first and combined samples, respectively.

Table 3.4 Multiple sampling plan AQL

Sample number	Sample size (n_i)	Acceptance number (c_i)	Rejection number (d_i)
Sample 1	$n_1 = 50$	$c_1 = $ NA	$r_1 = 4$
Sample 2	$n_2 = 50$	$c_2 = 1$	$r_2 = 5$
Sample 3	$n_3 = 50$	$c_3 = 2$	$r_3 = 6$
Sample 4	$n_4 = 50$	$c_4 = 3$	$r_4 = 7$
Sample 5	$n_5 = 50$	$c_5 = 5$	$r_5 = 8$
Sample 6	$n_6 = 50$	$c_6 = 7$	$r_6 = 9$
Sample 7	$n_7 = 50$	$c_7 = 9$	$r_7 = 10$

Like the DSP, the MSP also to each sample must possess an acceptance number (c_i) and rejection number (d_i).

Now, it is time to illustrate the MSP: The lot size is 5000. Here, we also set G-II so that the SSCL is L, and we set the AQL of 1.0. Table 3.4 gives the MDP.

Using the MSP, we will have the following decisions.

1. From the first sample, if we find a minimum of 4 (i.e., $d_1 = 4$ are rejection number) nonconformance (defected part, item, or material), then we will reject the lot. Else,
2. The first two samples (i.e., $50 + 50 = 100$), find a maximum of 1 (i.e., $c_2 = 1$ is acceptance number) nonconformance (defected part, item, or material), then we will accept the lot. Furthermore, find a minimum of 5 (i.e., $d_2 = 5$ is rejection number) nonconformance (defected part, item, or material), then we will rejection the lot.
3. The first three samples (i.e., $50 + 50 + 50 = 150$), find a maximum of 1 (i.e., $c_3 = 2$ is acceptance number) nonconformance (defected part, item, or material), then we will accept the lot. Furthermore, find a minimum of 6 (i.e., $d_3 = 6$ is rejection number) nonconformance (defected part, item, or material), then we will rejection the lot.
4. The k samples (i.e., $50 + 50 + 50 + 50 + 50 + 50 + 50 = 350$), find a maximum of 9 (i.e., $c_9 = 9$ is acceptance number) nonconformance (defected part, item, or material), then we will accept the lot. Furthermore, find a minimum of 10 (i.e., $d_{10} = 10$ is rejection number) nonconformance (defected part, item, or material), then we will rejection the lot.

Compared to DSP or SSP, the MSP has a small ASN (average sampling number). In general, in terms of ASN, MSP > DSP > SSP.

3.3.4 Sequential Sampling Plan (SqSP)

The Z1.4 standard lets the quality inspector choose amongst SSP, DSP, or MSP. Furthermore, the Z1.4 standard contributes to the quality inspector in setting up the sampling systems and schemes. We have already discoursed the distinctions of the SSP, DSP, and MSP. Now, it is time to discuss the sequential sampling plans.

The sequential sampling plan (SqSP, also called PRST (probability ratio sequential tests), or SPRT (sequential probability ratio tests)) is the advancement of multiple sampling plans (MSP). In the SqSP, the sampling process until further notice that the whole lot inspected. In the SqSP, an acceptance or rejection decision is made after inspecting a sample at a time.

For "too little" or "very high" quantities (value) of defectives, in SqSP, make available for a smaller ANI (average sampling number) than any equivalent sampling plans. However, suppose the quantity of defective units lies among the acceptable quality level (AQL) and the Lot Tolerance Percent Defective (LTPD). In that case, the SqSP could have a bigger ANI than any other sampling plans (i.e., SSP, DSP, or MSP). In general, the SqSP may decrease the ANI up to 50% of that needed by the equivalent SSP.

Therefore, the SqSP is cost-effective than the other acceptance sampling plans.

Several lots have an unidentified (unknown) amount of nonconformance. Furthermore, suppose we suspect that the proportion of defects is not homogenous among the lots. In that case, the SqSP is the best sampling plan. With suitable power and confidence, the acceptable at AQL of 5 (5 out of 100, or only 5%), we apply the SqSP set to determine.

Consider the following information to illustrate the SqSP:

- Manufacturer acceptance quality level (AQL), $p_1 = 1\%$ with Producer Risk (PR), $\alpha = 0.05$, which brings 95% confidence.
- Customer acceptance quality level (AQL), $p_2 = 5\%$ with Consumers Risk (CR), $\beta = 0.10$, which brings 90% power.

3.3.4.1 Decision Measures Calculation

- Calculation of $h_1 = \dfrac{\log\,[(1-\alpha)/\beta\,]}{\left[\log\,\left(\frac{p_2}{p_1}\right) + \log\,[\,(\,1-p_1)-\log\,(1-p_2\,)]\,\right]}$

- Calculation of $h_2 = \dfrac{\log\,[(1-\beta)/\alpha\,]}{\left[\log\,\left(\frac{p_2}{p_1}\right) + \log\,[(1-p_1)/(1-p_2)]\right]}$

- Calculation of $s = \dfrac{\log\,(\,1-p_1)-\log\,(1-p_2)}{\left[\log\,\left(\frac{p_2}{p_1}\right) + \log\,[(1-p_1)/(1-p_2)]\right]}$

Suppose k is the number of samples taken, then acceptance and rejection lines, respectively, given as

$$y_1 = ks - h_1$$

And

$$y_1 = ks + h_2$$

In the kth sample, let us let $d_k = \sum_{i=1}^{k} d_i$ is the aggregate number of nonconforming, the decision rule becomes

1. Accept the lot if $d_k \leq y_1 = ks - h_1$
2. Reject the lot if $d_k \geq y_2 = ks + h_2$
3. Keep sampling if $y_1 < d_k < y_2$. That is $ks - h_1 < d_k < ks + h_2$.

Suppose the quality inspector set $\alpha = 10\%$ and $\beta = 20\%$, the confidence and power decrease to 90% and 80%, respectively. In this case, the inspector should sample 37 items. If the inspector finds no nonconforming item, then we accept the lot. Else, if the inspector samples a maximum of 29 items and finds a minimum of 2 nonconforming items, we reject the lot.

3.4   General Inspection Levels of Sampling and Sample Size

As can be seen from Table 3.5, the ANSI standard has three GIS (general inspection levels, i.e., G-I, G-II, and G-III) with four SPs (special inspections, i.e., S-1, S-2, S-3, and S-4). The GIS classifies the inspection rule based on the need and requirements of the sampling plan. However, the SPs classification helps us when we have a deficiency in getting the required number of sample points for inspection.

Typically, we emphasize the GIS primarily within the ANSI standard. However, naturally, samples whose sample size is small due to large sampling may risk the manufacturer and consumer. The SPs plan necessity comes with a valuation and acceptance of the possible risks.

From Table 3.5, we observe that as the GIL (general inspection level) moves from I → II → III the SSCL (sample size code letter) changes. That means the number of sample points related to the sampling plan.

For instance, in Table 3.6, suppose the lot size of inspection is 1000 items (units). The SSCL changes as follows

1. G-I interprets the SSCL, L, which brings the sample size of 200.
2. G-II interprets the SSCL, N, which brings the sample size of 500.
3. G-III interprets the SSCL, P, which brings the sample size of 800.

From Table 3.6, we learn that having the same quality acceptance level (AQL), the different general inspection levels (GILs) lead to having different sample sizes. We know that customer's risk (CR) will be reduced if we need to inspect more samples. Therefore, the GIL is a very good indicator for selecting proper sample sizes that need to be inspected before the lot moves to the customer. G-I relatively higher, and G-III have a lower customer risk (β).

Table 3.5 ANSI standard: GIL and SPs

Lot or batch size			Special inspection levels				General inspection levels		
			S-1	S-2	S-3	S-4	I	II	III
2	to	8	A	A	A	A	A	A	B
9	to	15	A	A	A	A	A	B	C
16	to	25	A	A	B	B	B	C	D
26	to	50	A	B	B	C	C	D	E
51	to	90	B	B	C	C	C	E	F
91	to	150	B	B	C	D	D	F	G
151	to	280	B	C	D	E	E	G	H
281	to	500	B	C	D	E	F	H	J
501	to	1200	C	C	E	F	G	J	K
1201	to	3200	C	D	E	G	H	K	L
3201	to	10000	C	D	F	G	J	L	M
10001	to	35000	C	D	F	H	K	M	N
35001	to	150000	D	E	G	J	L	N	P
150001	to	500000	D	E	G	J	M	P	Q
	>	500001	D	E	H	K	N	Q	R

Table 3.6 General inspection levels of sampling and sample size

Lot or batch size			General inspection levels				Sample size Code letter	Sample size
			I	II	III			
							K	125
							L	200
							M	315
							N	500
35001	to	150000	L	N	P		P	800
150001	to	500000	M	P	Q		Q	1250
	>	500001	N	Q	R		R	2000

3.5 Sampling Scheme

A sampling scheme is a sort of sampling plans. The sampling plan is based on ISO standards (ISO 3951 and ISO 2859). It sets specific criteria for defining rules whether the lot is accepted or rejected based on the sample inspected.

3.5.1 Sampling Inspection Categories: Reduced, Normal, and Heightened

During the sampling process, a normal inspection is the starting point of the sampling schemes. Then if necessary, a reduced inspection follows. The reduced inspection process is a sampling plan that produces smaller sample sizes than the normal inspection.

Essentially, we should remark that reduced inspection (i.e., reducing the sample size) intensifies the consumer risk, β. Nevertheless, that risk may be tolerated because of the strong quality performance of the vendor.

Tightened inspection is a sort of sampling plan that fixes the sample size. However, it decreases the acceptance number (c) and rejection number (d) to modify and upgrade the consumer risk, β.

Suppose let us set the acceptance quality level (AQL) to 1 (i.e., 1 out of 100), let us compare the heightened, normal, and reduced sampling plans. Table 3.7 shows the sample sizes (n), and acceptance number (c) reject number (d) for each plan.

In Table 3.7, the difference between the sample sizes the normal and reduce inspection (sampling plan) painted in purple. Moreover, the difference between the acceptance number (c) and rejection number (d) the normal and reduce inspection (sampling plan) is painted in yellow.

From Table 3.7, we observe that for the same SSCL, for instance, K:

- The sample size varies with reduced and normal inspections. The table showed that the sample sizes are 50 and 125, respectively, for reduced and normal inspections.
- The sample size is the same for normal and highlighted inspections.
- The acceptance numbers are different for the different inspection methods. The acceptance number is tighter for heightened (i.e., $d = 2$ out of 125, which allows tolerating a maximum of 1.6% of nonconfining) and softer (weaker) for reduced (i.e., $d = 1$ out of 50, which allows tolerating up to 2% nonconfining).

The rejection numbers are different for the different inspection methods. The rejection number is tighter for heightened (i.e., $c = 3$ out of 125, which is 2.4% nonconfining) and weaker for reduced (i.e., $d = 4$ out of 50, 8%). An important point about the reduced inspection is that if the real number of rejections is within the acceptance and rejection numbers, the lot will be accepted. However, in that case, the reduced inspection plan back to normal inspection.

Table 3.7 Comparing the reduced, normal, and heightened (tightened) sampling plans

Reduced inspection				Normal inspection				Heightened inspection			
Sample size Code letter	Sample size	1 Ac	Re	Sample size	Sample size	1 Ac	Re	Sample size	Sample size	1 Ac	Re
A	2			A	2			A	2		
B	2			B	3			B	3		
C	2			C	5			C	5		
D	3	↓		D	8	↓		D	8		
E	5	0	1	E	13	0	1	E	13	↓	
F	8	↑		F	20	↑		F	20	0	1
G	13	↓		G	32	↓		G	32		
H	20	0	2	H	50	1	2	H	50	↓	
J	32	1	3	J	80	2	3	J	80	1	2
K	50	1	4	K	125	3	4	K	125	2	3
L	80	2	5	L	200	5	6	L	200	3	4
M	125	3	6	M	315	7	8	M	315	5	6
N	200	5	8	N	500	10	11	N	500	8	9
P	315	7	10	P	800	14	15	P	800	12	13
Q	500	10	13	Q	1250	21	22	Q	1250	18	19
R	800	↑		R	2000	↑		R	2000	↑	

The other condition for the switching scheme is that value revealing is discontinued sampling. Suppose we perform the tightened inspection on the incoming lot (batch) and find five subsequent rejection. In that case, the whole sampling methodological process will be terminated. Instead, that moves 100% of the incoming material.

The sampling record states that the vendor has problems delivering the items (products, parts, or material). The customer needs to report the problem to the vendor. Any sampling plan cannot be applied to the incoming items unless the vendor provides effective and efficient corrective and preventive actions (CAPA) for those specific problems. The customer verified and approved the CAPA.

3.6   Cost of Rejection

Successful supplier relationships necessitate two-way right information at the right time to improve the quality of the product.

Cost reduction is one of the goals of process improvement practices. Cost reduction strategies can reduce operations costs while increasing productivity, allowing for strategic reallocation of resources. Cost reduction strategies are operative approaches for upgrading operations efficiency. The cost of rejection is one of the leading causes of indirect costs. So, we need to estimate the cost as follows.

Whenever we do an incoming inspection, there are costs to perform the inspection. These costs are:

1. *Labor Cost (LC)*

   To calculate the labor cost (LC), we use the following formula:

   - $LC = TH \times HR$

     where TH = Total hours taken for inspecting the rejected parts
     HR = Hourly rate for the technician

2. *Calibration Cost*

   - Yearly tool calibration Cost = YCC
   - Daily tool calibration cost (DCC) = YCC/365
   - Hourly tool calibration cost (HCC) = YCC/(8 × 365) = YCC/2920

   Therefore, to calculate the tool calibration cost (TCC) for inspection of defective parts, we use the following formula:

   - $TCC = [(YCC/2920)] \times TH = [YCC \times TH]/2920$

3. *Machine Cost*

   Suppose the machine's purchase price is PMP, and the machine gives a total service for years of the machine (TS) in years.

   - The yearly service cost becomes = PMP/TS
   - The Daily service cost becomes = PMP/(365 × TS)
   - The Hourly service cost becomes = PMP/(8 × 365 × TS) = PMP/(2920 × TS)

   Therefore, to calculate the machine service cost (MSC) for inspection of defective parts, we use the following formula:

   - $MSC = [PMP/(2920 \times TS)] \times TH = [PMP \times TH]/[2920 \times TS]$
     where TH = Total hours taken for inspecting the rejected parts.

   Total rejection cost (TRC)

   - $TRC = LC + MSC + TCC = TH \times HR + PMP/(2920 \times TS) + YCC \times TH/2920$
   - $TRC = TH \times [HR + PMP/(2920 \times TS) + YCC/2920]$

where

TH = Total hours taken for inspecting the rejected parts
PMP = Purchase price of the machine
HR = Hourly rate for the technician
TS = Total service for years of the machine in years
YCC = Yearly tool calibration cost

3.7 Switching Rules

We have elaborated above section; the sampling scheme is a sort of sampling plan. In this section, we will learn to see how the sampling scheme will be applied for switching rules. The switching rules shift back-and-forth among the three-sampling plans, which have different operating characteristic curves (OC curve). That is, the reduced, normal, and heightened (tightened) plans, with each plan has its unique OC curve. For instance, switching from the normal plan to the reduced plan brings the sampling plan not as good as the consumers' point.

The purpose of the ANSI Z1.4 standard is to be applied as a structure that customs the switching rules. These sampling systems are meant to recompence vendors who have established a stable and high-quality process. With reduced sampling, poorly performing vendors can encourage continuous improvement programs to apply heightened sampling. That means the best performing vendor can apply the reduced sampling plan, whereas the vendor who has problems may push its sampling plan to the heightened one.

These various sampling plans (i.e., the normal sampling, heightened sampling, and reduced sampling) are used per the recommendation of the ANSI Z1.4 standard. Many quality inspectors use any of the sampling plans with no-switching rules. However, if we do not apply the switching rules, we do not perform inspections based on the ANSI Z1.4 standard. Therefore, hence, switching rules are essential to the standard; we emphasize the applications.

Consider Fig. 3.1 to illustrate the switching rules.

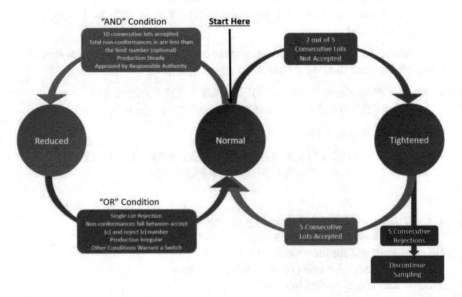

Fig. 3.1 Switching rules

We always start our sampling plan from normal. Then, based on Fig. 3.1, we observe that

1. *Switching from normal sampling to tightened sampling (stringent sampling)*
   Switching from the normal sampling to tightened sampling should happen if we reject 2 lots out of 5 successive lots. That is a signal of the poor performance of the that the vendor. Therefore, the vendor is expected to improve the outgoing inspection level and show verifications on the process capability requirements as a part of corrective and preventive actions (CAPA).
2. *Switching from tightened sampling to normal sampling (right back to normal)*
   Switching from the tightened sampling to normal sampling should happen if 5 successive lots are accepted. That is a good signal of the process improvement of the vendor. Therefore, the vendor acknowledged and recognized as a suitable vendor via the approved corrective and preventive actions (CAPA).
3. *Switching from normal sampling to reduced sampling (soft sampling)*
   Switching from the normal sampling to reduced sampling should happen if all the following preconditions must fulfill.

   Precondition 1: Manufacturing style should be continuous with a stable rate.
   Precondition 2: The total number of nonconformances in the preceding 10 lots should be less than the "Limit Numbers" in the ANSI/ASQ Z1.4 standard.
   Precondition 4: If DSP or MSP is used, then all the samples should be inspected.
   Precondition 5: 10, successive lots should be accepted.

   Then, the reduced sampling should be approved by the customer.

4. *Switching from reduced sampling to normal sampling (left back to normal)*
   Switching from the reduced sampling to normal sampling should happen if at least one of the following conditions fulfills.

   (a) One lot is rejected.
   (b) If the number of nonconformances is between the acceptance number (c) and the rejection number (d). In that situation, the lot can be accepted. Nevertheless, the reduced sampling should be withdrawn.
   (c) The vendor's production schedule is infrequent or irregular.
   (d) Additional situations permit the switch.

3.8 ANSI/ASQ Z1.9: The Variable Sampling Plan

The ANSI/ASQ Z1.9 standard is and initiated as Military Standard 414.
   The ANSI/ASQ Z1.9 standard is an adequate sampling for variable data (to estimate the lot percentage nonconformance). The benefit of applying variable data is that without disturbing the producer risk and customer risk, the sample sizes (n) are much lesser than attribute sampling. The disadvantage is that the ANSI/ASQ Z1.9 standard uses advanced mathematics than the attribute sampling plan.

The ANSI/ASQ Z1.9 standard is established based on the assumption that the data are normally distributed. Usually, the ANSI/ASQ Z1.9 standard uses descriptive statistics. Such as central tendencies (i.e., mean, mode, and median) and dispersion measures (i.e., variance, standard deviation, and range) make inferences about the overall population.

Like the ANSI/ASQ Z1.4 standard, the ANSI/ASQ Z1.9 standard is founded on the switching scheme. That is based on the performance of the incoming quality of the vendor. The inspection can switch among reduced, normal, or tightened inspection. The switching principle within ANSI/ASQ Z1.9 virtually reflects the ANSI/ASQ Z.14 switching principle.

Moreover, ANSI/ASQ Z1.9 standard also the SSP, DSP, and MSP. Variable sampling habitually necessitates a contrast of sample statistics (i.e., mean, mode, median, variance, standard deviation, and range) against the upper and lower specification limits. In that way, the ANSI/ASQ Z1.9 standard helps evaluate the overall percentage of nonconforming items in the lot. That intensified the complexity of the process and increased the cost of sampling.

The ANSI/ASQ Z1.4 (attribute sampling) standard allows categorizing many quality characteristics into a given sampling plan. The ANSI/ASQ Z1.4 standard is appropriate solitary to the valuation of a single quality characteristic. The sampling costs related to the variable plan upsurges when many quality features are evaluated.

The ANSI/ASQ Z1.4 standard provides three different calculation phases for sampling grounded on three different estimations of the lot variability.

These are:

- If the population standard deviation (σ) is given
- Suppose we know the population standard deviation (σ). In that case, the average range (R-bar) is used to estimate the population standard deviation
- If the standard deviation (σ) of the population is not given, then the sample standard deviation (s) will be used to estimate the population standard deviation.

Note that the ANSI/ASQ Z1.4 standard also delivers five different inspection levels. Three levels of the general (G-I, G-II, and G-III) and the last two levels of special (S3 and S4) inspections. Like the ANSI/ASQ Z1.4 standard, the defaulting inspection level is for the general inspection level is G-II.

3.9   Dodge–Romig Sampling Tables (DRST)

We have seen that the ANSI/ASQ Z1.4 and ANSI/ASQ Z1.9 standards deliver sampling plans grounded on AQL (acceptance quality level). However, the DRST (Dodge–Romig sampling tables) establish sampling plans based on the required AOQL (average outgoing quality limit) or based on LTPD (lot tolerance percent defective).

The Dodge–Romig sampling tables are detailed to attribute data for only SSP and DSP. The concept of AOQL (average outgoing quality limit) undertakes that the entire rejected lots need to be sorted (i.e., 100% inspected). The nonconforming items need to be replaced by conforming items.

The LTPD (lot tolerance percent defective) is the incoming quality level related to a small likelihood of acceptance. LTPD is related to consumer risk (β) is typically set at the 10% likelihood of acceptance. The plans are generated so that the LTPD links to a precise incoming quality level. For instance, 1% of nonconforming. That means the plan considers a 10% probability of accepting (or 90% of rejection) the lot with the quality level is 1% of nonconforming.

The AOQL (average outgoing quality limit) plans are intended to verify that the AOQL is less than the number indicated in the sampling plan. To use the DRST (Dodge–Romig sampling table), we essentially need to know two-parameters. The number of items in the incoming batch (lot) and the expected value of the process average of nonconforming items related to the sample being inspected.

One of the main advantages of the DRST (Dodge–Romig sampling tables) is that they tend to minimalize the average number of items inspected. Nevertheless, that is solitary accurate if the average number of nonconformance related to the process has inspected the sample. If the process's average number of nonconformance proportions lies, the subsequent lot rejections and 100% sorting of the sample. That results in minimizing the offset linked with the lower inspection quantity.

3.10   Sample Integrity

The sample points (i.e., sample) we take to perform acceptance sampling are essential. Therefore, it is very essential to be careful when we pick the sample points for inspection. Hence, the sample should be effective and efficient, enabling us to make perdition about the population. Suppose the sample is biased or has a sampling error. In that case, the consequence is costly and a lot of dangerous implications in the subsequent processes.

Precisely, the sampling process is exceptionally critical randomness. Randomness certifies that the sample is a typical representative of the population parameters (i.e., proportion, central tendencies, and measures of dispersion or any other parameter). If the randomization process of removing systematic pattern or trend from the data. Thus, a decision (either rejection or acceptance) using the inspection of the biased sample leads to the lot's wrong disposition.

The primary assumption of random sampling is that every single item in the population has an equal (i.e., equal chance) chance of being included (selected) in the sample. For instance, suppose let us consider a secondary machining process. The sharpness of the knife for the secondary machining operation declines with the number of parts processed. If we take the sample during the initial (beginning) of production, then the sample will be biased.

That is because the items produced at last are not identical to the items produced at the beginning. Therefore, to control the sample's biasedness, we should take the sample from the beginning, the middle, and the end of the operation. Alternatively, we can use the top–middle–bottom approach.

Like randomization, sampling without a replacement is also another issue. That is because when we take one sample point from the population of size N, the probability of taking the first sample point is $\frac{1}{N}$. Then, since one sample point is already taken, the probability of taking the second sample point is $\frac{1}{N-1}$. Similarly, the probability of taking the third sample is $\frac{1}{N-2}$, and so on. Therefore, if we employ a sampling method without replacement, then the probability of the sample points to be included in the sample is not equal. In this case, sampling without replacement is the best to use. The DSP and MSP approach uses a sampling technique sampling without replacement. In this scenario, we should not return the inspected item to the lot.

Annex

Cumulative Binomial Probabilities

$$P(x \leq c) = \sum_{x}^{c} \binom{n}{x} p^x (1-p)^{n-x}$$

	p											
	c	0.05	0.10	0.20	0.30	0.40	0.50	0.60	0.70	0.80	0.90	0.95
$n = 1$	0	0.950	0.900	0.800	0.700	0.600	0.500	0.400	0.300	0.200	0.100	0.050
	1	1.000	1.000	1.000	1.000	1.000	1.000	1.000	1.000	1.000	1.000	1.000
$n = 2$	0	0.903	0.810	0.640	0.490	0.360	0.250	0.160	0.090	0.040	0.010	0.003
	1	0.998	0.990	0.960	0.910	0.840	0.750	0.640	0.510	0.360	0.190	0.098
	2	1.000	1.000	1.000	1.000	1.000	1.000	1.000	1.000	1.000	1.000	1.000
$n = 3$	0	0.857	0.729	0.512	0.343	0.216	0.125	0.064	0.027	0.008	0.001	0.000
	1	0.993	0.972	0.896	0.784	0.648	0.500	0.352	0.216	0.104	0.028	0.007
	2	1.000	0.999	0.992	0.973	0.936	0.875	0.784	0.657	0.488	0.271	0.143
	3	1.000	1.000	1.000	1.000	1.000	1.000	1.000	1.000	1.000	1.000	1.000
$n = 4$	0	0.815	0.656	0.410	0.240	0.130	0.063	0.026	0.008	0.002	0.000	0.000
	1	0.986	0.948	0.819	0.652	0.475	0.313	0.179	0.084	0.027	0.004	0.000
	2	1.000	0.996	0.973	0.916	0.821	0.688	0.525	0.348	0.181	0.052	0.014
	3	1.000	1.000	0.998	0.992	0.974	0.938	0.870	0.760	0.590	0.344	0.185
	4	1.000	1.000	1.000	1.000	1.000	1.000	1.000	1.000	1.000	1.000	1.000
$n = 5$	0	0.774	0.590	0.328	0.168	0.078	0.031	0.010	0.002	0.000	0.000	0.000
	1	0.977	0.919	0.737	0.528	0.337	0.188	0.087	0.031	0.007	0.000	0.000
	2	0.999	0.991	0.942	0.837	0.683	0.500	0.317	0.163	0.058	0.009	0.001
	3	1.000	1.000	0.993	0.969	0.913	0.813	0.663	0.472	0.263	0.081	0.023
	4	1.000	1.000	1.000	0.998	0.990	0.969	0.922	0.832	0.672	0.410	0.226
	5	1.000	1.000	1.000	1.000	1.000	1.000	1.000	1.000	1.000	1.000	1.000
$n = 6$	0	0.735	0.531	0.262	0.118	0.047	0.016	0.004	0.001	0.000	0.000	0.000
	1	0.967	0.886	0.655	0.420	0.233	0.109	0.041	0.011	0.002	0.000	0.000
	2	0.998	0.984	0.901	0.744	0.544	0.344	0.179	0.070	0.017	0.001	0.000
	3	1.000	0.999	0.983	0.930	0.821	0.656	0.456	0.256	0.099	0.016	0.002
	4	1.000	1.000	0.998	0.989	0.959	0.891	0.767	0.580	0.345	0.114	0.033
	5	1.000	1.000	1.000	0.999	0.996	0.984	0.953	0.882	0.738	0.469	0.265
	6	1.000	1.000	1.000	1.000	1.000	1.000	1.000	1.000	1.000	1.000	1.000
$n = 7$	0	0.698	0.478	0.210	0.082	0.028	0.008	0.002	0.000	0.000	0.000	0.000
	1	0.956	0.850	0.577	0.329	0.159	0.063	0.019	0.004	0.000	0.000	0.000

(continued)

	0.996	0.974	0.852	0.647	0.420	0.227	0.096	0.029	0.005	0.000	0.000
3	1.000	0.997	0.967	0.874	0.710	0.500	0.290	0.126	0.033	0.003	0.000
4	1.000	1.000	0.995	0.971	0.904	0.773	0.580	0.353	0.148	0.026	0.004
5	1.000	1.000	1.000	0.996	0.981	0.938	0.841	0.671	0.423	0.150	0.044
6	1.000	1.000	1.000	1.000	0.998	0.992	0.972	0.918	0.790	0.522	0.302
7	1.000	1.000	1.000	1.000	1.000	1.000	1.000	1.000	1.000	1.000	1.000

	p											
	c	0.05	0.10	0.20	0.30	0.40	0.50	0.60	0.70	0.80	0.90	0.95
$n=8$	0	0.663	0.430	0.168	0.058	0.017	0.004	0.001	0.000	0.000	0.000	0.000
	1	0.943	0.813	0.503	0.255	0.106	0.035	0.009	0.001	0.000	0.000	0.000
	2	0.994	0.962	0.797	0.552	0.315	0.145	0.050	0.011	0.001	0.000	0.000
	3	1.000	0.995	0.944	0.806	0.594	0.363	0.174	0.058	0.010	0.000	0.000
	4	1.000	1.000	0.990	0.942	0.826	0.637	0.406	0.194	0.056	0.005	0.000
	5	1.000	1.000	0.999	0.989	0.950	0.855	0.685	0.448	0.203	0.038	0.006
	6	1.000	1.000	1.000	0.999	0.991	0.965	0.894	0.745	0.497	0.187	0.057
	7	1.000	1.000	1.000	1.000	0.999	0.996	0.983	0.942	0.832	0.570	0.337
	8	1.000	1.000	1.000	1.000	1.000	1.000	1.000	1.000	1.000	1.000	1.000
$n=9$	0	0.630	0.387	0.134	0.040	0.010	0.002	0.000	0.000	0.000	0.000	0.000
	1	0.929	0.775	0.436	0.196	0.071	0.020	0.004	0.000	0.000	0.000	0.000
	2	0.992	0.947	0.738	0.463	0.232	0.090	0.025	0.004	0.000	0.000	0.000
	3	0.999	0.992	0.914	0.730	0.483	0.254	0.099	0.025	0.003	0.000	0.000
	4	1.000	0.999	0.980	0.901	0.733	0.500	0.267	0.099	0.020	0.001	0.000
	5	1.000	1.000	0.997	0.975	0.901	0.746	0.517	0.270	0.086	0.008	0.001
	6	1.000	1.000	1.000	0.996	0.975	0.910	0.768	0.537	0.262	0.053	0.008
	7	1.000	1.000	1.000	1.000	0.996	0.980	0.929	0.804	0.564	0.225	0.071
	8	1.000	1.000	1.000	1.000	1.000	0.998	0.990	0.960	0.866	0.613	0.370
	9	1.000	1.000	1.000	1.000	1.000	1.000	1.000	1.000	1.000	1.000	1.000
$n=10$	0	0.599	0.349	0.107	0.028	0.006	0.001	0.000	0.000	0.000	0.000	0.000
	1	0.914	0.736	0.376	0.149	0.046	0.011	0.002	0.000	0.000	0.000	0.000
	2	0.988	0.930	0.678	0.383	0.167	0.055	0.012	0.002	0.000	0.000	0.000
	3	0.999	0.987	0.879	0.650	0.382	0.172	0.055	0.011	0.001	0.000	0.000
	4	1.000	0.998	0.967	0.850	0.633	0.377	0.166	0.047	0.006	0.000	0.000

(continued)

	0.000	0.002	0.033	0.150	0.367	0.623	0.834	0.953	0.994	1.000	1.000	1.000
5	0.000	0.002	0.033	0.150	0.367	0.623	0.834	0.953	0.994	1.000	1.000	1.000
6	0.001	0.013	0.121	0.350	0.618	0.828	0.945	0.989	0.999	1.000	1.000	1.000
7	0.012	0.070	0.322	0.617	0.833	0.945	0.988	0.998	1.000	1.000	1.000	1.000
8	0.086	0.264	0.624	0.851	0.954	0.989	0.998	1.000	1.000	1.000	1.000	1.000
9	0.401	0.651	0.893	0.972	0.994	0.999	1.000	1.000	1.000	1.000	1.000	1.000
10	1.000	1.000	1.000	1.000	1.000	1.000	1.000	1.000	1.000	1.000	1.000	1.000
$n = 11$												
0	0.000	0.000	0.000	0.000	0.000	0.000	0.004	0.020	0.086	0.314	0.569	1.000
1	0.000	0.000	0.000	0.000	0.001	0.006	0.030	0.113	0.322	0.697	0.898	1.000
2	0.000	0.000	0.000	0.001	0.006	0.033	0.119	0.313	0.617	0.910	0.985	1.000
3	0.000	0.000	0.000	0.004	0.029	0.113	0.296	0.570	0.839	0.981	0.998	1.000
4	0.000	0.000	0.002	0.022	0.099	0.274	0.533	0.790	0.950	0.997	1.000	1.000
5	0.000	0.000	0.012	0.078	0.247	0.500	0.753	0.922	0.988	1.000	1.000	1.000
6	0.000	0.003	0.050	0.210	0.467	0.726	0.901	0.978	0.998	1.000	1.000	1.000
7	0.002	0.019	0.161	0.430	0.704	0.887	0.971	0.996	1.000	1.000	1.000	1.000
8	0.015	0.090	0.383	0.687	0.881	0.967	0.994	0.999	1.000	1.000	1.000	1.000
9	0.102	0.303	0.678	0.887	0.970	0.994	0.999	1.000	1.000	1.000	1.000	1.000
10	0.431	0.686	0.914	0.980	0.996	1.000	1.000	1.000	1.000	1.000	1.000	1.000
11	1.000	1.000	1.000	1.000	1.000	1.000	1.000	1.000	1.000	1.000	1.000	1.000

n	p											
	c	0.05	0.10	0.20	0.30	0.40	0.50	0.60	0.70	0.80	0.90	0.95
$n=12$	0	0.540	0.282	0.069	0.014	0.002	0.000	0.000	0.000	0.000	0.000	0.000
	1	0.882	0.659	0.275	0.085	0.020	0.003	0.000	0.000	0.000	0.000	0.000
	2	0.980	0.889	0.558	0.253	0.083	0.019	0.003	0.000	0.000	0.000	0.000
	3	0.998	0.974	0.795	0.493	0.225	0.073	0.015	0.002	0.000	0.000	0.000
	4	1.000	0.996	0.927	0.724	0.438	0.194	0.057	0.009	0.001	0.000	0.000
	5	1.000	0.999	0.981	0.882	0.665	0.387	0.158	0.039	0.004	0.000	0.000
	6	1.000	1.000	0.996	0.961	0.842	0.613	0.335	0.118	0.019	0.001	0.000
	7	1.000	1.000	0.999	0.991	0.943	0.806	0.562	0.276	0.073	0.004	0.000
	8	1.000	1.000	1.000	0.998	0.985	0.927	0.775	0.507	0.205	0.026	0.002
	9	1.000	1.000	1.000	1.000	0.997	0.981	0.917	0.747	0.442	0.111	0.020
	10	1.000	1.000	1.000	1.000	1.000	0.997	0.980	0.915	0.725	0.341	0.118
	11	1.000	1.000	1.000	1.000	1.000	1.000	0.998	0.986	0.931	0.718	0.460
	12	1.000	1.000	1.000	1.000	1.000	1.000	1.000	1.000	1.000	1.000	1.000
$n=13$	0	0.513	0.254	0.055	0.010	0.001	0.000	0.000	0.000	0.000	0.000	0.000
	1	0.865	0.621	0.234	0.064	0.013	0.002	0.000	0.000	0.000	0.000	0.000
	2	0.975	0.866	0.502	0.202	0.058	0.011	0.001	0.000	0.000	0.000	0.000
	3	0.997	0.966	0.747	0.421	0.169	0.046	0.008	0.001	0.000	0.000	0.000
	4	1.000	0.994	0.901	0.654	0.353	0.133	0.032	0.004	0.000	0.000	0.000
	5	1.000	0.999	0.970	0.835	0.574	0.291	0.098	0.018	0.001	0.000	0.000
	6	1.000	1.000	0.993	0.938	0.771	0.500	0.229	0.062	0.007	0.000	0.000
	7	1.000	1.000	0.999	0.982	0.902	0.709	0.426	0.165	0.030	0.001	0.000
	8	1.000	1.000	1.000	0.996	0.968	0.867	0.647	0.346	0.099	0.006	0.000
	9	1.000	1.000	1.000	0.999	0.992	0.954	0.831	0.579	0.253	0.034	0.003
	10	1.000	1.000	1.000	1.000	0.999	0.989	0.942	0.798	0.498	0.134	0.025
	11	1.000	1.000	1.000	1.000	1.000	0.998	0.987	0.936	0.766	0.379	0.135

(continued)

p	1.000	1.000	1.000	1.000	1.000	1.000	0.999	0.990	0.945	0.746	0.487
13	1.000	1.000	1.000	1.000	1.000	1.000	1.000	1.000	1.000	1.000	1.000
0	0.488	0.229	0.044	0.007	0.001	0.000	0.000	0.000	0.000	0.000	0.000
1	0.847	0.585	0.198	0.047	0.008	0.001	0.000	0.000	0.000	0.000	0.000
2	0.970	0.842	0.448	0.161	0.040	0.006	0.001	0.000	0.000	0.000	0.000
3	0.996	0.956	0.698	0.355	0.124	0.029	0.004	0.000	0.000	0.000	0.000
4	1.000	0.991	0.870	0.584	0.279	0.090	0.018	0.002	0.000	0.000	0.000
5	1.000	0.999	0.956	0.781	0.486	0.212	0.058	0.008	0.000	0.000	0.000
6	1.000	1.000	0.988	0.907	0.692	0.395	0.150	0.031	0.002	0.000	0.000
7	1.000	1.000	0.998	0.969	0.850	0.605	0.308	0.093	0.012	0.000	0.000
8	1.000	1.000	1.000	0.992	0.942	0.788	0.514	0.219	0.044	0.001	0.000
9	1.000	1.000	1.000	0.998	0.982	0.910	0.721	0.416	0.130	0.009	0.000
10	1.000	1.000	1.000	1.000	0.996	0.971	0.876	0.645	0.302	0.044	0.004
11	1.000	1.000	1.000	1.000	0.999	0.994	0.960	0.839	0.552	0.158	0.030
12	1.000	1.000	1.000	1.000	1.000	0.999	0.992	0.953	0.802	0.415	0.153
13	1.000	1.000	1.000	1.000	1.000	1.000	0.999	0.993	0.956	0.771	0.512
14	1.000	1.000	1.000	1.000	1.000	1.000	1.000	1.000	1.000	1.000	1.000

$n = 14$

	c	0.05	0.10	0.20	0.30	0.40	0.50	0.60	0.70	0.80	0.90	0.95
$n = 15$	0	0.463	0.206	0.035	0.005	0.000	0.000	0.000	0.000	0.000	0.000	0.000
	1	0.829	0.549	0.167	0.035	0.005	0.000	0.000	0.000	0.000	0.000	0.000
	2	0.964	0.816	0.398	0.127	0.027	0.004	0.000	0.000	0.000	0.000	0.000
	3	0.995	0.944	0.648	0.297	0.091	0.018	0.002	0.000	0.000	0.000	0.000
	4	0.999	0.987	0.836	0.515	0.217	0.059	0.009	0.001	0.000	0.000	0.000
	5	1.000	0.998	0.939	0.722	0.403	0.151	0.034	0.004	0.000	0.000	0.000
	6	1.000	1.000	0.982	0.869	0.610	0.304	0.095	0.015	0.001	0.000	0.000
	7	1.000	1.000	0.996	0.950	0.787	0.500	0.213	0.050	0.004	0.000	0.000
	8	1.000	1.000	0.999	0.985	0.905	0.696	0.390	0.131	0.018	0.000	0.000
	9	1.000	1.000	1.000	0.996	0.966	0.849	0.597	0.278	0.061	0.002	0.000
	10	1.000	1.000	1.000	0.999	0.991	0.941	0.783	0.485	0.164	0.013	0.001
	11	1.000	1.000	1.000	1.000	0.998	0.982	0.909	0.703	0.352	0.056	0.005
	12	1.000	1.000	1.000	1.000	1.000	0.996	0.973	0.873	0.602	0.184	0.036
	13	1.000	1.000	1.000	1.000	1.000	1.000	0.995	0.965	0.833	0.451	0.171
	14	1.000	1.000	1.000	1.000	1.000	1.000	1.000	0.995	0.965	0.794	0.537
	15	1.000	1.000	1.000	1.000	1.000	1.000	1.000	1.000	1.000	1.000	1.000
$n = 16$	0	0.440	0.185	0.028	0.003	0.000	0.000	0.000	0.000	0.000	0.000	0.000
	1	0.811	0.515	0.141	0.026	0.003	0.000	0.000	0.000	0.000	0.000	0.000
	2	0.957	0.789	0.352	0.099	0.018	0.002	0.000	0.000	0.000	0.000	0.000
	3	0.993	0.932	0.598	0.246	0.065	0.011	0.001	0.000	0.000	0.000	0.000
	4	0.999	0.983	0.798	0.450	0.167	0.038	0.005	0.000	0.000	0.000	0.000
	5	1.000	0.997	0.918	0.660	0.329	0.105	0.019	0.002	0.000	0.000	0.000
	6	1.000	0.999	0.973	0.825	0.527	0.227	0.058	0.007	0.000	0.000	0.000
	7	1.000	1.000	0.993	0.926	0.716	0.402	0.142	0.026	0.001	0.000	0.000
	8	1.000	1.000	0.999	0.974	0.858	0.598	0.284	0.074	0.007	0.000	0.000

(continued)

p	0.05	0.10	0.20	0.30	0.40	0.50	0.60	0.70	0.80	0.90	0.95
9	1.000	1.000	1.000	0.993	0.942	0.773	0.473	0.175	0.027	0.001	0.000
10	1.000	1.000	1.000	0.998	0.981	0.895	0.671	0.340	0.082	0.003	0.000
11	1.000	1.000	1.000	1.000	0.995	0.962	0.833	0.550	0.202	0.017	0.001
12	1.000	1.000	1.000	1.000	0.999	0.989	0.935	0.754	0.402	0.068	0.007
13	1.000	1.000	1.000	1.000	1.000	0.998	0.982	0.901	0.648	0.211	0.043
14	1.000	1.000	1.000	1.000	1.000	1.000	0.997	0.974	0.859	0.485	0.189
15	1.000	1.000	1.000	1.000	1.000	1.000	1.000	0.997	0.972	0.815	0.560
16	1.000	1.000	1.000	1.000	1.000	1.000	1.000	1.000	1.000	1.000	1.000

$n = 17$

p	0.05	0.10	0.20	0.30	0.40	0.50	0.60	0.70	0.80	0.90	0.95
c	0.05	0.10	0.20	0.30	0.40	0.50	0.60	0.70	0.80	0.90	0.95
0	0.418	0.167	0.023	0.002	0.000	0.000	0.000	0.000	0.000	0.000	0.000
1	0.792	0.482	0.118	0.019	0.002	0.000	0.000	0.000	0.000	0.000	0.000
2	0.950	0.762	0.310	0.077	0.012	0.001	0.000	0.000	0.000	0.000	0.000
3	0.991	0.917	0.549	0.202	0.046	0.006	0.000	0.000	0.000	0.000	0.000
4	0.999	0.978	0.758	0.389	0.126	0.025	0.003	0.000	0.000	0.000	0.000
5	1.000	0.995	0.894	0.597	0.264	0.072	0.011	0.001	0.000	0.000	0.000
6	1.000	0.999	0.962	0.775	0.448	0.166	0.035	0.003	0.000	0.000	0.000
7	1.000	1.000	0.989	0.895	0.641	0.315	0.092	0.013	0.000	0.000	0.000
8	1.000	1.000	0.997	0.960	0.801	0.500	0.199	0.040	0.003	0.000	0.000
9	1.000	1.000	1.000	0.987	0.908	0.685	0.359	0.105	0.011	0.000	0.000
10	1.000	1.000	1.000	0.997	0.965	0.834	0.552	0.225	0.038	0.001	0.000
11	1.000	1.000	1.000	0.999	0.989	0.928	0.736	0.403	0.106	0.005	0.000
12	1.000	1.000	1.000	1.000	0.997	0.975	0.874	0.611	0.242	0.022	0.001
13	1.000	1.000	1.000	1.000	1.000	0.994	0.954	0.798	0.451	0.083	0.009
14	1.000	1.000	1.000	1.000	1.000	0.999	0.988	0.923	0.690	0.238	0.050

(continued)

Cumulative binomial probability table (columns are values of p; rows are values of c).

Continuation rows (from the preceding table):

c	0.95	0.90	0.80	0.70	0.60	0.50	0.40	0.30	0.20	0.10	0.05
16	0.208	0.518	0.882	0.981	0.998	1.000	1.000	1.000	1.000	1.000	1.000
17	0.582	0.833	0.977	0.998	1.000	1.000	1.000	1.000	1.000	1.000	1.000

$n = 18$

c	0.95	0.90	0.80	0.70	0.60	0.50	0.40	0.30	0.20	0.10	0.05
0	0.000	0.000	0.000	0.000	0.000	0.000	0.000	0.002	0.018	0.150	0.397
1	0.000	0.000	0.000	0.000	0.000	0.000	0.001	0.014	0.099	0.450	0.774
2	0.000	0.000	0.000	0.000	0.000	0.001	0.008	0.060	0.271	0.734	0.942
3	0.000	0.000	0.000	0.000	0.001	0.004	0.033	0.165	0.501	0.902	0.989
4	0.000	0.000	0.000	0.000	0.006	0.015	0.094	0.333	0.716	0.972	0.998
5	0.000	0.000	0.000	0.001	0.020	0.048	0.209	0.534	0.867	0.994	1.000
6	0.000	0.000	0.000	0.006	0.058	0.119	0.374	0.722	0.949	0.999	1.000
7	0.000	0.000	0.001	0.021	0.135	0.240	0.563	0.859	0.984	1.000	1.000
8	0.000	0.000	0.004	0.060	0.263	0.407	0.737	0.940	0.996	1.000	1.000
9	0.000	0.000	0.016	0.141	0.437	0.593	0.865	0.979	0.999	1.000	1.000
10	0.000	0.001	0.051	0.278	0.626	0.760	0.942	0.994	1.000	1.000	1.000
11	0.000	0.006	0.133	0.466	0.791	0.881	0.980	0.999	1.000	1.000	1.000
12	0.000	0.028	0.284	0.667	0.906	0.952	0.994	1.000	1.000	1.000	1.000
13	0.002	0.098	0.499	0.835	0.967	0.985	0.999	1.000	1.000	1.000	1.000
14	0.011	0.266	0.729	0.940	0.992	0.996	1.000	1.000	1.000	1.000	1.000
15	0.058	0.550	0.901	0.986	0.999	0.999	1.000	1.000	1.000	1.000	1.000
16	0.226	0.850	0.982	0.998	1.000	1.000	1.000	1.000	1.000	1.000	1.000
17	0.603	0.982	0.998	1.000	1.000	1.000	1.000	1.000	1.000	1.000	1.000
18	1.000	1.000	1.000	1.000	1.000	1.000	1.000	1.000	1.000	1.000	1.000

$n = 19$

c	0.95	0.90	0.80	0.70	0.60	0.50	0.40	0.30	0.20	0.10	0.05
0	0.000	0.000	0.000	0.000	0.000	0.000	0.000	0.001	0.014	0.135	0.377
1	0.000	0.000	0.000	0.000	0.000	0.000	0.001	0.010	0.083	0.420	0.755

(continued)

p											
2	0.933	0.705	0.237	0.046	0.005	0.000	0.000	0.000	0.000	0.000	0.000
3	0.987	0.885	0.455	0.133	0.023	0.002	0.000	0.000	0.000	0.000	0.000
4	0.998	0.965	0.673	0.282	0.070	0.010	0.001	0.000	0.000	0.000	0.000
5	1.000	0.991	0.837	0.474	0.163	0.032	0.003	0.001	0.000	0.000	0.000
6	1.000	0.998	0.932	0.666	0.308	0.084	0.012	0.003	0.000	0.000	0.000
7	1.000	1.000	0.977	0.818	0.488	0.180	0.035	0.011	0.002	0.000	0.000
8	1.000	1.000	0.993	0.916	0.667	0.324	0.088	0.033	0.007	0.000	0.000
9	1.000	1.000	0.998	0.967	0.814	0.500	0.186	0.084	0.023	0.002	0.000
10	1.000	1.000	1.000	0.989	0.912	0.676	0.333	0.182	0.068	0.009	0.000
11	1.000	1.000	1.000	0.997	0.965	0.820	0.512	0.334	0.163	0.035	0.002
12	1.000	1.000	1.000	0.999	0.988	0.916	0.692	0.526	0.327	0.115	0.013
13	1.000	1.000	1.000	1.000	0.997	0.968	0.837	0.718	0.545	0.295	0.067
14	1.000	1.000	1.000	1.000	0.999	0.990	0.930	0.867	0.763	0.580	0.245
15	1.000	1.000	1.000	1.000	1.000	0.998	0.977	0.954	0.917	0.865	0.623
16	1.000	1.000	1.000	1.000	1.000	1.000	0.995	0.990	0.986	1.000	1.000
17	1.000	1.000	1.000	1.000	1.000	1.000	0.999	0.999	1.000	1.000	1.000
18	1.000	1.000	1.000	1.000	1.000	1.000	1.000	1.000	1.000	1.000	1.000
19	1.000	1.000	1.000	1.000	1.000	1.000	1.000	1.000	1.000	1.000	1.000
n = 20											
0	0.358	0.122	0.012	0.001	0.000	0.000	0.000	0.000	0.000	0.000	0.000
1	0.736	0.392	0.069	0.008	0.001	0.000	0.000	0.000	0.000	0.000	0.000
2	0.925	0.677	0.206	0.035	0.004	0.000	0.000	0.000	0.000	0.000	0.000
3	0.984	0.867	0.411	0.107	0.016	0.001	0.000	0.000	0.000	0.000	0.000
4	0.997	0.957	0.630	0.238	0.051	0.006	0.000	0.000	0.000	0.000	0.000
5	1.000	0.989	0.804	0.416	0.126	0.021	0.002	0.000	0.000	0.000	0.000
6	1.000	0.998	0.913	0.608	0.250	0.058	0.006	0.000	0.000	0.000	0.000
7	1.000	1.000	0.968	0.772	0.416	0.132	0.021	0.001	0.000	0.000	0.000

(continued)

p	0.05	0.10	0.20	0.30	0.40	0.50	0.60	0.70	0.80	0.90	0.95
8	1.000	1.000	0.990	0.887	0.596	0.252	0.057	0.005	0.000	0.000	0.000
9	1.000	1.000	0.997	0.952	0.755	0.412	0.128	0.017	0.001	0.000	0.000
10	1.000	1.000	0.999	0.983	0.872	0.588	0.245	0.048	0.003	0.000	0.000
11	1.000	1.000	1.000	0.995	0.943	0.748	0.404	0.113	0.010	0.000	0.000
12	1.000	1.000	1.000	0.999	0.979	0.868	0.584	0.228	0.032	0.000	0.000
13	1.000	1.000	1.000	1.000	0.994	0.942	0.750	0.392	0.087	0.002	0.000
14	1.000	1.000	1.000	1.000	0.998	0.979	0.874	0.584	0.196	0.011	0.000
15	1.000	1.000	1.000	1.000	1.000	0.994	0.949	0.762	0.370	0.043	0.003
16	1.000	1.000	1.000	1.000	1.000	0.999	0.984	0.893	0.589	0.133	0.016
17	1.000	1.000	1.000	1.000	1.000	1.000	0.996	0.965	0.794	0.323	0.075
18	1.000	1.000	1.000	1.000	1.000	1.000	0.999	0.992	0.931	0.608	0.264
19	1.000	1.000	1.000	1.000	1.000	1.000	1.000	0.999	0.988	0.878	0.642
20	1.000	1.000	1.000	1.000	1.000	1.000	1.000	1.000	1.000	1.000	1.000

$n = 25$

p	0.05	0.10	0.20	0.30	0.40	0.50	0.60	0.70	0.80	0.90	0.95
c											
0	0.277	0.072	0.004	0.000	0.000	0.000	0.000	0.000	0.000	0.000	0.000
1	0.642	0.271	0.027	0.002	0.000	0.000	0.000	0.000	0.000	0.000	0.000
2	0.873	0.537	0.098	0.009	0.000	0.000	0.000	0.000	0.000	0.000	0.000
3	0.966	0.764	0.234	0.033	0.002	0.000	0.000	0.000	0.000	0.000	0.000
4	0.993	0.902	0.421	0.090	0.009	0.000	0.000	0.000	0.000	0.000	0.000
5	0.999	0.967	0.617	0.193	0.029	0.002	0.000	0.000	0.000	0.000	0.000
6	1.000	0.991	0.780	0.341	0.074	0.007	0.000	0.000	0.000	0.000	0.000
7	1.000	0.998	0.891	0.512	0.154	0.022	0.001	0.000	0.000	0.000	0.000
8	1.000	1.000	0.953	0.677	0.274	0.054	0.004	0.000	0.000	0.000	0.000
9	1.000	1.000	0.983	0.811	0.425	0.115	0.013	0.000	0.000	0.000	0.000

(continued)

p	0.000	0.000	0.000	0.002	0.034	0.212	0.586	0.902	0.994	1.000	1.000
11	0.000	0.000	0.000	0.006	0.078	0.345	0.732	0.956	0.998	1.000	1.000
12	0.000	0.000	0.000	0.017	0.154	0.500	0.846	0.983	1.000	1.000	1.000
13	0.000	0.000	0.000	0.044	0.268	0.655	0.922	0.994	1.000	1.000	1.000
14	0.000	0.000	0.002	0.098	0.414	0.788	0.966	0.998	1.000	1.000	1.000
15	0.000	0.000	0.006	0.189	0.575	0.885	0.987	1.000	1.000	1.000	1.000
16	0.000	0.000	0.017	0.323	0.726	0.946	0.996	1.000	1.000	1.000	1.000
17	0.000	0.002	0.047	0.488	0.846	0.978	0.999	1.000	1.000	1.000	1.000
18	0.000	0.009	0.109	0.659	0.926	0.993	1.000	1.000	1.000	1.000	1.000
19	0.000	0.033	0.220	0.807	0.971	0.998	1.000	1.000	1.000	1.000	1.000
20	0.001	0.098	0.383	0.910	0.991	1.000	1.000	1.000	1.000	1.000	1.000
21	0.007	0.236	0.579	0.967	0.998	1.000	1.000	1.000	1.000	1.000	1.000
22	0.034	0.463	0.766	0.991	1.000	1.000	1.000	1.000	1.000	1.000	1.000
23	0.127	0.729	0.902	0.998	1.000	1.000	1.000	1.000	1.000	1.000	1.000
24	0.358	0.928	0.973	1.000	1.000	1.000	1.000	1.000	1.000	1.000	1.000
25	0.723	0.996	0.996	1.000	1.000	1.000	1.000	1.000	1.000	1.000	1.000

Cumulative Probability of Poisson Distribution

$$P(x \leq c) = \sum_{x}^{c} \frac{e^{-\lambda}\lambda^x}{x!}$$

λ										
X	0.1	0.2	0.3	0.4	0.5	0.6	0.7	0.8	0.9	1.0
0	0.90480	0.81870	0.74080	0.67030	0.60650	0.54880	0.49660	0.44930	0.40660	0.36790
1	0.09050	0.16370	0.22220	0.26810	0.30330	0.32930	0.34760	0.35950	0.36590	0.36790
2	0.00450	0.01640	0.03330	0.05360	0.07580	0.09880	0.12170	0.14380	0.16470	0.18390
3	0.00020	0.00110	0.00330	0.00720	0.01260	0.01980	0.02840	0.03830	0.04940	0.06130
4	0.00000	0.00010	0.00030	0.00070	0.00160	0.00300	0.00500	0.00770	0.01110	0.01530
5	0.00000	0.00000	0.00000	0.00010	0.00020	0.00040	0.00070	0.00120	0.00200	0.00310
6	0.00000	0.00000	0.00000	0.00000	0.00000	0.00000	0.00010	0.00020	0.00030	0.00050
7	0.00000	0.00000	0.00000	0.00000	0.00000	0.00000	0.00000	0.00000	0.00000	0.00010

λ										
X	1.1	1.2	1.3	1.4	1.5	1.6	1.7	1.8	1.9	2.0
0	0.33290	0.30120	0.27250	0.24660	0.22310	0.20190	0.18270	0.16530	0.14960	0.13530
1	0.36620	0.36140	0.35430	0.34520	0.33470	0.32300	0.31060	0.29750	0.28420	0.27070
2	0.20140	0.21690	0.23030	0.24170	0.25100	0.25840	0.26400	0.26780	0.27000	0.27070
3	0.07380	0.08670	0.09980	0.11280	0.12550	0.13780	0.14960	0.16070	0.17100	0.18040
4	0.02030	0.02600	0.03240	0.03950	0.04710	0.05510	0.06360	0.07230	0.08120	0.09020
5	0.00450	0.00620	0.00840	0.01110	0.01410	0.01760	0.02160	0.02600	0.03090	0.03610
6	0.00080	0.00120	0.00180	0.00260	0.00350	0.00470	0.00610	0.00780	0.00980	0.01200
7	0.00010	0.00020	0.00030	0.00050	0.00080	0.00110	0.00150	0.00200	0.00270	0.00340
8	0.00000	0.00000	0.00010	0.00010	0.00010	0.00020	0.00030	0.00050	0.00060	0.00090
9	0.00000	0.00000	0.00000	0.00000	0.00000	0.00000	0.00010	0.00010	0.00010	0.00020

λ										
X	2.1	2.2	2.3	2.4	2.5	2.6	2.7	2.8	2.9	3.0
0	0.12250	0.11080	0.10030	0.09070	0.08210	0.07430	0.06720	0.06080	0.05500	0.04980
1	0.25720	0.24380	0.23060	0.21770	0.20520	0.19310	0.18150	0.17030	0.15960	0.14940
2	0.27000	0.26810	0.26520	0.26130	0.25650	0.25100	0.24500	0.23840	0.23140	0.22400

(continued)

X										
3	0.22400	0.22370	0.22250	0.22050	0.21760	0.21380	0.20900	0.20330	0.19660	0.18900
4	0.16800	0.16220	0.15570	0.14880	0.14140	0.13360	0.12540	0.11690	0.10820	0.09920
5	0.10080	0.09400	0.08720	0.08040	0.07350	0.06680	0.06020	0.05380	0.04760	0.04170
6	0.05040	0.04550	0.04070	0.03620	0.03190	0.02780	0.02410	0.02060	0.01740	0.01460
7	0.02160	0.01880	0.01630	0.01390	0.01180	0.00990	0.00830	0.00680	0.00550	0.00440
8	0.00810	0.00680	0.00570	0.00470	0.00380	0.00310	0.00250	0.00190	0.00150	0.00110
9	0.00270	0.00220	0.00180	0.00140	0.00110	0.00090	0.00070	0.00050	0.00040	0.00030
10	0.00080	0.00060	0.00050	0.00040	0.00030	0.00020	0.00020	0.00010	0.00010	0.00010
11	0.00020	0.00020	0.00010	0.00010	0.00010	0.00000	0.00000	0.00000	0.00000	0.00000
12	0.00010	0.00000	0.00000	0.00000	0.00000	0.00000	0.00000	0.00000	0.00000	0.00000

λ

X	4.0	3.9	3.8	3.7	3.6	3.5	3.4	3.3	3.2	3.1
0	0.01830	0.02020	0.02240	0.02470	0.02730	0.03020	0.03340	0.03690	0.04080	0.04500
1	0.07330	0.07890	0.08500	0.09150	0.09840	0.10570	0.11350	0.12170	0.13400	0.13970
2	0.14650	0.15390	0.16150	0.16920	0.17710	0.18500	0.19290	0.20080	0.20870	0.21650
3	0.19540	0.20010	0.20460	0.20870	0.21250	0.21580	0.21860	0.22090	0.22260	0.22370
4	0.19540	0.19510	0.19440	0.19310	0.19120	0.18880	0.18580	0.18230	0.17810	0.17340
5	0.15630	0.15220	0.14770	0.14290	0.13770	0.13220	0.12640	0.12030	0.11400	0.10750
6	0.10420	0.09890	0.09360	0.08810	0.08260	0.07710	0.07160	0.06620	0.06080	0.05550
7	0.05950	0.05510	0.05080	0.04660	0.04250	0.03850	0.03480	0.03120	0.02780	0.02460
8	0.02980	0.02690	0.02410	0.02150	0.01910	0.01690	0.01480	0.01290	0.01110	0.00950
9	0.01320	0.01160	0.01020	0.00890	0.00760	0.00660	0.00560	0.00470	0.00400	0.00330
10	0.00530	0.00450	0.00390	0.00330	0.00280	0.00230	0.00190	0.00160	0.00130	0.00100
11	0.00190	0.00160	0.00130	0.00110	0.00090	0.00070	0.00060	0.00050	0.00040	0.00030
12	0.00060	0.00050	0.00040	0.00030	0.00030	0.00020	0.00020	0.00010	0.00010	0.00010
13	0.00020	0.00020	0.00010	0.00010	0.00010	0.00010	0.00000	0.00000	0.00000	0.00000
14	0.00010	0.00000	0.00000	0.00000	0.00000	0.00000	0.00000	0.00000	0.00000	0.00000

(continued)

X	λ									
	4.1	4.2	4.3	4.4	4.5	4.6	4.7	4.8	4.9	5.0
0	0.01660	0.01500	0.01360	0.01230	0.01110	0.01010	0.00910	0.00820	0.00740	0.00670
1	0.06790	0.06300	0.05830	0.05400	0.05000	0.04620	0.04270	0.03950	0.03650	0.03370
2	0.13930	0.13230	0.12540	0.11880	0.11250	0.10630	0.10050	0.09480	0.08940	0.08420
3	0.19040	0.18520	0.17980	0.17430	0.16870	0.16310	0.15740	0.15170	0.14600	0.14040
4	0.19510	0.19440	0.19330	0.19170	0.18980	0.18750	0.18490	0.18200	0.17890	0.17550
5	0.16000	0.16330	0.16620	0.16870	0.17080	0.17250	0.17380	0.17470	0.17530	0.17550
6	0.10930	0.11430	0.11910	0.12370	0.12810	0.13230	0.13620	0.13980	0.14320	0.14620
7	0.06400	0.06860	0.07320	0.07780	0.08240	0.08690	0.09140	0.09590	0.10020	0.10440
8	0.03280	0.03600	0.03930	0.04280	0.04630	0.05000	0.05370	0.05750	0.06140	0.06530
9	0.01500	0.01680	0.01880	0.02090	0.02320	0.02550	0.02800	0.03070	0.03340	0.03630
10	0.00610	0.00710	0.00810	0.00920	0.01040	0.01180	0.01320	0.01470	0.01640	0.01810
11	0.00230	0.00270	0.00320	0.00370	0.00430	0.00490	0.00560	0.00640	0.00730	0.00820
12	0.00080	0.00090	0.00110	0.00140	0.00160	0.00190	0.00220	0.00260	0.00300	0.00340
13	0.00020	0.00030	0.00040	0.00050	0.00060	0.00070	0.00080	0.00090	0.00110	0.00130
14	0.00010	0.00010	0.00010	0.00010	0.00020	0.00020	0.00030	0.00030	0.00040	0.00050
15	0.00000	0.00000	0.00000	0.00000	0.00010	0.00010	0.00010	0.00010	0.00010	0.00020

X	λ									
	5.1	5.2	5.3	5.4	5.5	5.6	5.7	5.8	5.9	6.0
0	0.00610	0.00550	0.00500	0.00450	0.00410	0.00370	0.00330	0.00300	0.00270	0.00250
1	0.03110	0.02870	0.02650	0.02440	0.02250	0.02070	0.01910	0.01760	0.01620	0.01490
2	0.07930	0.07460	0.07010	0.06590	0.06180	0.05800	0.05440	0.05090	0.04770	0.04460
3	0.13480	0.12930	0.12390	0.11850	0.11330	0.10820	0.10330	0.09850	0.09380	0.08920
4	0.17190	0.16810	0.16410	0.16000	0.15580	0.15150	0.14720	0.14280	0.13830	0.13390
5	0.17530	0.17480	0.17400	0.17280	0.17140	0.16970	0.16780	0.16560	0.16320	0.16060
6	0.14900	0.15150	0.15370	0.15550	0.15710	0.15840	0.15940	0.16010	0.16050	0.16060

(continued)

7	0.10860	0.11250	0.11630	0.12000	0.12340	0.12670	0.12980	0.13260	0.13530	0.13770
8	0.06920	0.07310	0.07710	0.08100	0.08490	0.08870	0.09250	0.09620	0.09980	0.10330
9	0.03920	0.04230	0.04540	0.04860	0.05190	0.05520	0.05860	0.06200	0.06540	0.06880
10	0.02000	0.02200	0.02410	0.02620	0.02850	0.03090	0.03340	0.03590	0.03860	0.04130
11	0.00930	0.01040	0.01160	0.01290	0.01430	0.01570	0.01730	0.01900	0.02070	0.02250
12	0.00390	0.00450	0.00510	0.00580	0.00650	0.00730	0.00820	0.00920	0.01020	0.01130
13	0.00150	0.00180	0.00210	0.00240	0.00280	0.00320	0.00360	0.00410	0.00460	0.00520
14	0.00060	0.00070	0.00080	0.00090	0.00110	0.00130	0.00150	0.00170	0.00190	0.00220
15	0.00020	0.00020	0.00030	0.00030	0.00040	0.00050	0.00060	0.00070	0.00080	0.00090
16	0.00010	0.00010	0.00010	0.00010	0.00010	0.00020	0.00020	0.00020	0.00030	0.00030
17	0.00000	0.00000	0.00000	0.00000	0.00000	0.00000	0.00010	0.00010	0.00010	0.00010

	λ									
X	**6.1**	**6.2**	**6.3**	**6.4**	**6.5**	**6.6**	**6.7**	**6.8**	**6.9**	**7.0**
0	0.00220	0.00200	0.00180	0.00170	0.00150	0.00140	0.00120	0.00110	0.00100	0.00090
1	0.01370	0.01260	0.01160	0.01060	0.00980	0.00900	0.00820	0.00760	0.00700	0.00640
2	0.04170	0.03900	0.03640	0.03400	0.03180	0.02960	0.02760	0.02580	0.02400	0.02230
3	0.08480	0.08060	0.07650	0.07260	0.06880	0.06520	0.06170	0.05840	0.05520	0.05210
4	0.12940	0.12490	0.12050	0.11620	0.11180	0.10760	0.10340	0.09920	0.09520	0.09120
5	0.15790	0.15490	0.15190	0.14870	0.14540	0.14200	0.13850	0.13490	0.13140	0.12770
6	0.16050	0.16010	0.15950	0.15860	0.15750	0.15620	0.15460	0.15290	0.15110	0.14900
7	0.13990	0.14180	0.14350	0.14500	0.14620	0.14720	0.14800	0.14860	0.14890	0.14900
8	0.10660	0.10990	0.11300	0.11600	0.11880	0.12150	0.12400	0.12630	0.12840	0.13040
9	0.07230	0.07570	0.07910	0.08250	0.08580	0.08910	0.09230	0.09540	0.09850	0.10140
10	0.04410	0.04690	0.04980	0.05280	0.05580	0.05880	0.06180	0.06490	0.06790	0.07100
11	0.02450	0.02650	0.02850	0.03070	0.03300	0.03530	0.03770	0.04010	0.04260	0.04520
12	0.01240	0.01370	0.01500	0.01640	0.01790	0.01940	0.02100	0.02770	0.02450	0.02640
13	0.00580	0.00650	0.00730	0.00810	0.00890	0.00980	0.01080	0.01190	0.01300	0.01420
14	0.00250	0.00290	0.00330	0.00370	0.00410	0.00460	0.00520	0.00580	0.00640	0.00710

Bibliography

T.W. Anderson, *Introduction to Multivariate Statistical Analysis*, 2nd edn. (Wiley, New York, NY, 1984)

ANSI/ASQ Z1.4, *Sampling Procedures and Tables for Inspection by Attributes* 2008

ANSI/ASQ Z1.9, *Sampling Procedures and Tables for Inspection by Variables for Percent Nonconforming* (2003)

Army Chemical Corps, *Master Sampling Plans for Single, Duplicate, Double and Multiple Sampling, Manual No. 2* (Chemical Corps Engineering Agency, Army Chemical Center, Arsenal, MD, 1953)

A.F. Bissell, How reliable is your capability index? Appl. Stat. **39**, 331–340 (1990)

C.W. Champ, W.H. Woodall, Exact results for Shewhart control charts with supplementary runs rules. Technometrics **29**, 393–399 (1987)

J.-P. Chen, C.G. Ding, A new process capability index for non-normal distributions. Int. J. Qual. Reliab. Mange. **18**(7), 762–770 (2001)

A.J. Duncan, *Quality Control and Industrial Statistics*, 5th edn. (Irwin, Homewood, IL, 1986)

M.A. Durivage, *Practical Engineering, Process, and Reliability Statistics* (ASQ Quality Press, Milwaukee, 2014)

H. Hotelling, Multivariate quality control, in *Techniques of Statistical Analysis*, ed. by C. Eisenhart, M. W. Hastay, W. A. Wallis, (McGraw-Hill, New York, NY, 1947)

R.A. Johnson, D.W. Wichern, *Applied Multivariate Statistical Analysis*, 4th edn. (Prentice Hall, Upper Saddle River, NJ, 1998)

J.M. Juran, Early SQC: A historical supplement. Qual. Prog. **30**(9), 73–81 (1997)

S. Kotz, N.L. Johnson, *Process Capability Indices* (Chapman & Hall, London, 1992)

C.A. Lowry, W.H. Woodall, C.W. Champ, S.E. Rigdon, A multivariate exponentially weighted moving average chart. Technometrics **34**, 46–53 (1992)

J.M. Lucas, M.S. Saccucci, Exponentially weighted moving average control schemes: Properties and enhancements. Technometrics **32**, 1–29 (1990)

D.C. Montgomery, *Introduction to Statistical Quality Control*, 4th edn. (Wiley, New York, NY, 2000)

E.R. Ott, E.G. Schilling, *Process Quality Control*, 2nd edn. (McGraw-Hill, New York, NY, 1990)

W.L. Pearn, Y.T. Tai, F. Hsiao, Y.P. Ao, Approximately unbiased estimator for non-normal process capability index CNpk. J. Test. Eval. **42**(6), 1–10 (2014)

C.P. Quesenberry, The effect of sample size on estimated limits for and X control charts. J. Qual. Technol. **25**(4), 237–247 (1993)

T.P. Ryan, *Statistical Methods for Quality Improvement*, 2nd edn. (Wiley, New York, NY, 2000)

T.P. Ryan, N.C. Schwertman, Optimal limits for attributes control charts. J. Qual. Technol. **29**(1), 86–98 (1997)

E.G. Schilling, *Acceptance Sampling in Quality Control* (Marcel Dekker, New York, NY, 1982)

N.K. Squeglia, *Zero Acceptance Number Sampling Plans*, 5th edn. (ASQ Quality Press, Milwaukee, 2008)

N.D. Tracy, J.C. Young, R.L. Mason, Multivariate control charts for individual observations. J. Qual. Technol. **24**(2), 88–95 (1992)

W.H. Woodall, Control charting based on attribute data: Bibliography and review. J. Qual. Technol. **29**, 172–183 (1997)

W.H. Woodall, B.M. Adams, The statistical design of CUSUM charts. Qual. Eng. **5**(4), 559–570 (1993)

N.F. Zhang, A. Stenback, D.M. Wardrop, Interval estimation of the process capability index. Commun. Stat. Theory Methods **19**(21), 4455–4470 (1990)

Chapter 4
Statistical Process Control (SPC)

4.1 Introduction

The International Organization for Standardization 9000 (abbreviated as ISO 9000) emphasizes the world's manufacturing companies' agreement on the quality requirements. Suppose the company is certified with ISO 9000. In that case, it means that the company comprehends documentation, demonstrates the effectiveness of quality stems, and can implement the ISO well-defined quality programs.

ISO 9000 guiding principles are basic and useful for users to establish their quality systems. The ISO 13485 aimed for manufacturing medical devices, ISO 9100 for the aerospace industry suppliers, and several standards for other industries. Statistical process control (SPC) is the authoritative instrument in accomplishing the ISO 9000 certification via structuring their quality systems.

In the ecosphere of engineering, designs are determined, calculations are precise, and dimensions set exact. The engineering designing process is just as one plus one is two. However, in the manufacturing ecosphere, outputs show (significant) variations against the product's original engineering design.

In manufacturing, variation is an intrinsic nature of processes that produce the product. The variations that have an insignificant effect on the quality of the product are called acceptable variations. Else, suppose the variations have a significant effect on the quality and the functionality of the product. In that case, the manufacturing process is characterized as incapable of producing the product.

Here, the most important thing is quantifying the variation and characterizing whether the variation is acceptable. Therefore, to assist the manufacturing's efficiency and effectiveness, analyzing the variation is a critical question. The clear answer is that a statistics expert will lead to the analysis. Statisticians are data engineers who can quantify the variability in the process.

In the 1920s, at Bell LAB, Walter Shewhart introduced Statistical Process Control (SPC). Since then, step-by-step, statistical process control (SPC) brings a

Y. Y. Tesfay, *Developing Structured Procedural and Methodological Engineering Designs*, https://doi.org/10.1007/978-3-030-68402-0_4

set of tools to measure and analyze process data. That assists in characterizing our process's behavior and to achieve process control.

Today, statistical process control (SPC) is a generic tool and significantly applied and helps different successful manufacturing operations and service sector industries worldwide. Statistical process control (SPC) is an assortment of statistical tools that allow to ensure and confirm that the operational process is in control to produce a quality product or not.

Stereotypically, variability is the main cause of quality problems. That is, non-conformities are mainly the result of process variability. The ongoing process[1] variability is called inherent (short term) variability, measured by the process measurement's standard deviation. Hence, statistical tools, models, and techniques, together with statistical process control (SPC), are essential for quality engineering applications. "If you are not managing our processes, you are not going to like the result you get."

This chapter intended to introduce and give clarification about the objectives and benefits of SPC. Then we clarify the "WHY?" behindhand SPC, which is variation, and the two types of variation (common cause variation and special cause variation) that the entire processes experience. Subsequently, we will assess the process of generating control charts. During generating control charts, selecting the right variable is the key to monitor the process.

An item's quality is contingent on our personal and frame of preference for a product or service. Beforehand we can talk about how to control quality; we must define and know the term quality from the engineering context. In engineering, quality can be defined and then analyzed based on the eight quality dimensions (performance, durability, reliability, serviceability, conformance, features, aesthetics, and perceived quality). Furthermore, quality can be defined based on the three components (quality of design, quality of conformance, and reliability). See Fig. 4.1 to understand the process and quality control (incoming inspection, process quality control, and outgoing inspection).

4.2 Benefits of Using Control Charts

A control chart is one of the finest tools to help control the quality of the process in real time. The control chart gives visual (graphic) information about the process (in real time). Hence, the control chart shows the performance and behavior of the ongoing process. It is a vital tool to understand whether the process is performing based on the acceptable quality requirement or not.

Having the visual tool that shows the real-time process data that characterize the process's fitness allows the engineer or operation manager to control the process

---

[1]A process is not a machine. It is a set of activities (materials, machines, tools, procedures, personnel, environment, etc.) that turn the input into an output.

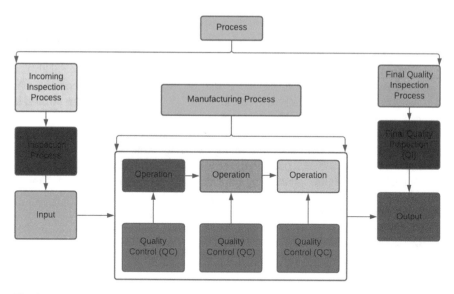

Fig. 4.1   Typical manufacturing product development process

proactively. Therefore, the control chart helps to decide on the process of whether the process requires corrective actions, adjustment, or not. That is the primary objective of the control chart.

Depending on the type of process (or operation), using control charts have several benefits. So, let us discuss these benefits.

A given process's performance can be labeled by the average measurement (nominal value) of the process and the variation associated with the process. The control chart is used to show these parameters of the process. Knowing these parameters is essential to control the process.

Every process has a common variation, which is common to the process. Such variations are natural and can be caused by the environment, equipment, people, etc. However, there is another type of variation which is not caused by natural factors. That variation is called special cause variation. Hence, the variation is not common (natural or normal) to the process. It is attributed to some specific cause. So, each quality expert, manufacturing engineer, or operation manager aims to eliminate such variation.

The following are benefits of using control charts:

- The straightforward advantage of using control charts is to have process data over time, which brings information and facilitates measuring the process's capability.
- They are simple visual (graphical or figurative) tools that empower to evaluate and monitor process performance.
- The control chart's prime advantage is its exceptional ability to distinguish special cause variation from the natural variation within the process. The control chart highlights sections (areas) of performance that may be regarded as a special

cause variation for further examination. Distinguishing between these two types of variability means it will be easier to react to the problem and find the best corrective action to eliminate the special cause variations.

- Control charts can measure (quantify) the impact of process improvement efforts. For example, in the DMAIC process of the Six Sigma method, the second step is "Measure," and the last step is "Control." Therefore, control charts can be used in these critical steps of the DMAIC process.
- Traceability is the ability to recover some (historical) fact in operation via tracking and verifying data (performing research). Traceability can be used to find the cause of a defect, poor performance, etc. Traceability intended to answer "when?" and "where?" something (our target investigation) happened. Control charts can be used as a traceability analyzer tool.
- Last but not least is that control charts can be applied to predict many instances in the process. Used in many instances to predict when problems occur and prevent them.

4.3   Types of Within Process Variations

Walter Shewhart and Edwards Deming familiarize two types of variations within the process. These are (1) common cause variation and (2) special cause variation.

Common cause variations are natural patterns (usual and historical), which are quantifiable (predictable) variations within the process. Special cause variations are unusual (not previously observed), nonquantifiable (unpredictable) variations within the process. Let us explain in detail and demonstrate these variations types.

4.3.1   Common Cause Variation

Shewhart recognized and identified that some variations are common for every single process. No other external factor is the cause of such variations within each process. They are random; hence they occur just by chance. However, such variations can be measured (estimated).

In 1947, Harry Alpert introduced the common cause variation (also called natural problems, noises, or simply random cause). He defined common cause variation as the variabilities that are natural but also quantifiable in the process. Common cause variation is instability triggered by uncontrolled factors resulting in a stable (steady) random distribution of output around the data's average. It is a measure of the process's potential capability of meeting customer requirements. Alternatively, how well the process can perform when special cause variation is removed.

The natural variations are the causes of the problem. They are an intrinsic part of the process itself. Therefore, any specific action cannot be taken to avoid failure from happening. The following are essential characteristics of common cause variations:

Fig. 4.2 Stable process

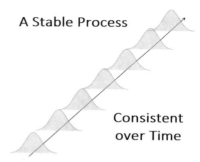

- The variation is statistically predictable.
- The variation is continuously active in the process.
- Variation within experience or knowledge base is not consistent.
- Nonexistence of significance or any implication in individual (high and low) values.

These categories of variations (i.e., common cause variations) require engineering management involvement to make some actions because there are no immediate processes to correct it. For instance, if the engineering team wants to reduce the common cause variations, then engineering design may need to be changed. Therefore, engineering management involvement is crucial for such decisions.

Assuming a normally distributed process data, 99.73% of observations (values) lie within three standard deviations from the process mean. Therefore, the common cause variations lie within three standard deviations (3-sigma) from the mean. On the control chart, they are represented by random points within the lower and upper control limits. The process is said to be statistically stable if it contains only common cause variations.

Hence common cause variations are unavoidable, and the ideal consistent manufacturing process looks like Fig. 4.2. Such a process is called a stable process. So, the goal of statistical process control (SPC) is to achieve such a stable process. The important characteristic of a stable process is its predictability. That is, manufacturing with a stable process is predictable.

The control charts can help to visualize whether the process is stable or not. A stable process may show unexpected variation. In this scenario, the only way to reduce the variation is by making major changes in the process. These changes may be buying new equipment, design change, maintenance cycle change, calibration cycle change, material change, etc.

4.3.1.1 Illustrations of Common Causes of Variance

An athlete may wish to finish a 100-m race within 9 s. However, because of several natural constraints, he may complete the race by 9.58 s. On August 16, 2009, the Jamaican athlete named Usain Bolt broke the men's world record of the 100 m by

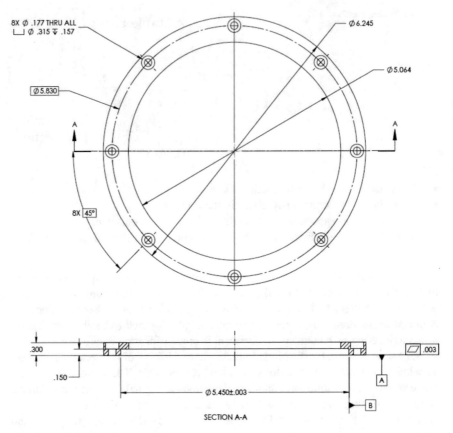

SECTION A-A

Fig. 4.3 Drawing for illustration

finishing the race with 9.58 s. Since 2009 no athlete can finish the 100-m race within lower than 9.58 s.

When we bring a similar analogy with engineering, we find the following example. Consider the engineering drawing specifications in Fig. 4.3.

Suppose the product engineer design the part with total flatness speciation of 0.003. As shown in Fig. 4.4, the product is produced in the following vertically integrated process.

During the inspection of the final product, we have the following measurement for the surface flatness:

Part 1	Part 2	Part 3	Part 4	Part 5	Part 6	Part 7	Part 8	Part 9	Part 10
0.0032	0.0033	0.0032	0.0031	0.0031	0.0033	0.0032	0.0033	0.0034	0.0032

As we have seen from the sample observations, the minimum value of the surface flatness that the process capably produces is 0.0031. The sample information

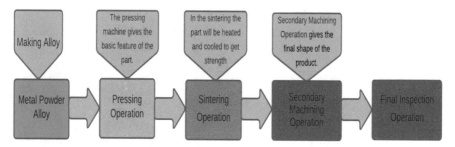

Fig. 4.4 Production process

suggests that the process cannot produce the part with the maximum surface flatness of 0.003.

Furthermore, the process-average output of surface flatness is found to be 0.00323 with a standard deviation of 0.0001. The cause of such variability on the surface flatness of the product is called common cause variation.

Therefore, the sample information suggests that the engineer cannot meet the requirement of the design plan. So, if production is necessary, then the engineering team must change the original design. That is why the common cause variations needed engineering involvement for decision.

The following are examples of common cause variations

- Inadequate procedure
- Poor engineering design
- Computer response time
- The measurement capability of precision
- Work conditions: vibration, humidity, temperature, etc.
- Raw material natural variation: impurity
- Quality control capability
- Voltage variations in microprocessors
- Vibration in industrial processes

4.3.2   Special Cause Variation

In the previous section, we discussed the common cause variations. The normal process can have a variation that results in some problems. However, another type of variation, called the special cause variation, is much severe.

The common cause variations are generated randomly without any external or internal effect within the process. However, some variations are caused by the changes in the given process's external factor(s). The factors can be environmental (weather change), machine failure, nonconforming materials, which have identifiable causes such as human error. Such types of variations within the process are called special (assignable) cause variations.

Fig. 4.5 Unstable process

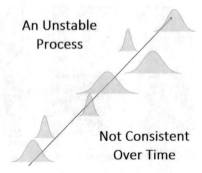

In contrast to the common cause variations, the special cause variations are caused by identified and known factors. That is, special cause variations are unusual to the processes and are nonquantifiable variations. When the process is faced with special cause variations, the distribution of output is not random. That is why the special cause variations are also called Also referred to as exceptional (or assignable) variations.

Special cause variation is the least of any variation that possibly will be attributed to a situation affecting the process. These variations are sporadic, resulting from a specific change in the process that causes the problem. On the control chart, the observations (process measured points) lie outside the preferred control limits. As seen in Fig. 4.5, the effect can be instability of the process and results in process unpredictability. These special causes affect the process in undesirable ways. There-fore, the special causes need to be identified swiftly and should be controlled to restabilize the process.

These variations are sporadic, resulting from a specific change in the process that causes the problem.

Characteristics of special causes of variations are:

- An unanticipated (new or previously neglected) incident within the process.
- Usually unpredictable and even more problematic than the common cause variations.
- The control chart with the special cause variation shows a significant discrepancy than the usual process.
- They happen either eventually or through time.

The control charts can identify when the process is being affected by special causes of variations. Therefore, as a tool, control charts may play a significant role in identifying the problem's root causes. Knowing these, special cause variations allow the manufacturing engineer to choose and then implement the best corrective and preventive actions (CAPA). Furthermore, distinguishing common cause and special cause variations play a significant role in developing corrective actions.

4.3.2.1   Instances of Special Causes of Variance

In the above section, we mentioned that Usain Bolt holds the 100-m world record by finishing the race with 9.58 s. So, let us further assume that Usain Bolt (at his prime) can finish the 100-m race in under 10 s. One day, suppose Usain Bolt finishes the race within 10.30 s. Therefore, the athlete's poor performance may come from external factors such as a lack of training, bad weather conditions (higher temperature), the athlete's health problems, mental preparedness, etc. The factors that cause the athlete not to finish the race under 10 s are called special cause variations.

When we come to the engineering problem, recall the above example. In the pressing operation, during maintenance of the pressing machine (compressor), the upper compressor misaligned (inclined some degree from the horizontal axis, see Fig. 4.6). Then during the production, the parts found nonconformance (defective) to meet the surface flatness specification (max 0.003).

Therefore, the cause of defect (nonconformance) parts in the production is the machine's poor maintenance and setup. Such kinds of variations are special cause variations. So, the potential corrective and preventive action for this problem must involve maintenance and setup of the machine.

The following are examples of special cause variations:

- Machine malfunctions, breakdowns, errors, etc.
- Power cut and the consequences
- Faulty or inadequate maintenance (adjustment) of equipment
- Inadequate preventive maintenance cycles
- Instant weather changes
- Faulty PLC programming of robots
- Operator skill and error

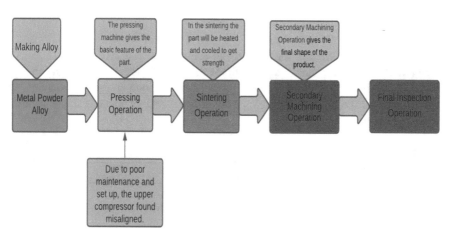

Fig. 4.6   The root cause of misspecification of surface flatness of the products

- Nonconforming incoming materials
- Insufficient system awareness
- Missing operation

4.3.3   Process States

If the process is stable (controlled), it shows only common cause variations. The variation is intrinsic to the process. On the other hand, if the process is unstable (uncontrolled), it shows the special cause variation. External factor(s) disturbs the process.

Based on their output conformance characteristics, a process can further be divided into four groups.

1. *Ideal state*: A process that shows only common cause variations, which produces non-defective parts.
2. *The threshold*: A process that does not show any special cause variations but occasionally produces some (acceptable amount of) defective parts.
3. *The brink of chaos*: A process with special cause variations (not under statistical control), but still, it meets requirements. Do not be surprised at any time if the process produces defective parts.
4. *The state of chaos*: A process with special cause variations produces a significant number of defective parts.

4.4   The Three Key Elements of Control Charts

Control charts are a visual displayer (graphical device) of process performance. These charts can tell whether the given manufacturing (operation or process) is going well based on specifications or not. So, the term control is associated with the ability to have the ability to know whether an engineer or operation manager has a statistical control tool on the given process. For both variable and attribute data, the control chart must have three elements.

Fig. 4.7 The three key elements of the control chart

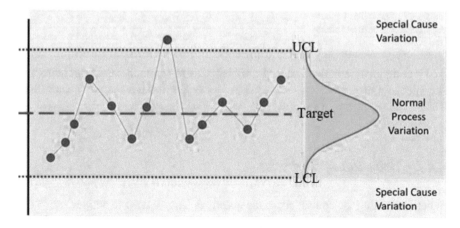

Fig. 4.8   The control chart shows common cause and special cause process variations

As it can be shown from Fig. 4.7, control charts must have three basic components: (1) the centerline (target), (2) upper and lower statistical control limits (simply control limits), and (3) time-series process data points.

1. *Centerline (target)*: This is a line drawn parallel to the *x*-axis representing the specification's target (nominal value).
2. *Upper and lower statistical control limits (simply control limits)*: These are lines parallel to the *x*-axis representing the specification's acceptable limits.
3. *Time series process data points*: These are the data points measured from the ongoing process. Note that for both variable and attribute data, the control chart must have three elements.

After we have recognized and built the control chart, we need to analyze the control chart. We will evaluate the different rules that help us to determine whether the process is in control or not. The upper and lower control limits create boundaries (margins) that the process is in statistical control or not. Suppose all observations (process measurements) lie within the upper and lower control limits (boundaries). In that case, the process has only common cause variations. Such a process is called a normal process. Otherwise, if some of the observations of the process measurements out of the control limits. Then the process has special cause variations. See Fig. 4.8 for figurative clarification.

The target value also characterizes the process mean. If the process can perform based on the design, then the target and the process mean are the same. Usually, the control limits are set within three-standard deviations from the target, which is the process's nominal value. That shows that if the process is stable (controlled), about 99.73% of the entire production falls within the upper and lower control limits.

The statistical upper limit and lower limit may or may not be the upper and the lower specification limits, respectively, of the part specification (customer requirement).

Table 4.1 Some lists of variable charts and attribute charts

Variable charts	I-MR Chart, X-bar Chart, R Chart, S chart
Attribute charts	p-chart, np-chart, c-chart, and u-chart

To completely comprehend and differentiate the common cause variations and the special cause variation, we need to assess the different control charts (see Table 4.1).

4.4.1 Selection of Variables

When generating the control chart, the questions that we need to answer are:

1. What sort of data is accessible for the process?
2. Which variable within the process needs to be controlled?

Concerning the data, we need to assess the difference between discrete data and continuous data. After characterizing the data type, we will determine and select the best control chart. Furthermore, we will see the difference between a defect and a defective one.

To determine and select the variable within the process to be controlled, we will apply the Pareto principle. Using data, the Pareto analysis (also known as the 80/20 rule, Pareto Diagram, Pareto Principle) makes a vibrant representation of the most important variable among the candidate variables.

4.4.1.1   Discrete Data and Continuous

Definition 1 A data type is said to be discrete if and only if it satisfies (1) the dataset is finite (2) countable (maps with the set of natural numbers). Precisely, discrete data is the form of data that takes a clear separation between successive values.

Discrete data is (any) data that is restricted to an exact range of data. Discrete data can be quantitative. Usually, discrete data takes values in natural numbers (i.e., 1, 2, 3, 4). Alternatively, qualitative data, for instance, are conformance (pass or fail).
Examples of discrete data are:

- Number of defective parts
- Categorical choices (Disagree, Agree, Strongly Agree)
- Number of malfunctions of an equipment
- Number of operators
- Number of product lines
- Number of production batches
- Number of operator errors
- Number of operations in the process
- Number of distribution channel for a production

Note that based on the analysis's objective, one can categorize the data to apply discrete data (categorical data). For instance, if you want to study the failure rates of the machines in the plant. You can categorize the ages of the machines as:

- New (machines whose age less than a year)
- Moderate (machines whose age is between 1 and 5 years)
- Old (machines whose age greater than 5 years)

Definition 2 Data are said to be continuous if and only if satisfies (1) the dataset is infinite (2) measurable (maps with the set of real or rational numbers). Precisely, continuous data are the form of data that take infinite values between two distinct values.

Suppose we intended to measure the temperature inside a furnace. The supply air temperature at a register should be from 100 to 110 °F. That means during the given measurement, the temperature can take any value between 100 and 110 °F. Thus, the quantity temperature is continuous data (please note that such interpretation is based on classical mechanics. Hence, quantum mechanics may contradict this idea).
Examples of continuous data are:

- Dimensions (length)
- Weight and mass
- Density
- Viscosity
- Discharge rate
- Time

Sometimes, the data are considered discrete data, but (actually) they are not. Instances of such data are:

- Age of the machine (a year, 2 years, 3 years). Here a year (31,536,000 s) is a measure of time, and physics still did not prove that time is discrete.
- Color of a mobile (Black, Gold, White). In physics, we do not still have a proof that color spectrums are discrete.

4.4.1.2   Defect and Defective

Definition 3 Defective part (or product) is a part (or a product) that does not work properly. Precisely, defective is a part or product that is unable to meet specifications.

Definition 4 Defect on a part (or product) is the undesirable (mis-specified) condition on the part (or a product). Precisely, a defect means a shortcoming or fault.

From Definitions 3 and 4, we can see that defect and defective are different concepts than being the same. The part is defective if it has at least one defect. Therefore, one or more defects are the causes of the part to be defective. Distinguishing the difference between a defect and a defective one is very vital to problem descriptions.

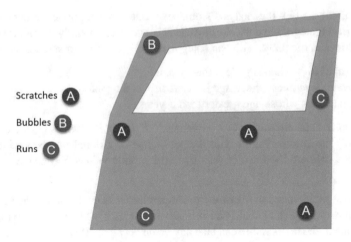

Fig. 4.9 Demonstration of a defect and a defective

To distinguish an important difference between defects and defectives, let us see the illustration in Fig. 4.9.

Suppose a quality superintendent inspects a part. He found a defective part with defects of scratches (out of spec scratch), bubbles (out of spec bubbles), and runs (out of spec runs). That is, the single defective part has found three different defects.

Another example to illustrate defect, let us see the following instance in the semiconductor industry. Assume a quality engineer needs to know problems that affect the quality of water. One of the quality problems in the silicon carbide (SiC) water is micropipes. Micropipes are also called micropores or microtube. Temperature, impurities, vapor-phase stoichiometry, and polarity of the seed (crystal surface) supersaturation are the major causes of micropipes formations, which is one of the defects in the water. Here, engineers quantify the micropipe density is in terms of percentage.

Statistical process control (SPC), like control charts alone, will not enhance and improve quality. Statistical process control (SPC) delivers management awareness into what is occurring on the manufacturing floor. SPC can only show the result so that skilled personnel use the information to improve quality. The management is responsible for adjusting the process's rules so employees can use their knowledge and skills. Certain degrees of changes that employees can do; however, the essential improvement hails from management decisions. Therefore, management necessity sets the phase for employees to perform to do the job at the required level.

Control charts are overwhelming for the process control. A part may have several quality variables (parameters or dimensions). The selection of all the variables leads to an uncontrollable number of control charts in the process. Therefore, we should efficiently select the variable(s) to be included in the control chart(s).

When we set the variables (parameters) critical to the process, we should consider the following essential points.

- The statistical correlation among those variables and the process.
- Knowledge of the important variable to the customer.
- Knowledge of risk to the process and the customer (this includes analysis of revenue and other interrelated penalties).
- The potential financial benefits or consequences associated with a variable.
- The technical difficulty related to executing the control chart.
- An additional common endorsement is to prefer a variable by asking which variable is sensitive to the special cause variations.

At last, we apply the Pareto to select the variables that critically quality parameters to the process.

4.5 Rational Subgrouping

Rational subgrouping is systematically arranging the given data into groups of objects produced under the same conditions to measure the variation among the subgroups in its place among individual data points. Commonly, the arrangement of the subgroups based on homogenous portions of the population.

During creating control charts, the concept of a rational subgroup is essential. The chart's control limits are determined on the basis of the product specification and the variability within the sample. Through the control charts, we take samples from the process. We will measure the sample data, and then we will determine whether the process is in statistical control or not.

Taking the samples has an enormous influence on the overall characteristics and sensitivity of the control chart. Therefore, we should determine the rational subgroup of the samples to consider.

A rational subgroup is a collection of elements that are produced within identical conditions. These samples must be as standardize into homogenous groups as possible. A little variation within the samples must solitary comprise of the normal and intrinsic process variation. Typically, the rational subgroup is the reflection of the process at any instant in time. The chart's control limits are determined on the basis of the product specification and the variability within those samples taken.

By including the common cause process variation in the rational subgroup, we can recognize that the control chart's control limits are adequately sensitive to the special causes of variation.

The control limits of the process regulate the variation within the rational subgroup of the samples. Here we aim to minimize the control limits within the group variation. That leads the control chart to be sensitive only to the special causes of variation. For instance, in Fig. 4.10, we see that the rational subgroup from five samples in the subgroup defines the within subgroup variation.

We aim to select a rational subgroup that diminishes the variation within the same subgroup. Such a process will magnify the chance for the variation between subgroups. Therefore, systematically capturing any small external variations (special

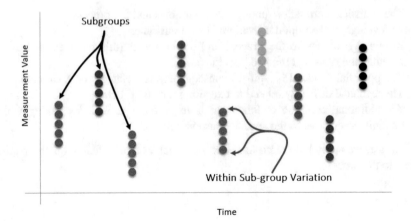

Fig. 4.10 Rational subgrouping

cause variation) will easily be detected in the process. Suppose we do not use the rational subgroup in the samples. In that case, some essential variabilities will be hidden not to detect the special cause variations.

What does the point subgroup mean in manufacturing? To answer that, assume that a given machine produces 200 parts per 6 h. A quality inspector may decide to take ten parts for a quality check every 6 h. Here the ten parts are characterizing the (first) subgroup. The second subgroup of ten parts will be taken after 6 h and so on.

4.6   The Eight Rules for Statistical Process Control (SPC)

If the process is in statistical control, most of the data points concentrate near the average (centerline, target value). Furthermore, the controlled process does not have any data points that are plotted outside the control limits. However, when we say the target's data points, it does not give quantifiable information. To have a piece of quantifiable information, we need to have a systematic rule.

The eight rules of statistical process control are introduced to quantify the distribution of data points in the control chart. To comprehend the eight rules of statistical process control, let us consider Fig. 4.11.

From Fig. 4.11, we see the following zones:

- Zone A (Alert): the region between the two standard deviations and three standard deviations.
- Zone B (Caution): the region between the one standard deviation and two standard deviations.
- Zone C (Green Light): the region within one standard deviation from the target.

Operating the zones, we will have the following rules.

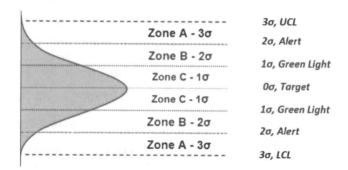

Fig. 4.11 The eight rules of statistical process control

- Rule 1: Any data point outside Zone A. This region is the identification region for the existence of special cause variation.
- Rule 2: 2 out of 3 (2/3) successive data points in Zone A (same side).
- Rule 3: 4 out of 5 (4/5) successive data points in Zone B (same side).
- Rule 4: More than 8 successive data points on any side of the target.
- Rule 5: 6 data points in a row, all decreasing or increasing trend.
- Rule 6: 14 successive data points in a row show systematic variation, alternating up and down.
- Rule 7: 15 points of successive data points within Zone C, which is a little variation.
- Rule 8: successive data points outside of Zone C on either side.

Violating at least one of the eight rules can be a solid signal that the process is under the effect of a special cause variation (SCV). Therefore, the process requires corrective and preventive action.

Bibliography

R.E. Barlow, T.Z. Irony, Foundations of statistical quality control, in *Current Issues in Statistical Inference: Essays in Honor of D. Basu*, ed. by M. Ghosh, P. K. Pathak, (Institute of Mathematical Statistics, Hayward, CA, 1992), pp. 99–112

B. Bergman, Conceptualistic Pragmatism: A framework for Bayesian analysis? IIE Trans. **41**, 86–93 (2009)

W.E. Deming, On probability as a basis for action. Am. Stat. **29**(4), 146–152 (1975)

W.E. Deming, *Out of the Crisis: Quality, Productivity and Competitive Position* (Massachusetts Institute of Technology, Center for Advanced Engineering Study, Cambridge, MA, 1982) ISBN 0-521-30553-5

E.L. Grant, *Statistical Quality Control* (McGraw-Hill Book Company, New York, NY, 1946) ISBN 0071004475

J. Oakland, *Statistical Process Control* (MPG Books Limited, Bodmin, UK, 2002) ISBN 0-7506-5766-9

T. Salacinski, *SPC - Statistical Process Control* (The Warsaw University of Technology Publishing House, Warsaw, 2015) ISBN 978-83-7814-319-2

W.A. Shewhart, *Economic Control of Quality of Manufactured Product* (D. Van Nostrand Company, Inc., New York, NY, 1931) ISBN 0-87389-076-0

W.A. Shewhart, *Statistical Method from the Viewpoint of Quality Control* (The Graduate School, The Dept. of Agriculture, Washington, DC, 1939) ISBN 0-486-65232-7

D.J. Wheeler, *Understanding Variation: The Key to Managing Chaos*, 2nd edn. (SPC Press, Knoxville, TN, 1999) ISBN 0-945320-53-1

D.J. Wheeler, *Normality and the Process-Behaviour Chart* (SPC Press, Knoxville, TN, 2000) ISBN 0-945320-56-6

D.J. Wheeler, D.S. Chambers, *Understanding Statistical Process Control* (SPC Press, Knoxville, TN, 1992) ISBN 0-945320-13-2

S.A. Wise, D.C. Fair, *Innovative Control Charting: Practical SPC Solutions for Today's Manufacturing Environment* (ASQ Quality Press, Milwaukee, WI, 1998) ISBN 0-87389-3859

S.L. Zabell, Predicting the unpredictable. Synthese **90**(2), 205 (1992). https://doi.org/10.1007/bf00485351

Chapter 5
Process Control Charts

5.1 Control Charts for Variable Data

Control charts of variable data are time series plots of data, which are characterized by three lines: the target (nominal value) line, the lower control limit (LCL), and the upper control limit (UCL).

Typically, variable control charts function in pairs. For instance, an $\overline{X}$ (also called X-Bar) and R chart consists of two-charts. The X-Bar chart monitors the mean (the average of the process), and the Range chart monitors the range of the process. The control charts of the variable data showed a tendency to be sensitive to the process changes. For instance, the $\overline{X}$ chart indicates whether the process is accurate or missed the process accuracy concerning the target (the nominal value of the specification).

Similarly, the R chart tells us the precision of the process. Therefore, if there is a significant change (i.e., the existence of special cause variation), either the trend of observations (data) in the $\overline{X}$ chart or in the R chart will show alteration. Furthermore, the S-chart and the Moving Range (MR) chart show process precision (variability).

Thus, the typical applications of control charts of variable data are:

- Determining whether the process is stable (contains only common cause variations, or we call it in statistical control) or not.
- Prediction tool for the process outcomes.
- Adjusting production problems as they happen.
- Used as a validation tool for whether the implemented corrective action is successful or not.
- To evaluate quality improvement programs.

The three important control charts of variable data are:

1. $\overline{X}$ and R chart (also called X-bar and R Chart, where $\overline{X}$ is for the mean and R is for the range).
2. $\overline{X}$ and S chart (also called X-bar and S Chart, where $\overline{X}$ is for the mean and S is for the standard deviation).

3. I-MR chart (individual and moving range chart).

5.1.1 $\overline{X}$ (X-Bar) and R Chart

$\overline{X}$ and R chart is the pillar of control charts. The $\overline{X}$ and R charts are the most common. They are extensively used as the control chart for variable data for process stability in various industries: manufacturing, hospitals, and other services sector industries.

Definition 5 $\overline{X}$-chart is a chart that presents the change of process mean (average) of the subgroups over time.

Definition 6 R-Chart is a chart that presents the change of process range of the subgroups over time.

Definition 7 $\overline{X}$ and R chart is a chart consists of the $\overline{X}$ chart and the R chart.

$\overline{X}$ and R charts are applied to display the variable's process characteristics and performance (continuous) data. The data for the $\overline{X}$ and R charts data should be collected within the rational subgroups (usually the subgroup size is between 2 and 9) overtime in the process.

It is essentially two plots to displace and monitor the process mean and the process precision (variation, difference, and a discrepancy of measurements) over time. The combination of the two charts helps comprehend the process's stability else, and they show the existence of special cause variation in the process.

If the process is perfect, the $\overline{X}$ coincides with the target (the nominal value of design), and the R-chart will always zero. However, due to the unavoidable common cause variations, such measurements are not achievable. However, the perfect process is our reference to control the process and to take any necessary action. A process that is close enough to the ideal approach is acceptable. Otherwise, the quality engineer, manufacturing engineer, or operation manager should identify the source of process incapability (variability). And we should fix the factors that cause the process incapability.

5.1.1.1 Steps of Building the $\overline{X}$ and R Charts Chart

1. Select the suitable control chart for the dataset.
2. Determine the sample size of the dataset and organize the data for analysis.
3. Calculations (e.g., X-Bar and R chart)

   (a) Sample mean $\overline{X} = \sum_{i=1}^{n} \frac{x_i}{n}$
   (b) Sample Range $R = \text{Max}\{x_i\} - \text{Min}\{x_i\}$
   (c) Assuming we consider k-samples, then, Grand-average $= \overline{\overline{X}} = \frac{\sum_{i=1}^{k} \overline{X}_i}{k}$
   (d) Grand-Average Range $\overline{R} = \sum_{i=1}^{k} \frac{R_i}{k}$

Table 5.1 Data for the X-bar and R Chart illustration

| | Sample number | | | | |
Subgroups	1	2	3	Subgroup average	Subgroup range
Subgroup 1	25.8	27.5	27.4	26.9	1.7
Subgroup 2	27.5	25.8	27.5	26.9	1.7
Subgroup 3	26.5	27.1	27.0	26.8	0.6
Subgroup 4	26.7	27.4	27.0	27.0	0.7
Subgroup 5	26.7	27.1	27.2	27.0	0.4
Subgroup 6	26.9	27.5	27.3	27.2	0.6
Subgroup 7	27.2	28.0	27.9	27.7	0.9
Subgroup 8	26.0	27.1	28.2	27.1	2.2
Subgroup 9	27.7	26.4	27.3	27.1	1.3
			Grand	27.08	1.122

4. Then, the Lower Control Limit is calculated as $LCL = \overline{\overline{X}} - A_2\overline{R}$ and the Upper Control Limit is calculated as $UCL = \overline{\overline{X}} + A_2\overline{R}$
   where A_2 is hooked on the sample size and can be obtained from statistical tables.
5. Observe the plots to analyze whether the data are inside the control limits or not. If we find data out of the control limit, then that is caused by special cause variations.
6. When one is identified, mark it on the chart and investigate the cause—document how we investigated, what we learned, the cause, and how it was corrected.

Consider the data in Table 5.1 to build the X-bar and R chart with nine subgroups with a sample size of 3 below. The data nominal (target) value of the data is 27 with USL = 28.5 and LSL = 25.5.
Now, let us calculate the important values to build the $\overline{X}$ and R chart.

Calculation of the Critical Values of the $\overline{X}$ Chart

The centerline (the grand average) of the $\overline{X}$ chart is the average of all the subgroups' averages. It is the best linear unbiased estimator of the population mean (μ, which is the value of the design nominal (target) value for our process.
The centerline the $\overline{X}$ chart is calculated as follows

$$\overline{\overline{X}} = \frac{\sum_{i=1}^{k}\overline{X_i}}{k} = \frac{26.9 + 26.9 + 26.8 + 27 + 27 + 27.2 + 27.7 + 27.1 + 27.1}{9}$$
$$= 27.08 = 27.1$$

The grand mean of the range is given as

Table 5.2 Constants for the $\overline{X}$ chart and R chart

X-Bar and R chart				
Subgroup sample size	X-Bar factor	Range factors		Variance factor
n	A_2	D_3	D_4	d_2
2	1.880	-	3.267	1.128
3	1.023	-	2.575	1.693
4	0.729	-	2.282	2.059
5	0.577	-	2.115	2.326
6	0.483	-	2.004	2.534
7	0.419	0.076	1.924	2.704
8	0.373	0.136	1.864	2.847
9	0.337	0.184	1.816	2.970
10	0.308	0.223	1.777	3.078
15	0.223	0.347	1.653	3.472
20	0.180	0.415	1.585	3.735
25	0.153	0.459	1.541	3.931

$$\overline{R} = \frac{\sum_{i=1}^{k} R}{k} = \frac{1.7 + 1.7 + 0.6 + 0.7 + 0.4 + 0.6 + 0.9 + 2.2 + 1.3}{9} = 1.12$$

As we have seen in the steps above, the upper control limit (UCL) and the lower control limit (LCL) are calculated utilizing the grand average ($\overline{\overline{X}}$) and the grand average range ($\overline{R}$), and the value of A_2 varies dependent on the size of the subgroup.

Using Table 5.2, we will calculate the control limits of the $\overline{X}$ chart as follows. The value of A_2 for the subgroup with a sample size of 3 is 1.023. Therefore,

$$\text{Lower control limit LCL} = \overline{\overline{X}} - A_2\overline{R} = 27.08 - 1.023 \times 1.122 = 25.952$$

$$\text{Upper control limit UCL} = \overline{\overline{X}} + A_2\overline{R} = 27.08 + 1.023 \times 1.122 = 28.248$$

Calculation of the Critical Values of the R Chart

The critical elements of the R chart are specified as follows:

The grand mean of the range is given as

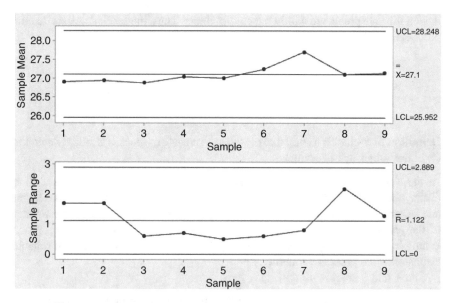

Fig. 5.1 $\overline{X}$ and R chart example

$$\overline{R} = \frac{\sum_{i=1}^{k} R}{k} = \frac{1.7 + 1.7 + 0.6 + 0.7 + 0.4 + 0.6 + 0.9 + 2.2 + 1.3}{9} = 1.12$$

One can multiply the average range ($\overline{R}$) by two range factors (i.e., D_4 and D_3) to calculate the lower and upper control limits intended for the R chart. The range factors (i.e., D_4 and D_3) vary with the subgroup size, which is specified as follows.

The lower control limit of the R-chart is calculated as $LCL_R = D_3\overline{R}$ and the upper control limit of the R-chart is calculated as $UCL_R = D_4\overline{R}$. As we can see from Table 5.2. The value of D_3 and D_4 for the subgroup with the sample size of 3 are 0 and 2.575, respectively. Therefore,

Lower Control Limit $LCL_R = D_3\overline{R} = 0$ and
Upper Control Limit $UCL_R = D_4\overline{R} = 2.575 \times 1.12 = 2.889$

As the grand average is the best linear unbiased (BLU) estimator of the population mean (μ), correspondingly, just adjusting by a constant (i.e., $1/d_2$, the inverse of the variance factor) of the grand average-range ($\overline{R}$) is the best linear unbiased (BLU) estimator of the standard deviation of the population. Therefore, the estimate of the population standard deviation $s = \overline{R}/d_2$.

From Table 5.2, we see that the value of d_2 for the subgroup with the sample size of 3 is 1.693, that results in the standard deviation, $s = 0.721$ (Fig. 5.1).

Both the $\overline{X}$ and R charts have the centerline (Target), the lower control limit (LCL), and the upper control limits (UCL). To interpret the control chart, first, we start looking at the R-chart. The R-chart shows that all the data points are plotted within the control limits. Our next step is to look at the $\overline{X}$-chart. In the $\overline{X}$-chart, we see that all the data points are within the control limits. Thus, the result suggests that the process is stable.

5.1.2 $\overline{X}$ *Chart and S Chart*

Definition 8 S-chart is a chart that presents the change of process standard deviation (*s*) of the subgroups over time.

As the $\overline{X}$ and R chart, the $\overline{X}$ and S Chart (also called X-bar and S Chart) is used to display the data's central tendency. If the sample size of the subgroup becomes larger and larger, the range estimate of the subgroup turns out to be an inefficient estimator of the process variability. Studies show that as the sample size of the subgroup getting higher, the sample standard deviation becomes a more efficient and reliable estimator of the process variability.

The $\overline{X}$ and S chart is used when

(a) The sampling process is the same for the individual sample and is accepted reliably
(b) The data are distributed normally
(c) The subgroup size is more than 10. Thus, to build the chart, we need to collect more data than the $\overline{X}$ and R chart
(d) The data are continuous
(e) The inspection and the sampling process should be the same for each subsample, and is carried out consistently from sample to sample

5.1.2.1   Steps of Building the $\overline{X}$ and S Charts Chart

1. Select the suitable control chart for the dataset.
2. Determine the sample size of the dataset and organize the data for analysis.

   (a) Calculations (e.g., X-Bar and S chart)
   (b) Sample mean $\overline{X} = \sum_{i=1}^{n} \frac{x_i}{n}$
   (c) Sample standard deviation $s = \sqrt{\frac{(x-\overline{x})^2}{n-1}}$

3. The disadvantage of switching range by the standard deviation is that it is more problematic to implement and preserve due to the calculations needed to be intended for the standard deviation value. Like the $\overline{X}$ and R chart, in the $\overline{X}$ and S chart, we calculate the grand average as follows. Assuming we consider *k*-samples, then, Grand-average $= \overline{\overline{X}} = \frac{\sum_{i=1}^{k} \overline{X}_i}{k}$.
4. Average standard deviation $\overline{s} = \sum_{i=1}^{k} \frac{s_i}{k}$.

Table 5.3 Constants for the $\overline{X}$ Chart and S chart

X-Bar and S chart				
Subgroup sample size	X-Bar factor	Standard deviation factors		Variance factor
n	A_3	B_3	B_4	C_4
2	2.659	-	3.267	0.7979
3	1.954	-	2.568	0.8862
4	1.628	-	2.266	0.9213
5	1.427	-	2.089	0.9400
6	1.287	0.030	1.970	0.9515
7	1.182	0.118	1.882	0.9594
8	1.099	0.185	1.815	0.9650
9	1.032	0.239	1.761	0.9693
10	0.975	0.284	1.716	0.9727
15	0.789	0.428	1.572	0.9823
20	0.680	0.510	1.490	0.9869
25	0.606	0.565	1.435	0.9896

5. Then, the lower control limit of $\overline{X}$ chart is calculated as LCL $= \overline{\overline{X}} - A_3\overline{s}$ and the upper control limit is calculated as UCL $= \overline{\overline{X}} + A_3\overline{s}$, where A_3 is hooked on the sample size and can be obtained from statistical tables.
6. The lower control limit of S chart is calculated as LCL $= \overline{s} - B_3\overline{s}$ and the upper control limit is calculated as UCL $= \overline{s} - B_4\overline{s}$, where B_3 and B_4 is hooked on the sample size and can be obtained from statistical Table 5.3.
7. Observe the plots to analyze whether the data are inside the control limits or not. If you find data out of the control limit, then that is caused by special cause variations.
8. Plot new data and continue checking the point that is listed in step 5. When one is identified, mark it on the chart and investigate the cause—document how you studied, what you learned, the reason, and how it was corrected.

Furthermore, with a correction factor $(1/c_4)$, the best linear unbiased (BLU) estimator of the population standard deviation is given below.

$$\hat{\sigma} = s = \frac{\overline{s}}{c_4}$$

As we can see the values in Table 5.3, the values B_3, B_4, and c_4 vary with the subgroups.

Table 5.4 Data to build the $\overline{X}$ and S Chart

Subgroups	Sample number			Subgroup average	Subgroup standard deviation
	1	2	3		
Subgroup 1	27.4	27.6	27.4	27.467	0.115
Subgroup 2	27.4	27.1	27.5	27.333	0.208
Subgroup 3	27.4	27.6	27.6	27.533	0.115
Subgroup 4	25.8	26.9	27.3	26.667	0.777
Subgroup 5	27.5	27.4	27.4	27.433	0.058
Subgroup 6	26.4	26.4	27.5	26.767	0.635
Subgroup 7	27.1	25.8	27.3	26.733	0.814
Subgroup 8	27	27.9	27.1	27.333	0.493
Subgroup 9	26.7	28.1	27.1	27.300	0.721
Subgroup 10	27.4	26.8	26.6	26.933	0.416
Subgroup 11	27	27.4	27.1	27.167	0.208
Subgroup 12	26.7	27.4	27.2	27.100	0.361
Subgroup 13	27.1	27.3	25.5	26.633	0.987
Subgroup 14	27.1	27.6	27.4	27.367	0.252
Subgroup 15	26.9	27.4	27.1	27.133	0.252
Subgroup 16	27.4	27.5	28.3	27.733	0.493
Subgroup 17	27.3	26.9	26.6	26.933	0.351
Subgroup 18	27.1	27.4	27.6	27.367	0.252
Subgroup 19	28	26.3	27.1	27.133	0.850
Subgroup 20	27.9	27	26.8	27.233	0.586
Subgroup 21	26	25.4	27.3	26.233	0.971
Subgroup 22	27.1	26.9	27.1	27.033	0.115
Subgroup 23	28.2	27	27.2	27.467	0.643
Subgroup 24	27.6	27.5	27.1	27.400	0.265
Subgroup 25	26.4	26.9	27.1	26.800	0.361
Subgroup 26	27.3	27.3	26.3	26.967	0.577
Subgroup 27	26.3	27.1	28.3	27.233	1.007
Subgroup 28	27.3	27.6	27.9	27.600	0.300
Subgroup 29	27.8	27.3	27.2	27.433	0.321
Subgroup 30	27.4	27.1	25.9	26.800	0.794
Subgroup 31	27.1	27	27.2	27.100	0.100
Subgroup 32	26.6	27.6	27.8	27.333	0.643
Subgroup 33	27.4	27.2	27.1	27.233	0.153
Subgroup 34	27.3	26.8	27.8	27.300	0.500
			Grand	27.154	0.462

Using the data in Table 5.3, we will calculate all the necessary values in constructing for $\overline{X}$ and S Chart in Table 5.4.

The centerline the $\overline{X}$ chart is calculated as follows:

$$\bar{\bar{X}} = \frac{\sum_{i=1}^{k} \bar{X}_i}{k} = \frac{27.467 + 27.333 + \cdots + 27}{34} = 27.15$$

The grand mean of the standard deviation is given as:

$$\bar{s} = \frac{\sum_{i=1}^{k} s}{k} = \frac{0.115 + 0.208 + \cdots + 0.5}{34} = 0.462$$

Using Table 5.3, we will calculate the control limits of the $\bar{X}$ chart as follows. For the subgroup with a sample size of 3 the value of A_3 is 1.954. Therefore,

Lower control limit LCL $= \bar{\bar{X}} - A_3\bar{s} = 27.08 - 1.954 \times 0.462 = 26.252$
Upper control limit UCL $= \bar{\bar{X}} + A_2\bar{R} = 27.08 + 1.954 \times 0.462 = 28.056$

Calculation of the Critical Values of the S Chart

The critical elements of the S chart are specified as follows:

The grand mean of standard deviation is $\bar{s} = 0.462$. One can multiply the average standard deviation as $\bar{s} = 0.462$ by two range factors (i.e., $B_3 = 0$ and $B_4 = 2.568$) to calculate the lower and upper control limits, respectively, for the intended S chart.

The lower control limit of the S-chart is calculated as $\text{LCL}_s = B_3\bar{s}$ and upper control limit of the R-chart is calculated as $\text{UCL}_s = B_4\bar{s}$. Therefore,

Lower control limit $\text{LCL}_s = B_3\bar{s} = 0$ and
Upper control limit $\text{UCL}_s = B_4\bar{s} = 2.568 \times 0.462 = 1.185$

As the grand average is the best linear unbiased (BLU) estimator of the population mean (μ), correspondingly, just adjusting by a constant (i.e., $1/c_4$), the inverse of the variance factor) of the grand average-standard deviation ($\bar{s}$) is the best linear unbiased (BLU) estimator of the standard deviation of the population. Therefore, the estimate of the population standard deviation $s = \bar{s}/c_4$.

Table 5.3 shows that the value of c_2 for the subgroup with the sample size of 3 is 0.8862. That results in $s = 0.520$.

Finally, we have the $\bar{X}$ and S chart in Fig. 5.2.

Both the $\bar{X}$ and S charts have the centerline (target), the lower control limit (LCL), and the upper control limits (UCL). To interpret the control chart, first, we start looking at the S-chart. The S-chart shows that all the data points are plotted within the control limits. Our next step is to observe the $\bar{X}$-chart. In the $\bar{X}$-chart, we see that the 21st observation is out of control. Thus, the result suggests that the process is unstable. The cause of process instability is the indication that the process is affected by the special cause variations. So, we need to figure out the special cause variation (s) and fix them.

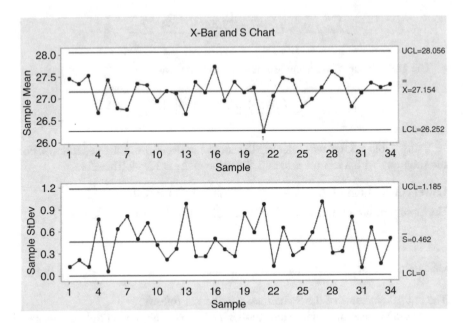

Fig. 5.2 $\overline{X}$ and S chart example

Often range is greater than the standard deviation. This leads that in substituting from the range of the $\overline{X}$ and R chart into the $\overline{X}$ and S chart becomes more powerful to detect unobservable special change variations in the process.

Once the S chart is in control, then review $\overline{X}$ chart and interpret the points against the control limits. All the points are to be interpreted against the control limits (not specification limits). If any point out of control in $\overline{X}$ chart. Identify the special cause and address the issue. $\overline{X}$ and S sensitive to slight deviations in the mean of the subgroup.

Thus, once the S chart is in control, review the chart, and interpret these points against the control limits. The entire points to be interpreted against the control limits (not specification limits). If any point out of control in the chart. Recognize the special cause and address the issue. Both the $\overline{X}$ and S sensitive to slight deviations in the mean of the subgroup.

Interpretation of the $\overline{X}$ and S Chart

(a) At all times, analyze the S chart before interpreting the whole chart (i.e., the $\overline{X}$ and S chart.

(b) The $\overline{X}$ chart control limits are resulting from the average standard deviation values. Therefore, if there are observations in the S-chart, which are out of the control limits, then the $\overline{X}$ chart will have inaccurate control limits. For instance, in Fig. 5.2, we see that one observation is out of the S-chart's control limits.

However, all the observations in the $\overline{X}$ chart falls within the control limits. However, the $\overline{X}$ chart information is inaccurate, and the process is not in statistical control. Hence, the S-chart already indicated that there are special cause variations in the process.

(c) If the S-chart shows that all the observations are within the control limits, we will analyze the $\overline{X}$ chart. Here, the process is in statistical control if all the observations are within the control limits of the $\overline{X}$ chart. Otherwise, we conclude that there are special cause variations in the process.

5.1.2.2   Individual and Moving Range (I-MR or X-MR) Charts

It is known that the contribution of Walter Shewart in statistical process control, especially introducing control charts for industrial applications, is significant. An Individual and Moving Range (I-MR or X-MR) control chart was one of the most important contributions of Walter Shewart.

Definition 9 Individual chart (I-Chart) is a chart that shows the data points collected over regular time intervals and displays the process mean.

Definition 10 Moving Range Chart (MR-Chart) is a chart that shows the unit data points collected over regular time intervals and displays the process range of the current from the previous measurement.

Definition 11 Individual and moving range char (I-MR, or X-MR) is a chart that consists of the I-chart and MR-chart.

I-MR charts comprehensive statistics on the process performances. If the interest of measurement is individuals rather than subgroups, then the I-MR chart is the best control chart.

Individual chart (I-Chart) plots each individual data point on a definite period. It is suitable to detect several tendencies and changes that are apparent in the process. It aids to picture common causes variations and any special cause variations if they are existing in the process. To infer the tendencies related to the time performance, for data I-Chart must be successive and accessible in the identical order against the time-axis.

The Moving Range Chart (MR-Chart) is technologically advanced by plotting values attained from the successive and timely ordered data. The moving range data point is calculated as $x_i - x_{i-1}$. So, if the I-Chart has k data points, then the MR-Chart will generate $k - 1$ data points.

Graphical I-MR control chart aid to recognize when the process is out of statistical control and signposts the origin of the special cause variation. Thus, the following are the most important advantages of I-MR control charts.

The I-MR control charts are very good when the sample size of the subgroup is a unit. These control charts are advantageous when in situations difficult to collect large sample sizes.

(a) There are situations where data collection is costly, time-consuming, and difficult to collect. For instance, calibration data.
(b) When the observation is continuous data, and there is no subgrouping. Or, we
can say that the subgrouping has a unit element.
(c) The chart assists in recognizing the existence of both the common cause and
special cause variations.
(d) To display the process performance of corrective and preventive actions
(CAPA).
(e) Typically, used for monitoring the batch process.

The Individual chart (I-chart or X-chart) displays the mean and the Moving Range
chart (MR-chart) to quantify the process variation. The benefit of an individual chart
is that it is simple to use. The individual chart needs a strong normality assumption
on the data. If the data are non-normal, then the graphical information from the chart
is extremely inefficient to show the process's characteristics. That hides some
important process variations so that the number of defective parts may be high.
The moving range (MR) is the difference between n (at least $n = 2$) consecutive data
points.

5.1.3 Statistical information for constructing the individual and moving range (I-MR or X-MR) chart

1. Select the suitable control chart for the dataset.
2. Calculate the centerline (mean) of individual observations

   (a) Sample mean $\overline{X} = \dfrac{\sum x_i}{n}$

3. Calculate the moving range from individual observations

   (a) Moving Range $MR_i = |x_{i+1} - x_i|$

4. Typically, for the moving range (MR) chart, specifically if $n = 2$, we are watching
the absolute difference between the last two successive data points. Likewise, if
$n = 3$, then the moving range will be calculated based on the previous three data
points. If the moving range (MR) is based on $n = 4$, we will calculate the last four
data points' moving range. The moving range (MR) value is calculated by
computing the difference between n successive data points. The centerline of
the moving range ($\overline{MR}$) is calculated by taking the moving ranges' averages as
follows.

$$\overline{MR} = \frac{\sum MR_i}{n-1}$$

5. Calculate the control limits of the centerline

   (a) Lower control limit $LCL_{\overline{X}} = \overline{X} - E_2\overline{MR}$

Table 5.5 Statistical information for constructing the individual and moving range (I-MR, or X-MR) chart

I-MR chart			
Subgroup sample size	Individual factor	Moving range factors		Variance factor
n	E_2	D_3	D_4	d_2
2	2.660	-	3.267	1.128
3	1.772	-	2.575	1.693
4	1.457	-	2.282	2.059
5	1.290	-	2.115	2.326
6	1.184	-	2.004	2.534
7	1.109	0.076	1.924	2.704
8	1.054	0.136	1.864	2.847
9	1.010	0.184	1.816	2.970
10	0.975	0.223	1.777	3.078

(b) Upper control limit $\text{UCL}_{\overline{X}} = \overline{X} + E_2\overline{\text{MR}}$

6. Furthermore, using constant coefficients (i.e., D_4 and D_3), we will calculate the upper and lower control limits of the moving range (MR) chart. Calculate the control limits of the moving average as follows.

(a) Lower control limit $\text{LCL}_{\overline{\text{MR}}} = D_3\overline{\text{MR}}$
(b) Upper control limit $\text{UCL}_{\overline{\text{MR}}} = D_4\overline{\text{MR}}$

Consider the data in Table 5.5 give important statistical information to contract the I-MR chart.

To illustrate how to build the I-MR chart, let us consider the data in Table 5.6. Hourly unit data are collected to monitor the temperature in the sintering process. Using the I-MR chart, let us examine that data to comprehend whether the process is in statistical control or not.

To proceed with the calculation, first, we should figure out the subgroup sample size. Here a single temperature measurement at a time is recorded in the process. Therefore, to calculate the moving range (MR), we use $n = 2$. Thus, the moving range (MR) will be the absolute value of the difference between the current and the previous temperature measurements. See the moving range (MR) calculated values above in Table 5.6. For instance, the moving range at hour 2 is measured as the absolute value of $370 - 355 = 15$.

Using the statistical information in Table 5.5, let us calculate the centerline and upper and lower control limits of the individual chart (I-Chart) and the moving range chart (MR-Chart).

Table 5.6 Hourly temperature data of the process

Sample	Temperature	Moving range
1	355	
2	370	15
3	371	1
4	375	4
5	370	5
6	372	2
7	368	4
8	367	1
9	369	2
10	370	1
11	375	5
12	374	1
13	374	0
14	373	1
15	364	9
16	357	7
17	360	3
18	375	15
19	372	3
20	373	1
21	370	3
22	367	3
23	369	2
24	370	1
25	375	5
26	374	1
27	374	0
28	373	1
29	364	9
30	357	7
31	370	13
32	372	2
33	368	4
34	367	1
35	369	2
36	370	1
37	375	5
Average	369.405	3.889

The centerline is calculated by

- Sample mean $\overline{X} = \frac{\sum x_i}{n} = \frac{355+370+\cdots+375}{37} = 369.405$

The moving range is calculated in the last column of Table 5.6.

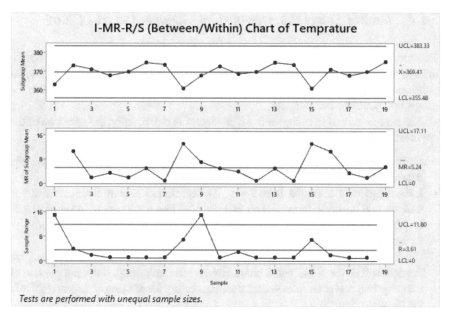

Fig. 5.3 Individual and moving range (I-MR) of temperature

The moving ranges as follows.

- $\overline{MR} = \frac{\sum MR_i}{n-1} = \frac{15+1+\cdots+5}{36} = 3.889$

The centerline ($\overline{MR}$) for the moving range (MR) chart is calculated by taking the average moving range (MR) values. Furthermore, the upper and lower and the control limits of the moving range (MR) chart are calculated using the constant coefficients D_4 and D_3 factors. To calculate the control limits of the centerline.

- Lower control limit $LCL_{\overline{X}} = \overline{X} - E_2\overline{MR} = 369.405 - 3.58 \times 3.889 = 355.48$
- Upper control limit $UCL_{\overline{X}} = \overline{X} + E_2\overline{MR} = 369.405 + 3.58 \times 3.889 = 383.33$

To calculate the control limits of the moving range

- Lower control limit $LCL_{\overline{MR}} = D_3\overline{MR} = 0 \times 3.889 = 0$
- Upper control limit $UCL_{\overline{MR}} = D_4\overline{MR} = 4.4 \times 3.889 = 17.11$

Now, let us construct our Individual and Moving Range (I-MR) chart to monitor the process's temperature. Figure 5.3 presents the required I-MR chart.

Figure 5.3 shows that the entire data points of the individual temperature measurements are within the upper and lower control limits. Furthermore, the moving range (MR) chart shows that all the successive range values of the process's temperature are between the lower and upper control limits. However, observations 1 and 9 are out of the control limits in the range chart. The result confirmed that the process is not stable, showing the existence of special cause variations.

5.1.4   How to Select the Appropriate Variable Control Chart

The subgroup is a set of observations that are generated under the identical conditions of the process circumstances. The subgroup needs to characterize the process performance. To get the best representative sample, the sample data points within the subgroup should be independent.

The sample size of the subgroup is a key factor in determining the selection of the appropriate statistical control chart to use.

- If the subgroup sample size is 1, the best statistical control chart to analyze the data will be the I-MR (individual and moving range) chart.
- If the subgroup size ranges from 2 to 9, the best statistical control chart to analyze the data will be the X-Bar and R chart. Hence, if the sample size of the subgroup is less than 10, the range (R) of the sample is an efficient estimator of the process variation than the sample standard deviation (s).
- If the subgroup size is greater than 9, the best statistical control chart to analyze the data will be the X-bar and S chart. Hence, if the subgroup, if the sample size of the subgroup is greater than 9, then the finest estimator of the process variability is the standard deviation.

5.2   Attribute Statistical Control Charts

In Sect. 5.1, we see that variables control charts are suitable for monitoring continuous data. Continuous data are characterized by the ability to measure in a set of real numbers. Data types such as mass, length, height, width, area, surface roundness, viscosity, density, temperature, pressure, humidity, and so on, are examples of continuous data.

On the other way around, some data types are countable, called attribute data. For instance, one may be interested to know the number of spelling errors that he/she committed per page. The number of errors per page takes only a whole number. Another instance is that an insurance company is interested to know whether the customer had accidents or not. Furthermore, the insurance company also wants to see the number of accidents the customer had before issuing an insurance policy. The number of accidents takes only a whole number. Such data types are not continuous.

Data types such as pass/fail, defective/nondefective, go/no-go, number of defects, number of quality parameters, product color type, and so on, are attribute (discrete) data types. For attribute data types, we use attribute statistical control charts.

In the subsequent sections, we will deliberate about the four most common attribute (discrete) data control charts (p-chart, np-chart, c-chart, and u-chart). As we have studied in the previous section, the variable data's control charts have their interpretations. For instance, in the X-bar and S chart, we have seen that if the S-chart shows data points outside its control limits, then whether the data points in the X-bar chart are inside the control limits or not, we conclude that the process is not in statistical control. However, contrasting the control chart of variable data (the X-bar and R chart, the X-bar and S chart, and the I-MR chart), the attribute (discrete) data control charts presented only a single chart.

Normally, attribute control charts are easier to build and execute. However, attribute control charts are not as sensitive to detect small but important changes in the process variation.

5.2.1   How to Select the Appropriate Attribute Control Chart

The four most common attribute data control charts are p-chart, np-chart, u-chart, and c-chart. The essential factors that differentiate the applications of these charts are the concepts of defective and defect. Therefore, to comprehend such statistical control charts, we need to recall some definitions from Chap. 2.

Definition 3 Defective part is a part that does not work properly. Precisely, defective means a part or product that is unable to meet specifications.

Definition 4 Defect on a part (or product) is the undesirable (misspecified) condition on the part. Precisely, a defect means a shortcoming or fault.

As seen in Table 5.6, the p-chart and np-chart help us evaluate, analyze, and control defectives over time. The c-chart and u-charts help us evaluate, analyze, and control defects over time.

The other essential factor that helps us differentiate these charts' applications is the characteristics of sample sizes, constant and variable. The np-chart and c-chart are applicable when the sample size is constant, whereas the p-chart and u-chart are applicable when the sample size varies.

From Table 5.7, we can see that:

- np-chart is appropriate when the data points are taken from a "constant" sample size with a category of "Defectives."
- p-chart is appropriate when the data points are taken from a "variable" sample size with a category of "Defectives."
- c-chart is appropriate when the data points are taken from a "constant" sample size with a category of "Defect."

Table 5.7 The matrix of attribute control charts

		Problem category	
		Defective	Defect
Sample size	Constant	np	c-chart
Characteristics	Variable	p	u-chart

- p-chart is appropriate when the data points are taken from a "constant" sample size with a category of "Defect."

Therefore, the appropriate control chart will be selected based on analyzing the problem category (i.e., defectives or defects), and the characteristics of the sample size (i.e., constant or variable).

5.2.2   p-Chart and np-Chart

Definition 12 A binomial distribution is a discrete probability distribution (also called a probability mass function) of the likelihood of successful trials from n-independent and successive trials.

The successful trials can be the observation of defective parts from the sample.
   Assumptions of the binomial random variable

1. There are n-successive trials.
2. The trials are mutually independent.
3. The trials are fixed.
4. In each trial, the probability of success (p) is constant.

Definition 13 Suppose p is the probability of success in each trial, and we have n-successive trials (sample size), then the binomial distribution is given as:

$$\Pr(X = x; p, n) = \binom{n}{x} p^x (1 - p)^{n-x}$$

where X is a binomial random variable, and $x = 0, 1, 2, \ldots, n$

Hence, the inspected item is either accepted or rejected (pass or fail, good or bad) when the p-chart and np-chart are developed based on the binomial distribution.

5.2.3   Proportion Chart (p-chart)

The proportion chart (p-chart) is applied to characterize and monitor the percentage of defective (or nonconformance) parts (items, products, or materials) over time. The p-chart is applied when the sampling size fluctuates over time. Among the other attribute control charts, the p-chart is utmost sensitive to changes in the process.
   As an alternative of proportion defective, we call this the fraction defective. Often people use the term defective to represent nonconformance and vice versa. However, as we have studied in Chap. 1, we should know that the word defective and nonconformance are not the same. To avoid confusion, as an alternative of proportion defective, let us call it fraction defective.

Now, let us calculate the centerline and the upper and lower control limits of the p-chart. Here, our interest is to characterize the proportion of defectives in the lot. The appropriate probability distribution becomes the Bernoulli distribution, which is the special case of Binomial distribution with $n = 1$.

$$\Pr(X = x; p) = p^x(1 - p)^{1-x} \quad \text{where } x = 0, 1$$

The mean of a Bernoulli distribution is computed as

$$E(x) = \sum_{x=0}^{1} x\Pr(X = x; p) = \sum_{x=0}^{1} xp^x(1 - p)^{1-x}$$

$$= (0)(1 - p) + (1)(p) = p$$

Therefore, the centerline of the p-chart is estimated by the average of proportions $\bar{p}$. The centerline is the average proportion of defective of the lot. We can calculate by taking the ratio of the sum of defective items to the sum of the number of parts in the subgroups.

$$\text{Centerline} = \bar{p} = \frac{\sum np}{\sum n}$$

When the sample size changes, the control limits of the p-chart changes. Fixing the proportion p, as the sample size of the subgroup increases, the width of the control limits gets narrower. Nevertheless, a communal method generates the control limits of the p-chart based on the average sample size of the subgroups ($\bar{n}$). Therefore, once we calculate the centerline $\bar{p}$, we will calculate the lower and the upper control limits as follows.

$$\text{Lower control limit } (\text{LCL}_{\bar{p}}) = \bar{p} - 3\sqrt{\frac{\bar{p}(1 - \bar{p})}{\bar{n}}}$$

$$\text{Upper control limit } (\text{UCL}_{\bar{p}}) = \bar{p} + 3\sqrt{\frac{\bar{p}(1 - \bar{p})}{\bar{n}}}$$

where $\bar{n} = \frac{\sum n}{k}$, k is the number of the subgroups

Now let us see an illustration for the application of the p-chart. Suppose we have measured 15 subgroups (lots). Furthermore, as we have seen in the data in Table 5.8, each subgroup has a different sample size. We aim to quantify the percentage of defective parts in the 15 batches.

From Table 5.10, we see that:

- The total number of parts in all the 15 lots is 3060.
- The total number of defectives in all the 15 lots is 140.

Table 5.8 Data for p-chart illustration

Lot	Lot size	Count of defective units	Percentage defective
1	200	8	4.0%
2	220	7	3.2%
3	180	5	2.8%
4	240	9	3.8%
5	200	11	5.5%
6	180	12	6.7%
7	200	9	4.5%
8	220	14	6.4%
9	220	7	3.2%
10	240	11	4.6%
11	160	7	4.4%
12	200	5	2.5%
13	180	10	5.6%
14	180	10	5.6%
15	240	15	6.3%
15	3060	140	4.58%

These leads:

$$\text{The centerline} = \overline{p} = \frac{\sum np}{\sum n} = \frac{140}{3060} = 0.0458.$$

$$\overline{n} = \frac{\sum n}{k} = \frac{3060}{15} = 204.$$

Therefore, the lower and upper control limits are calculated as follows

Lower Control Limit $\left(\text{LCL}_{\overline{p}}\right) = \overline{p} + 3\sqrt{\frac{\overline{p}(1-\overline{p})}{\overline{n}}} = 0.0458 + 3\sqrt{\frac{0.0458(1-0.0458)}{204}} = 0.0019$

Upper Control Limit $\left(\text{UCL}_{\overline{p}}\right) = \overline{p} - 3\sqrt{\frac{\overline{p}(1-\overline{p})}{\overline{n}}} = 0.0458 - 3\sqrt{\frac{0.0458(1-0.0458)}{204}} = 0.089$

At last, we construct the p-chart as presented in Fig. 5.4.

The p-chart shows that our process is in statistical control and, therefore, it is a stable process. Here, our process has a defective rate of 4.58%. The defective rate comes from common cause variations. However, as we have seen from Fig. 5.4, we cannot see any special cause variations in the process.

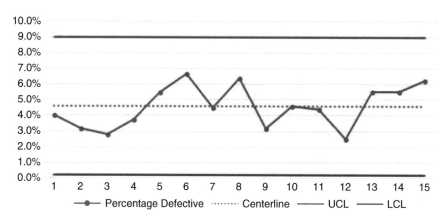

Fig. 5.4 p-chart

5.2.4 np Chart

Like the p-chart, the np-chart is also used to characterize and monitor the percentage of defective (or nonconformance) parts (items, products, or materials) over time. The difference is that unlike the p-chart, the np-chart is applied when the sampling size is fixed. Among the other attribute control charts, the p-chart is utmost sensitive to changes in the process.

When the sample size (n) is constant, then the average number of defectives in the subsample is calculated by np. Therefore, the centerline of the np-chart is calculated as follows.

$$\text{The centerline} = n\bar{p} = \frac{\sum np}{k}$$

where k is the number of subgroups.

Furthermore, we know that the standard deviation of $\bar{p}$ is $\sqrt{\frac{\bar{p}(1-\bar{p})}{n}}$. Therefore, the standard deviation of $n\bar{p}$ becomes $\sqrt{n}$. $\sqrt{\frac{\bar{p}(1-\bar{p})}{n}} = \sqrt{\bar{p}(1-\bar{p})}$. Using that information, we can calculate the lower and upper control limits, respectively, of the np-chart as follows.

Lower Control Limit $\left(\text{LCL}_{n\bar{p}}\right) = n\bar{p} - 3\sqrt{n\bar{p}(1-\bar{p})}$
Upper Control Limit $\left(\text{UCL}_{n\bar{p}}\right) = n\bar{p} + 3\sqrt{n\bar{p}(1-\bar{p})}$

Now let us see an illustration for the application of the np-chart. Suppose we have measured 15 subgroups (lots). Furthermore, as we have seen in the data in Table 5.9, each subgroup has the same sample size (n) that equals 120 parts. Like the p-chart, we aim to quantify the percentage of defective parts in the 15 batches in the np-chart.

Table 5.9 Data for np-chart illustration

Lot	Lot size	Count of defective units
1	120	12
2	120	20
3	120	20
4	120	6
5	120	14
6	120	6
7	120	17
8	120	18
9	120	5
10	120	11
11	120	11
12	120	15
13	120	11
14	120	13
15	120	20
15	1800	199

Now, we can calculate the centerline of the np-chart as centerline $= n\bar{p} = \sum \frac{np}{k} = \frac{199}{15} = 13.27$.

Furthermore, the estimate of the population defective part is calculated as $\bar{p} = \frac{\sum np}{\sum n} = \frac{199}{1800} = 0.1106$. The percentage of average defective ($\bar{p}$) is 11.06%. We can calculate the same result by taking the ratios of the average of the defective parts to the sample size of the subgroup as $\frac{13.27}{120} = 0.1106$.

The lower and upper control limits, respectively, of the np-chart, are calculated as follows:

- Lower control limit

$$\text{LCL}_{n\bar{p}} = n\bar{p} - 3\sqrt{n\bar{p}(1-\bar{p})} = 13.27 - 3\sqrt{120 \times 0.1106 \times (1 - 0.1106)} = 2.96$$

- Upper control limit

$$\text{UCL}_{n\bar{p}} = n\bar{p} - 3\sqrt{n\bar{p}(1-\bar{p})} = 13.27 + 3\sqrt{120 \times 0.1106 \times (1 - 0.1106)}$$
$$= 22.58$$

At last, we construct the np-chart as presented in Fig. 5.5.

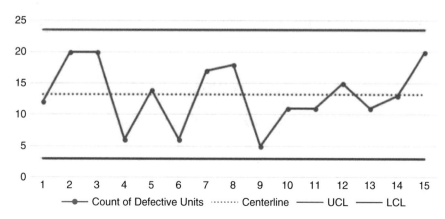

Fig. 5.5 np-chart

The np-chart shows that our process is in statistical control and, therefore, it is a stable process. Here, our process has a defective rate of 11.06%. That defective rate comes from common cause variations. However, as we have seen from Fig. 5.5, we cannot see any special cause variations in the process.

Based on this information, our process appears to be in control and stable, with no single subgroup having a count of defective items greater than our upper control limit of 23.58 (23) or less than the lower control limit of 2.96.

5.2.5 c-Chart and u-Chart

Like the p-chart and np-chart, the c-chart and u-chart are charts for attribute data. The p-chart and np-chart are used to analyze defectives, whereas the c-chart and u-chart are used to analyze defects in the subgroup.

As an illustration, in plastic production, one of the quality problems of production is that bubble, which is the product's defect. Here the bubbles are countable. Therefore, the c-chart and u-chart helps to visualize such defects in the process.

The p-chart and np-chart used the binomial distribution. In contrast, the c-chart and u-chart are used in the Poisson distribution. So, we need to define a Poisson distribution.

Definition 15 A Poisson distribution is a discrete probability distribution that gives the likelihood of independent events happening in a fixed time.

Assumptions of the Poisson distribution

- x is the number of success (event) occurs in the time interval
- Success (events) occurs independently
- The average event (λ) is independent time

- Two events cannot happen at an identical instant

Definition 15 Suppose λ is the mean of the success, Poisson distribution is given as:

$$f(X = x; \lambda) = \frac{e^{-\lambda} \lambda^x}{x!}$$

where X is a Poisson random variable and $x = 0, 1, 2, \ldots$.

5.2.6   Count Chart (c-Chart)

The c-chart is an unpretentious and single chart that displays defects of a part over time. The letter c represents the count of defects. The c-chart is an attributes (discrete) control chart applied to analyze defects for subgroups with the same sample size—c-charts display how the process via the number of defects per part over time. Nonconformities describe as any representative character existing but then must not be in the part of the production. For instance, nonconformities can be dents, scratches, bubbles, missing buttons, blemishes, and so on.

The c-chart is applied to analyze the number of nonconformities (or defects) per unit of the product. The sample size of all the subgroups is constant. In this circumstance, the process centerline is merely the average number of defects per inspection. The subgroup is samples that are taken during the inspection. Assuming we have k-subgroups, the centerline of the c-chart is calculated as follows.

$$\text{Centerline} = \bar{c} = \frac{\sum c}{k}$$

The lower and upper control limits, respectively, of the c-chart, is computed as follows.

Lower control limit $(\text{LCL}_{\bar{c}}) = \bar{c} - 3\sqrt{\bar{c}}$
Upper control limit $(\text{UCL}_{\bar{c}}) = \bar{c} + 3\sqrt{\bar{c}}$

Let us see the following example to clarify the application of the c-chart.

Suppose we have 20 lots (subgroups) that contain 1500 parts each. The number of defects in each lot is given in Table 5.10.

Let us start by calculating the center (centerline) of the process:

$$\text{Centerline} = \bar{c} = \frac{\sum c}{k} = \frac{1365}{20} = 68.25$$

The lower and upper control limits, respectively, of the c-chart, are computed as follows.

Table 5.10   Data for c-chart illustration

Lot	Lot size	Count of defects
1	1500	80
2	1500	70
3	1500	52
4	1500	75
5	1500	74
6	1500	75
7	1500	52
8	1500	78
9	1500	69
10	1500	50
11	1500	75
12	1500	67
13	1500	61
14	1500	54
15	1500	77
16	1500	66
17	1500	64
18	1500	66
19	1500	78
20	1500	82
	30000	1365

Lower control limit $(\text{LCL}_{\bar{c}}) = \bar{c} - 3\sqrt{\bar{c}} = 68.25 - 3\sqrt{68.25} = 43.47$
Upper control limit $(\text{UCL}_{\bar{c}}) = \bar{c} + 3\sqrt{\bar{c}} = 68.25 + 3\sqrt{68.25} = 93.03$

Therefore, the required c-chart is given in Fig. 5.6.

Since all the observations, the counts of defects per subgroup never fall outside of our control limits. The process appears to be in a state of statistical control. The c-chart shows that the process is in statistical control and, therefore, it is a stable process. The defect rate comes from common cause variations. However, as we have seen from Fig. 5.6, we cannot see any special cause variations in the process.

5.2.7   u-Chart

Just like the c-chart, the u-chart is modest and a single chart, which displays defects of a part over time. The c-chart is an attribute (discrete) control chart applied to analyze defects for subgroups that are different sample sizes. The u-chart stabilizes

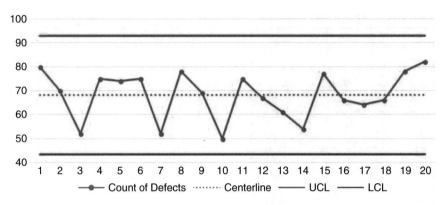

Fig. 5.6 c-Chart

the number of nonconformities (or defects) via the sample size of the subgroup. That brings the number of nonconformities (or defects) per subgroup. u-Charts display in what way the process is going with respect to the number of nonconformities (or defects) per part (item, product, material), over time.

Assuming, we have k-subgroups, the centerline of the u-chart is calculated as follows.

$$\text{Centerline} = \bar{c} = \frac{\sum c}{\sum n} = \frac{\sum \text{number of defects in each subgroup}}{\sum \text{sample size of each subgroup}}$$

Average sample size of subgroups $(\bar{n}) = \frac{\sum n}{k} = \frac{\sum \text{sample size of each subgroup}}{k}$. Then, the lower and upper control limits, respectively, of the u-chart are computed as follows.

Lower control limit $(LCL_{\bar{u}}) = \bar{c} - 3\sqrt{\frac{\bar{c}}{\bar{n}}}$

Upper control limit $(\text{UCL}_{\bar{u}}) = \bar{c} + 3\sqrt{\frac{\bar{c}}{\bar{n}}}$

Let us see the following example to clarify the application of the u-chart.

Suppose we have 15 lots (subgroups) that contain variable lot sizes, as seen in Table 5.11.

The centerline of the process is calculated as follows

$$\text{Centerline} = \bar{c} = \frac{\sum c}{\sum n} = \frac{\sum \text{number of defects in each subgroup}}{\sum \text{sample size of each subgroup}} = \frac{117}{1560}$$
$$= 0.075\%$$

Reminisce that the u-chart has the suppleness to compute a different control limit for each subgroup. That is based on the sample size of the subgroups. Nevertheless,

Table 5.11 Data for u-chart illustration

Lot	Lot size	Defects observed	Percentage defects
1	100	8	8.0%
2	120	10	8.3%
3	80	9	11.3%
4	140	6	4.3%
5	100	10	10.0%
6	80	5	6.3%
7	100	10	10.0%
8	120	10	8.3%
9	120	4	3.3%
10	140	7	5.0%
11	60	7	11.7%
12	100	5	5.0%
13	80	9	11.3%
14	80	8	10.0%
15	140	9	6.4%
15	1560	117	7.50%

several times n-bar $(\overline{n})$ helps to make the calculation simpler. The average sample size of subgroups $(\overline{n}) = \frac{\sum n}{k} = \frac{\sum \text{sample size of each subgroup}}{k} = \frac{1550}{15} = 104$. Then, the lower and upper control limits, respectively, of the u-chart is computed as follows.

Lower control limit $(\text{LCL}_{\overline{u}}) = \overline{c} - 3\sqrt{\frac{\overline{c}}{\overline{n}}} = 0.075 - 3\sqrt{\frac{0.075}{104}} = -0.0056$

Upper control limit $((\text{UCL}_{\overline{u}}) = \overline{c} + 3\sqrt{\frac{\overline{c}}{\overline{n}}} = 0.075 + 3\sqrt{\frac{0.075}{104}} = 0.156$

Therefore, the required c-chart is given in Fig. 5.7.

As we see from Fig. 5.7, the process gives the picture of statistical control, as not a single data point is outside the control limits.

5.3 Systematic Flow Chart to Select the Best Control Chart for the Given Process Dataset

See Fig. 5.8.

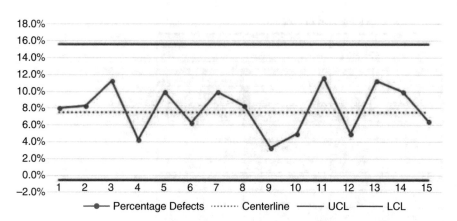

Fig. 5.7 u-Chart

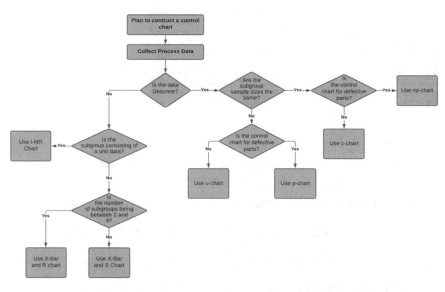

Fig. 5.8 Flow chart for chart selection

5.4   Analyzing and Interpreting Control Charts

The control chart analysis aims to guide us in recognizing special cause variations. Subsequently, process improvement comes. Nevertheless, we need a systematic analysis to determine the presence of special cause variations in the process. If the process is controlled, then it means it does not consist of special cause variations.

5.4.1   Controlled Process

A random oscillation (variation, fluctuation) around the centerline (target value)—if the curve connecting adjacent observations randomly distributed around the target value and within the control limits, then it is confirmed that the process contains only common cause variation. If the dataset lies within the control limits, but if the curve connecting the adjacent observations is shown any trend, then there is a problem in the process.

5.4.2   Unhealthy Process and Special Cause Variations

The process may show the following characteristics, which are signs of illness:

1. *Bias (a lower or upper trend)*
    If the control chart shows more than a few successive data points below or above the average, this process has a reliability problem.
2. *Observable either increasing or decreasing trend*
    If the control chart shows more than a few successive data points with either increasing or decreasing trend, then the process is not trustable, reliable, or accurate.
3. *Data points on the control limits*
    If the process shows some data points on control limits, you need to add additional observations to rely on the process.
4. *Data points outside the control limit*
    If the process shows some data points outside the control limits, the process consists of special cause variations. If that was the case, the process needs corrective and preventive action.

5.5   Common Control Chart Analysis Errors

Control charts are often essential tools we can apply to improve quality by making a stable process.

   However, if used inadequately, control charts can cause costly issues. Below are the four upmost common errors that we make when analyzing control charts are:

1. *Re-Sampling Fallacy*
    The error occurs when we do not trust an out-of-control measurement. Instead of analyzing the process and adjusting, we resample from the process, resulting in missing the original signal of recognizing the uncontrol process.

2. *Underadjustment*

Underadjustment happens mostly due to a lack of attention to the control charts. It can also include a lack of caring for the process data. These mistakes may lead to the out-of-control process to proceed with production without proper corrective action.

3. *Overadjustment*

That is possibly the most common mistake that occurs. Often, it results from a well-intentioned operator who adjusts anytime a process is not perfectly on center. In many cases, people will adjust the process when the right decision concludes. Any process variation is considered normal to the process. However, the adjustment may not be effective, and it does not improve the process.

4. *Not starting our analysis with the range chart*

Analysis stating from the mean is the starting point for many inspectors. However, when special cause variation exists, it can often affect the variability chart. That is either range or standard Deviation. Inspectors usually start by observing at the average value. However, the proper place to start is with the range or standard deviation chart, then move to the average chart.

Note that the range values are used to calculate the control limits on the average chart. Therefore, if the range data are out of control, our limitations on the average chart can often be wrong.

Annex: X-Bar and R Chart and X-Bar and S Chart constants

Sample size	X-bar Chart constants		For sigma estimate	R Chart constants		S Chart constants	
	A_2	A_3	d_2	D_3	D_4	B_3	B_4
2	1.8800	2.6590	1.1280	0.0000	3.2670	0.0000	3.2670
3	1.0230	1.9540	1.6930	0.0000	2.5740	0.0000	2.5680
4	0.7290	1.6280	2.0590	0.0000	2.2820	0.0000	2.2660
5	0.5770	1.4270	2.3260	0.0000	2.1140	0.0000	2.0890
6	0.4830	1.2870	2.5340	0.0000	2.0040	0.0300	1.9700
7	0.4190	1.1820	2.7040	0.0760	1.9240	0.1180	1.8820
8	0.3730	1.0990	2.8470	0.1360	1.8640	0.1850	1.8150
9	0.3370	1.0320	2.9700	0.1840	1.8160	0.2390	1.7610
10	0.3080	0.9750	3.0780	0.2230	1.7770	0.2840	1.7160
11	0.2850	0.9270	3.1730	0.2560	1.7440	0.3210	1.6790
12	0.2660	0.8860	3.2580	0.2830	1.7170	0.3540	1.6460
13	0.2490	0.8500	3.3360	0.3070	1.6930	0.3820	1.6180
14	0.2350	0.8170	3.4070	0.3280	1.6720	0.4060	1.5940
15	0.2230	0.7890	3.4720	0.3470	1.6530	0.4280	1.5720
16	0.2120	0.7630	3.5320	0.3630	1.6370	0.4480	1.5520
17	0.2030	0.7390	3.5880	0.3780	1.6220	0.4660	1.5340

(continued)

Sample size	X-bar Chart constants		For sigma estimate	R Chart constants		S Chart constants	
	A_2	A_3	d_2	D_3	D_4	B_3	B_4
18	0.1940	0.7180	3.6400	0.3910	1.6080	0.4820	1.5180
19	0.1870	0.6980	3.6890	0.4030	1.5970	0.4970	1.5030
20	0.1800	0.6800	3.7350	0.4150	1.5850	0.5100	1.4900
21	0.1730	0.6630	3.7780	0.4250	1.5750	0.5230	1.4770
22	0.1670	0.6470	3.8190	0.4340	1.5660	0.5340	1.4660
23	0.1620	0.6330	3.8580	0.4430	1.5570	0.5450	1.4550
24	0.1570	0.6190	3.8950	0.4510	1.5480	0.5550	1.4450
25	0.1530	0.6060	3.9310	0.4590	1.5410	0.5650	1.4350

Bibliography

R.E. Barlow, T.Z. Irony, Foundations of statistical quality control, in *Current Issues in Statistical Inference: Essays in Honor of D. Basu*, ed. by M. Ghosh, P. K. Pathak, (Institute of Mathematical Statistics, Hayward, CA, 1992), pp. 99–112

B. Bergman, Conceptualistic Pragmatism: A framework for Bayesian analysis? IIE Trans. **41**, 86–93 (2009)

W.E. Deming, On probability as a basis for action. Am. Stat. **29**(4), 146–152 (1975)

W.E. Deming, *Out of the Crisis: Quality, Productivity and Competitive Position* (Massachusetts Institute of Technology, Center for Advanced Engineering Study, Cambridge, MA, 1982) ISBN 0-521-30553-5

E.L. Grant, *Statistical Quality Control* (McGraw-Hill Book Company, New York, NY, 1946) ISBN 0071004475

J. Oakland, *Statistical Process Control* (MPG Books Limited, Bodmin, UK, 2002) ISBN 0-7506-5766-9

T. Salacinski, *SPC - Statistical Process Control* (The Warsaw University of Technology Publishing House, Warsaw, 2015) ISBN 978-83-7814-319-2

W.A. Shewhart, *Economic Control of Quality of Manufactured Product* (D. Van Nostrand Company, Inc., New York, NY, 1931) ISBN 0-87389-076-0

W.A. Shewhart, *Statistical Method from the Viewpoint of Quality Control* (The Graduate School, The Dept. of Agriculture, Washington, DC, 1939) ISBN 0-486-65232-7

D.J. Wheeler, *Understanding Variation: The Key to Managing Chaos*, 2nd edn. (SPC Press, Knoxville, TN, 1999) ISBN 0-945320-53-1

D.J. Wheeler, *Normality and the Process-Behaviour Chart* (SPC Press, Knoxville, TN, 2000) ISBN 0-945320-56-6

D.J. Wheeler, D.S. Chambers, *Understanding Statistical Process Control* (SPC Press, Knoxville, TN, 1992) ISBN 0-945320-13-2

S.A. Wise, D.C. Fair, *Innovative Control Charting: Practical SPC Solutions for Today's Manufacturing Environment* (ASQ Quality Press, Milwaukee, WI, 1998) ISBN 0-87389-3859

S.L. Zabell, Predicting the unpredictable. Synthese **90**(2), 205 (1992). https://doi.org/10.1007/bf00485351

Chapter 6
Process Capability Analysis

6.1 Introduction

In Chap. 5, we have seen that control charts are established based on referring to the process's statistical stability. The essential advantage of employing such tools (i.e., control charts) is that it is adequate and easy to implement and can help keep our process on center. Control charts are also a boundless way to make certain that the process is set up appropriately beforehand it starts. However, the shortcoming of using these charts is that the charts may not figure out understated changes in the process, and they do not show (reflect) the natural variation in the process. Furthermore, these control charts will not be applicable if the process is already unstable and incapable. Here, we need to distinguish that the control charts may not reflect the specification's engineering tolerance. Hence, we need to understand the core concept of the pre-control charts.

6.2 Pre-control Charts

The pre-control chart is usually different from the control charts. In fact, there is no absolute chart, but we compare process measurements against the part specification limits. The pre-control people have a habit of viewing every product within requirement specification as equally good. The entire lot classify as either conforming or nonconforming. Correspondingly, the classification can be defective and nondefective, or good or bad. The separation line for conformance and nonconformance is sharp. The part (item, product, material) that scarcely meets the requirement specification is as good (conforming, or nondefective) as the part that impeccably centered on the target value. Manufacturing products tighter than the requirement specification limits observed as a superfluous expense.

© The Author(s), under exclusive license to Springer Nature Switzerland AG 2021   187
Y. Y. Tesfay, *Developing Structured Procedural and Methodological Engineering Designs*, https://doi.org/10.1007/978-3-030-68402-0_6

Fig. 6.1 Specification
boundaries

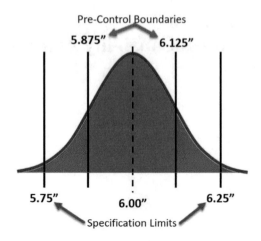

Pre-control charts are, to some extent, debatable in that they are generated employing the requirement specification limits. Therefore, it does not show the actual process. The pre-control chart displays and monitors the center of the process to make it unquestionable on target. The method is extremely sensitive to the normality assumption of the data and operates if it has adequate process capability (a good process has a process capability (Cpk) of at least 1). The system's mode to generate the pre-control chart is about to take the requirement specification tolerance (range) and create the boundaries of the pre-control. Every so often, the pre-control boundaries are 7%, 15%, or 25% of the overall requirement specification tolerance (range).

Let us analyze an illustration. Assume we are generating a widget whose length dimension is $6.00 \pm 0.25''$. Which is $6.00''$, and the tolerance is $\pm 0.25''$. The lower specification limit (LSL) is $5.75''$. The upper specification limit (USL) is $6.25''$.

In Fig. 6.1, let us generate the pre-control boundaries at 25% of the total tolerance (range) on both sides. Thus, if the total tolerance (range) is 0.50, then the pre-control lines are set at 25% of that. That results in a tolerance of $0.5 \times 0.25 = 0.125''$ on both sides, which leads to the lower specification limit (LSL) as $5.875''$, and the upper specification limit (USL) as $6.125''$.

As seen from Fig. 6.1, the pre-control and target specification limits and boundaries produce three different zones. That is the Green, Yellow, and Red Zones.

- The Red Zone is any data point that falls out of the specification limits. Suppose the process is in the Red Zone. In that case, we need to stop production and fix the problem by implementing the best corrective and preventive action.
- The Green Zone (i.e., the middle of process data point distribution) is the adequate (safe) area. If successive parts (items, products, material) are produced inside the Green Zone, the process is good to go.
- The Yellow Zone (Alter) is the area between the specification and limits the pre-control line. When the process is starting up, these lines assist in making

appropriate decisions. If we have successive parts in the Yellow Zone, we need to act and adjust to push to the green zone (back to the center).

As we have seen above, the pre-control charts are noticeably different from the control charts. In fact, "there is no real-chart-at-all," we are merely comparing and relating measurements versus the required specification limits. The benefit of using such types of tools is that it is laid-back to apply and implement, which will keep the process in the center (i.e., Green Zone). The tool gives different ways to make sure whether the process setup is correctly configured or not.

However, these charts may miss indirect changes in the process. They do not show the true (natural) variation in the process. Furthermore, if the process is not capable and stable, the tool may not have significant value.

6.3 Terms and Definitions

- *Capable process*: The process where almost all the measurements fall inside the specification limits is capable.
- *Ideal process*: The process centered over the Nominal, with a narrow distribution, leaving plenty of white space for the LSL and USL. Even if the process starts to drift, it is unlikely to produce defective parts.
- *Shifted process*: The process shifted toward either the lower (LSL) or the upper (USL). The small process shift could result in a defect.
- *Too broad process*: The process is too broad; the small process shift, toward either the lower (LSL) or upper (USL), could result in a defect.
- *Truncated distribution*: A truncated distribution may indicate that out-of-spec parts have been removed from the dataset.
- *Inadequate resolution*: Inadequate gage resolution may result in misleading calculations of the average and standard deviation.
- *Multi-modal*: Multiple operating conditions may cause a distribution with more than one peak (i.e., mode).
- *Quality records (QR)*: In the ISO Standard (specifically, in the ISO 9000), quality records (OR) are held in reserve, demonstrating conformance to definite requirements and the Quality Management System (QMS).
- *Part history record (PHR)*: PHR is a section of quality records (QR). This document stored the measurement data for a specific part.
- *Statistical process control (SPC)*: SPC is a statistical technique of quality control mechanisms.

6.4   Short-Run Statistical Process Control (SR-SPC)

Short-runs are an inevitability in a low-volume, high-mix production environment. The tendency in production has been in the direction of less important production runs. Production runs and products are tailor-made to customer needs. Although that reduces inventory and advances reaction to the customer requirement, it confounds the use of statistical process control (SPC).

An important point about the control charts is solitary operative and suitable for long-term and continuous production runs. These charts are not operational for short, rare production-runs. Providentially, short-run statistical process control (SPC) can help us regulate and control short (rare) processes.

Short-run statistical process control (SPC) charts are designed for both variable and attribute data. One of the utmost common methods in short-run SPC is via standardizing the control chart. The average of the subgroups is standardized to calculate the z-score (z-value) related to the subgroups. Below is an illustration of the short-run type of the u-chart. We calculate the z-score as follows.

$$Z_i = \frac{u_i - \bar{u}}{\sqrt{\bar{u}}}$$

In the above equation, we observe that Z_i transforms into the standard normal distribution. The difference is that the standard deviation is substituted by $\sqrt{\bar{u}}$. Otherwise, u_i is the observation and $\bar{u}$ is the process mean. In this case, the centerline (i.e., the target) of this kind of short-run SPC chart needs to be zero. If a process is stable (i.e., on target), the individual observation (u_i) is identical with the process mean (i.e., $\bar{u}$). That gives that Z_i to be set to zero.

In that scenario, we will plot the transformed values Z_i with the upper and lower control limits, respectively, are +3 and −3. That is the process will set to be controlled under 3-standard deviation, which enables the process to produce 99.73% of the parts that are conforming (nondefective).

6.5   The Fundamental Concept Process Capability

Before beginning a process capability analysis, we must check to ensure the process is stable. If the process is stable, then the short-term behavior of the process (during the initial run) will be a nice predictor of the process's long-term behavior. That is, we can predict future performance with confidence. For both the short and long-term, the process behavior is characterized by the average, range, and standard deviation. The process is considered stable when its mean and its standard deviation are both constants over time (Figs. 6.2 and 6.3).

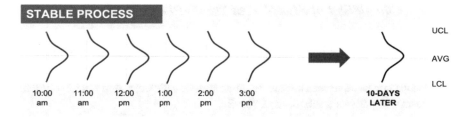

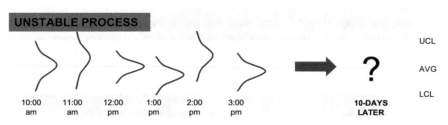

Fig. 6.2 Stable and unstable process

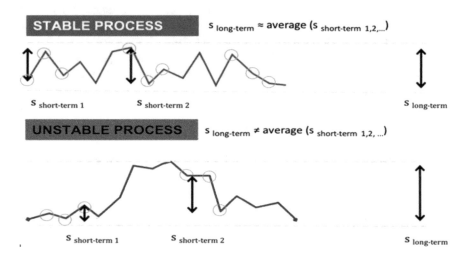

Fig. 6.3 Graphical representation of stable and unstable process

Short-term and long-term process behavior: For the stable process, the run chart must look relatively flat without an upward or downward trend and periodic fluctuations.

6.5.1 Calculation of Short-Term Standard Deviation

To calculate the short-term average and standard deviation, we create the data subgroups. Subgroups are produced in two ways.

1. Record consecutive measurements on an individual's chart and treat every two successive parts as a subgroup of size 2.
2. Record measurements for three to five samples at a fixed interval (e.g., take 5 parts every hour) on the $\overline{X}$ (X-bar) chart.

These subgroup measurements' average and standard deviation are called the short-term-average and short-term-standard-deviation or within-subgroup average and standard deviation.

6.5.2 Calculation of Long-Term Standard Deviation

Calculating the long-term mean and standard deviation is much simpler. We take all the measurements from the individual charts or the X-bar charts and calculate the entire dataset's mean and standard deviation (Fig. 6.4).

Process capability analysis applies data from an original run of parts. That aims to predict whether a manufacturing process can repeatably produce the products based

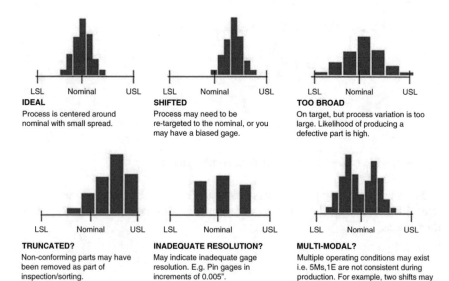

Fig. 6.4 Interpreting process capability histograms

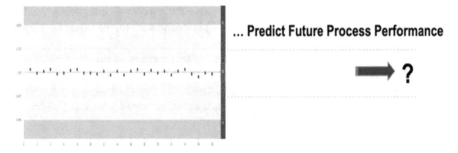

... Predict Future Process Performance

?

Fig. 6.5 Purpose of process capability analysis

on customer specifications. The interpretation is like thinking of how forecasting applies: take historical data, analyze it with a sound model, then perform prediction.

During the prediction process, we should answer the question, "can we rely on this process to deliver good parts?"

Even customers may require a process capability study as part of a Production Part Approval Process (PPAP). They will do this to ensure that the supplier's manufacturing processes can consistently produce good parts. When the manufacturing process is being defined, our goal is to ensure that the manufacturing product should be within the lower and upper specification limits (i.e., LSL and USL). Process Capability measures how consistently a manufacturing process can produce parts within specifications (Fig. 6.5).

6.5.3   Process Capability Index

Process capability index (Cpk) is a well-known statistical measure to predict the ability of the given process to produce products based on the customer's specific requirements or not. In simple words, Cpk measures the producer's capability to deliver the product within the customer's tolerance range.

Cpk is used to evaluate how close the manufacturing meets the target and how consistent it is about average performance.

Cpk provides the best-case scenario for the existing process against the required performance. It can also estimate future process performance, assuming performance is consistent over time.

The objectives of capability analysis are:

1. Producing fewer defective parts.
2. If not, guides to improve your process performance.

As we have defined earlier, Cpk is the index to measure the state of the given process's capability. The higher the Cpk value implies, the more capable to produce the product per the product's design specifications. Similarly, the lower the Cpk value suggests, the process is incapable of producing the product per the product's design specifications. The negative Cpk implies the process is completely inadequate. However, we should note that the Cpk interpreted for the specific single design specification of the product.

The Cpk is also applicable to compare the performances of different machines performing the same job. For instance, machine 1 has a Cpk of 1.7, and machine 2 has a Cpk of 1.1. From the Cpk value, one can derive that machine 1 is better than 2. Since Cpk uses specification limits and parts variation (sigma), we can also arrive at the machine's yield processed and losses.

6.5.4 Calculation of Cp and Cpk

Suppose

- Upper specification limit (USL)
- Lower specification limit (LSL)
- Process mean (μ)
- Process standard deviation (σ) (Fig. 6.6)

The formula to calculate Cpk is given as $Cpk = Min\left\{\frac{USL-\mu}{3\sigma}, \frac{\mu-LSL}{3\sigma}\right\}$.

The formula to calculate Cp is given as $Cp = \frac{USL-LSL}{6\sigma}$

Process capability index (Cpk) measures how consistently the manufacturing process can produce products within specifications. For instance, if Cpk is 1, then the statistical explanation when the curve stretches from +3 to -3. That is considered to occupy 99.73%, and here the process is producing 99.73% good parts.

In general, now it is clear that higher Cpk implies a better process. The Cpk value of less than 1.0 assumes that the process is poor. Or the process produces many defective parts. If the Cpk value is between 1.0 and 1.33, the process is moderately adequate and capable. If the Cpk is above 1.33, the process is quite a good process. However, if we aim to achieve a Cpk of above 2, we aim to make the process to have the best performance.

When we define the manufacturing process, our goal is to ensure that the products produced fall within the lower and upper specification limits (i.e., USL, LSL).

Fig. 6.6 Actual process spread versus ideal spread

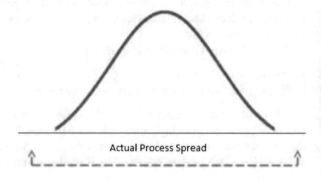

Actual Process Spread

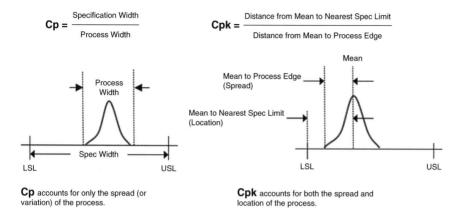

Fig. 6.7 Cp and Cpk

The fundamental idea is too simple. If we want the manufacturing process:

1. To be centered around the nominal, which is desired by the product engineer
2. Is within a spread narrower than the specification tolerance

   (a) We should distinguish the Cp and Cpk. The Cp measures the capability of the centered process. That measures whether the given process meets the specification.
   (b) However, Cpk measures the capability of uncentered of the process (Fig. 6.7).

6.5.5 Difference Between Cp, Cpk and Pp, Ppk

6.5.5.1 Potential Process Capability Analysis (Cp, Cpk)

The process capability analysis collects data from the sample to forecast the manufacturing process's capability, which produces products to meet the specifications. The prediction enables us to measure the manufacturing process as being fit for use in production. The process capability index Cp provides the measure of process capability. How well the process can perform the task if there is no change in the given process inputs. Cpk uses the short-term prediction of the behavior of the process.

6.5.5.2 Actual Process Performance Analysis (Pp, Ppk)

The process performance study is used to EVALUATE a manufacturing process and answers the question: "how did the process actually perform over a period of time?"

Table 6.1 The difference between Cp, Cpk and Pp, Ppk?

	Stable process under statistical control	Stable process NOT under statistical control
Process is centered	Cp	Pp
Process is NOT centered	Cpk	Ppk

This is a historical analysis rather than a predictive analysis but can still be used to drive process improvements. Ppk uses to evaluate the long-term behavior of the process.

If the process is stable, Ppk = Cpk, that is, the actual performance will match the predicted potential performance. However, if the process is unstable—that is, if it shifts or drifts over time—you will find Ppk ≪ Cpk.

Cp and Cpk are considered short-term potential capability measures for a process. In Six Sigma, we want to describe processes quality in terms of sigma because this gives us an easy way to talk about how capable different processes are using a common mathematical framework. In other words, it allows us to compare apple processes to orange processes.

The process performance analysis (Ppk) is used to evaluating the manufacturing process. The analysis answers the question "how did the process perform over a long period?"

That is a historical analysis rather than a predictive analysis but can still drive process improvements. Ppk uses to evaluate the long-term characteristics of the process.

Considering no external changes applied to the process, then if the Ppk = Cpk, then the process is stable. That is, the real performance matches the predicted potential performance of the process. Nevertheless, if the Ppk is lower than the Cpk, the process is unstable. That showed that the process faced some possible special cause variations that drift the process over time.

Precisely, Cp and Cpk are the short-term process capability, which measures the concurrent process's ability, whether it meets the specification or not.

For instance, in the Six Sigma approach, we intended to describe processes in terms of sigma. Hence, it is easy to communicate the discussion about the process capability.

The Six Sigma approach is a well-known tool for assessing process performance. The model applies a mathematical framework to analyze the process.

In other words, the Six Sigma analysis allows us to make basic comparisons of the process performance (Table 6.1).

6.5.6 Notes on Cpk

- Cpk quantifies how close the process performs compared to its specification limits and accounts for the process's natural variability.
- The larger Cpk suggests, the less likely that it is that any product fails the lower and upper specification limits.
- If the Cpk is negative, then the process is completely inadequate. That occurs if the average of the process is outside the specification limits.
- If the Cpk value is higher than 1, then the design tolerance is greater than the actual process spread. So, the process can be capable (depending on process centering).
- Generally, in most industries, it is acceptable that if we achieve a minimum Cpk of 1.33. That is, the process meets a 4 Sigma level.
- Suppose the process spread is greater than the design tolerance. In that case, the process variability cannot fit tolerance. Then even if the process is centered appropriately, we confirm that the process is incapable (Table 6.2).

Example 1 Figure 6.8 represents the capability report of big data on the torque measurement of the screwdriver. The specification of torque is 27 ± 1.5 psi. The

Table 6.2 Cpk, Sigma value, Z-score, nonconforming parts per million

Cpk value	Sigma value	Area under normal curve	Non conforming ppm
0.1	0.3	0.235822715	764177.2851
0.2	0.6	0.451493870	548506.1299
0.3	0.9	0.631879817	368120.1835
0.4	1.2	0.769860537	230139.4634
0.5	1.5	0.866385542	133614.4576
0.6	1.8	0.928139469	71860.531
0.7	2.1	0.964271285	35728.7148
0.8	2.4	0.983604942	16395.0577
0.9	2.7	0.993065954	6934.0461
1.0	3.0	0.997300066	2699.9344
1.1	3.3	0.999033035	966.9651
1.2	3.6	0.999681709	318.2914
1.3	3.9	0.999903769	96.231
1.333	3.999	0.999936360	63.6403
1.4	4.2	0.999973292	26.7082
1.5	4.5	0.999993198	6.8016
1.6	4.8	0.999998411	1.5887
1.666	4.998	0.999999420	0.5802
1.7	5.1	0.999999660	0.3402
1.8	5.4	0.999999933	0.0668
1.9	5.7	0.999999988	0.012
2.0	6.0	0.999999998	0.002

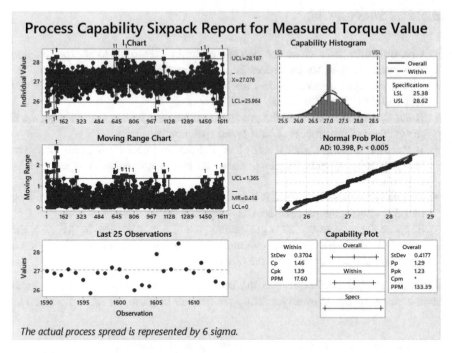

Fig. 6.8 Capability report of Torque measurement of screwdrivers

target of the torque spec measurement is 27 psi, with the lower and upper specifications are 25.5 psi and 28.5 psi, respectively.

From Fig. 6.7, we can infer the following points:

- The capability report showed that the Cpk of the process is 1.39.
- The $\overline{X}$ (X-Bar) chart reveals that the centerline is found 27.076 with the UCL and LCL of the process is found 28.187 psi and 25.964 psi, respectively. That showed that the tolerance 3 psi (i.e., 28.5 − 25.5) is wider than the range, 2.233 psi (i.e., 28.187 − 25.964) of the control chart. Defected parts out of a million parts.
- The range chart showed that there are a lot of measurements that were out of the control limit.

In general, the process has a good performance. However, it needs improvement.

Example 2 A manufacturer has three screw machines. However, due to poor performance, the manufacturer wants to replace one of the poorly performed machines.

The machine's replacement is based on the incapability of meeting the specification requirement of a steel pin with a diameter of 0.125 ± 0.003 in. The target of the steel pin is 0.125″ and the lower and upper specification limits are 0.122″ and 1.28″, respectively. Based on the data available at https://www.itl.nist.gov/div898/handbook/ppc/section5/ppc521.htm#data (Table 6.3).

Table 6.3 Data for steel pin diameter for the three machines

Machine	Time	Sample	Diameter	Machine	Time	Sample	Diameter	Machine	Time	Sample	Diameter
I	1	1	0.1247	II	1	1	0.1239	III	1	1	0.1255
I	1	2	0.1264	II	1	2	0.1239	III	1	2	0.1215
I	1	3	0.1252	II	1	3	0.1239	III	1	3	0.1219
I	1	4	0.1253	II	1	4	0.1231	III	1	4	0.1253
I	1	5	0.1263	II	1	5	0.1221	III	1	5	0.1232
I	1	6	0.1251	II	1	6	0.1216	III	1	6	0.1266
I	1	7	0.1254	II	1	7	0.1233	III	1	7	0.1271
I	1	8	0.1239	II	1	8	0.1228	III	1	8	0.1209
I	1	9	0.1235	II	1	9	0.1227	III	1	9	0.1212
I	1	10	0.1257	II	1	10	0.1229	III	1	10	0.1249
I	2	1	0.1271	II	2	1	0.122	III	2	1	0.1228
I	2	2	0.1253	II	2	2	0.1239	III	2	2	0.126
I	2	3	0.1265	II	2	3	0.1237	III	2	3	0.1242
I	2	4	0.1254	II	2	4	0.1216	III	2	4	0.1236
I	2	5	0.1243	II	2	5	0.1235	III	2	5	0.1248
I	2	6	0.124	II	2	6	0.124	III	2	6	0.1243
I	2	7	0.1246	II	2	7	0.1224	III	2	7	0.126
I	2	8	0.1244	II	2	8	0.1236	III	2	8	0.1231
I	2	9	0.1271	II	2	9	0.1236	III	2	9	0.1234
I	2	10	0.1241	II	2	10	0.1217	III	2	10	0.1246
I	1	1	0.1251	II	1	1	0.1247	III	1	1	0.1207
I	1	2	0.1238	II	1	2	0.122	III	1	2	0.1279
I	1	3	0.1255	II	1	3	0.1218	III	1	3	0.1268
I	1	4	0.1234	II	1	4	0.1237	III	1	4	0.1222
I	1	5	0.1235	II	1	5	0.1234	III	1	5	0.1244

(continued)

Table 6.3 (continued)

Machine	Time	Sample	Diameter	Machine	Time	Sample	Diameter	Machine	Time	Sample	Diameter
I	1	6	0.1266	II	1	6	0.1229	III	1	6	0.1225
I	1	7	0.125	II	1	7	0.1235	III	1	7	0.1234
I	1	8	0.1246	II	1	8	0.1237	III	1	8	0.1244
I	1	9	0.1243	II	1	9	0.1224	III	1	9	0.1207
I	1	10	0.1248	II	1	10	0.1224	III	1	10	0.1264
I	2	1	0.1248	II	2	1	0.1239	III	2	1	0.1224
I	2	2	0.1235	II	2	2	0.1226	III	2	2	0.1254
I	2	3	0.1243	II	2	3	0.1224	III	2	3	0.1237
I	2	4	0.1265	II	2	4	0.1239	III	2	4	0.1254
I	2	5	0.127	II	2	5	0.1237	III	2	5	0.1269
I	2	6	0.1229	II	2	6	0.1227	III	2	6	0.1236
I	2	7	0.125	II	2	7	0.1218	III	2	7	0.1248
I	2	8	0.1248	II	2	8	0.122	III	2	8	0.1253
I	2	9	0.1252	II	2	9	0.1231	III	2	9	0.1252
I	2	10	0.1243	II	2	10	0.1244	III	2	10	0.1237
I	1	1	0.1255	II	1	1	0.1219	III	1	1	0.1217
I	1	2	0.1237	II	1	2	0.1243	III	1	2	0.122
I	1	3	0.1235	II	1	3	0.1231	III	1	3	0.1227
I	1	4	0.1264	II	1	4	0.1223	III	1	4	0.1202
I	1	5	0.1239	II	1	5	0.1218	III	1	5	0.127
I	1	6	0.1266	II	1	6	0.1218	III	1	6	0.1224
I	1	7	0.1242	II	1	7	0.1225	III	1	7	0.1219
I	1	8	0.1231	II	1	8	0.1238	III	1	8	0.1266
I	1	9	0.1232	II	1	9	0.1244	III	1	9	0.1254
I	1	10	0.1244	II	1	10	0.1236	III	1	10	0.1258

I		n	x	II		n	x	III		n	x
1	2	1	0.1233	II	2	1	0.1231	III	2	1	0.1236
1	2	2	0.1237	II	2	2	0.1223	III	2	2	0.1247
1	2	3	0.1244	II	2	3	0.1241	III	2	3	0.124
1	2	4	0.1254	II	2	4	0.1215	III	2	4	0.1235
1	2	5	0.1247	II	2	5	0.1221	III	2	5	0.124
1	2	6	0.1254	II	2	6	0.1236	III	2	6	0.1217
1	2	7	0.1258	II	2	7	0.1229	III	2	7	0.1235
1	2	8	0.126	II	2	8	0.1205	III	2	8	0.1242
1	2	9	0.1235	II	2	9	0.1241	III	2	9	0.1247
1	2	10	0.1273	II	2	10	0.1232	III	2	10	0.125
								III	2	8	0.1242
								III	2	9	0.1247
								III	2	10	0.125

Table 6.4 Analysis of variance

Source	DF	Adj SS	Adj MS	F-Value	P-Value
Machine	2	0.000111	0.000056	30.46	0.000
Error	180	0.000328	0.000002		
Total	182	0.000439			

Table 6.5 Mean, standard deviation estimates, and 95% CI for the three machines

Machine	N	Mean	StDev	95% CI
Machine I	60	0.124887	0.001146	(0.124543, 0.125231)
Machine II	60	0.122968	0.000924	(0.122624, 0.123312)
Machine III	63	0.124051	0.001797	(0.123715, 0.124386)

Table 6.6 Tukey simultaneous tests for differences of means

Difference of levels	Difference of means	SE of difference	95% CI	t-Value	Adjusted P-value
Machine II − Machine I	−0.001918	0.000246	(−0.002500, −0.001336)	−7.78	0.000
Machine III − Machine I	−0.000836	0.000244	(−0.001411, −0.000261)	−3.43	0.002
Machine III − Machine II	0.001082	0.000244	(0.000507, 0.001658)	4.45	0.000

To get a comprehensive solution, we will do the following analysis:

1. Analysis of Variance (ANOVA): ANOVA is an advancement of the t-test, which is used to test different groups' means
2. Process Capability Analysis

The results of the ANOVA are given below. We have three-factor machine levels (i.e., machine I, machine II, machine III). Levels of the factor machine are three. Table 6.4 gives the ANOVA result.

From Table 6.4, we see that the P value is 0.000. At a 5% level of significance, we see that at least one of the machines performs differently than the others. In Table 6.5, the estimates of the process averages and means are presented.

Table 6.5 shows that the process averages of the three machines are 0.1249, 0.1229, and 0.1240, respectively, for machines I, II, and III. In terms of process accuracy (i.e., attaining the target), machine I, followed by machine III, is the most accurate. In terms of process precision, the minimum process variance is observed at machine II, and the maximum process variance is observed at machine III. Table 6.6 gives the tests of process mean differences between the three machines.

Grouping information using the Fisher LSD method and 95% confidence level. From Table 6.7, means that do not share a letter are significantly different.

Tables 6.6, 6.7, and 6.8, Figs. 6.9 and 6.10, we conclude that at a 5% level of significance, the process means of the three machines are different.

Table 6.7 Fisher pairwise comparisons

Machine	N	Mean	Grouping		
Machine I	60	0.124887	A		
Machine III	63	0.124051		B	
Machine II	60	0.122968			C

Table 6.8 Fisher individual tests for differences of means

Difference of levels	Difference of means	SE of difference	95% CI	t-Value	Adjusted P-value
Machine II − Machine I	−0.001918	0.000246	(−0.002405, −0.001432)	−7.78	0.000
Machine III − Machine I	−0.000836	0.000244	(−0.001316, −0.000355)	−3.43	0.001
Machine III − Machine II	0.001082	0.000244	(0.000602, 0.001563)	4.45	0.000

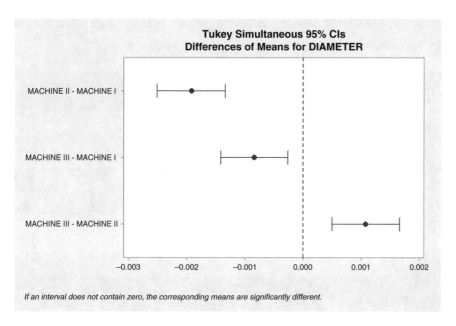

Fig. 6.9 Tukey simultaneous tests for differences of means

Figures 6.11, 6.12, and 6.13 give the interval plot, individual plots, and boxplots of the steel pin diameter of the processes in the three machines.

Figures 6.11, 6.12, and 6.13 conclude that the process variability is higher for machine III. Figure 6.14 gives the normal probability plot of the steel pin diameter. The plot suggests that the diameter is suitability approximated by a normal distribution.

The ANOVA analysis suggested that machine III, followed by machine I, perform poorly than machine II. To solidify our analysis, we need to analyze the process

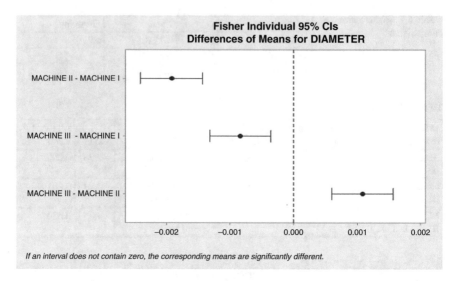

Fig. 6.10 Fisher 95% confidence level the means of steep pin diameter

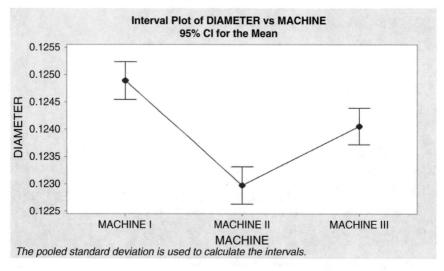

Fig. 6.11 Interval plots of the steel pin diameter of the processes in the three machines

capability of each machine. The process capability report of the machines, machine I, machine II, and machine III, is given in Figs. 6.15, 6.16, and 6.17, respectively.

From Fig. 6.15, we see that:

- The capability report showed that the Cpk of the process in machine I is 1.59.

  From Fig. 6.16, we see that:

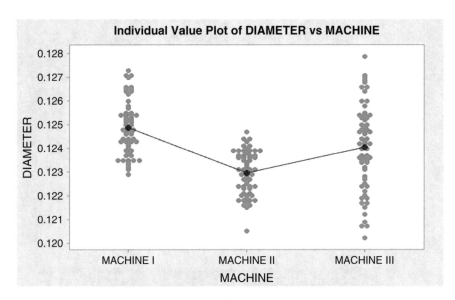

Fig. 6.12 Individual plots of the steel pin diameter of the processes in the three machines

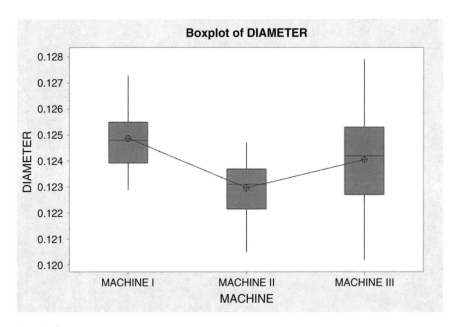

Fig. 6.13 Boxplots of the steel pin diameter of the processes in the three machines

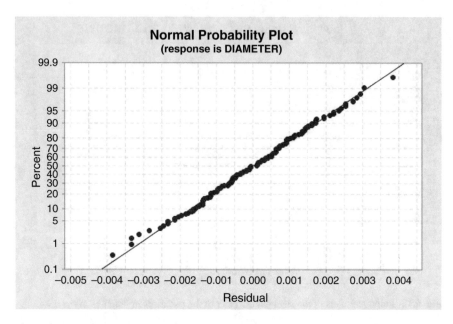

Fig. 6.14 Normal probability plot of the steel pin diameter

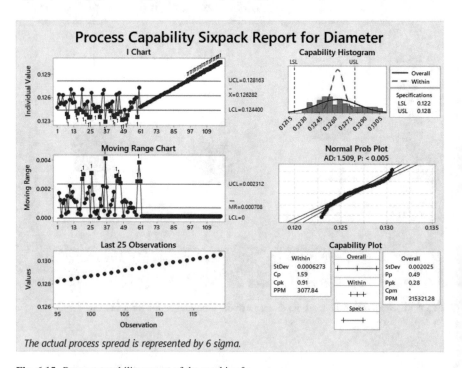

Fig. 6.15 Process capability report of the machine I

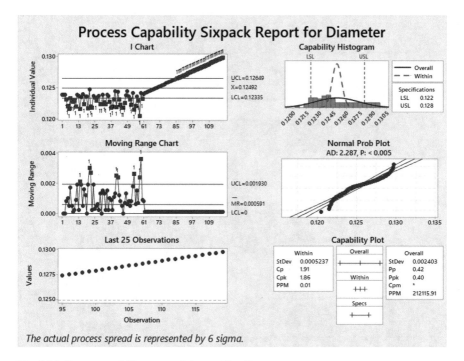

Fig. 6.16 Process capability report of the machine II

- The capability report showed that the Cpk of the process in machine II is 1.86. From Fig. 6.17, we see that:

- The capability report showed that the Cpk of the process in machine II is 0.62.

Like the ANOVA, the process capability analysis suggested that machine III has the lowest capability index, that is, Cpk of 0.62. Therefore, our analysis confirmed that machine III needs to be replaced.

Bibliography

R.E. Barlow, T.Z. Irony, Foundations of statistical quality control, in *Current Issues in Statistical Inference: Essays in Honor of D. Basu*, ed. by M. Ghosh, P. K. Pathak, (Institute of Mathematical Statistics, Hayward, CA, 1992), pp. 99–112

B. Bergman, Conceptualistic Pragmatism: A framework for Bayesian analysis? IIE Trans. **41**, 86–93 (2009)

J.M. Booker, M. Raines, K.G. Swift, *Designing Capable and Reliable Products* (Butterworth-Heinemann, Oxford, 2001) ISBN 978-0-7506-5076-2. OCLC 47030836

D.R. Bothe, *Measuring Process Capability* (Landmark Publishing Inc, Cedarburg, WI, 2001) ISBN 0-07-006652-3

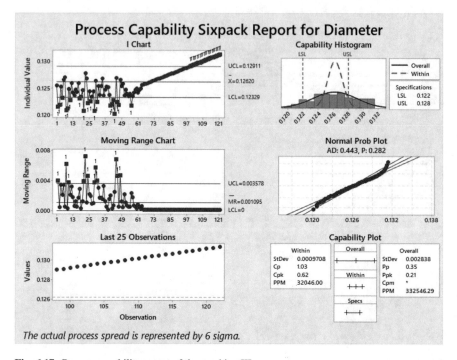

Fig. 6.17 Process capability report of the machine III

R. Boyles, The Taguchi Capability Index. J. Qual. Technol. **23**(1), 17–26 (1991) ISSN 0022-4065. OCLC 1800135

W.E. Deming, On probability as a basis for action. Am. Stat. **29**(4), 146–152 (1975)

W.E. Deming, *Out of the Crisis: Quality, Productivity and Competitive Position* (MIT Press, Cambridge, MA, 1982) ISBN 0-521-30553-5

A.B. Godfrey, *Juran's Quality Handbook* (McGraw-Hill, New York, NY, 1999) ISBN 007034003X

E.L. Grant, *Statistical Quality Control.* ISBN 0071004475 (McGraw-Hill, New York, NY, 1946)

D. Montgomery, *Introduction to Statistical Quality Control* (John Wiley & Sons, Inc, New York, NY, 2004), p. 776. ISBN 978-0-471-65631-9. OCLC 56729567. Archived from the original on 2008-06-20

J. Oakland, *Statistical Process Control* (MPG Books Limited, Bodmin, UK, 2002) ISBN 0-7506-5766-9

T. Pyzdek, *Quality Engineering Handbook* (CRC Press, Boca Raton, FL, 2003) ISBN 0-8247-4614-7

T. Salacinski, *SPC - Statistical Process Control* (The Warsaw University of Technology Publishing House, Warsaw, 2015) ISBN 978-83-7814-319-2

W.A. Shewhart, *Economic Control of Quality of Manufactured Product.* ISBN 0-87389-076-0 (D. Van Nostrand Company, Inc., New York, NY, 1931a)

W.A. Shewhart, *Statistical Method from the Viewpoint of Quality Control* (The Graduate School, The Dept. of Agriculture, Washington, DC, 1931b) ISBN 0-486-652327

D.J. Wheeler, *Understanding Variation: The Key to Managing Chaos*, 2nd edn. (SPC Press, Knoxville, TN, 1999) ISBN 0-945320-53-1

D.J. Wheeler, *Normality and the Process-Behaviour Chart* (SPC Press, Knoxville, TN, 2000) ISBN 0-945320-56-6

D.J. Wheeler, D.S. Chambers, *Understanding Statistical Process Control* (SPC Press, Knoxville, TN, 1992) ISBN 0-945320-13-2

S.A. Wise, D.C. Fair, *Innovative Control Charting: Practical SPC Solutions for Today's Manufacturing Environment* (ASQ Quality Press, Milwaukee, WI, 1998) ISBN 0-87389-3859

S.L. Zabell, Predicting the unpredictable. Synthese **90**(2), 205 (1992). https://doi.org/10.1007/bf00485351

Part III
Engineering Problem-Solving Tools and Continuous Improvement Techniques

Chapter 7
Concurrent Problem-Solving Models for Industrial Applications

7.1 Introduction

The eight disciplines (8D) model is a problem-solving engineering method characteristically applied by quality professionals. The eight disciplines (8D) model can be used in the automotive industry, aerospace industry, semiconductor industry, healthcare, retail, finance, government, and manufacturing. The 8D refers to the eight essential and critical steps to solving the problem. Although it initially encompassed eight stages (or disciplines), each step may have several substages later augmented by an initial planning stage. The eight disciplines (8D) problem-solving model is an exceedingly well-organized and productive scientific approach for solving enduring and recurrent problems.

The primary aims of the 8D methodology are to identify, correct, and eliminate recurrent problems, making them useful in product and process improvement. The model uses and practices team collaboration and make available excellent guidelines and instructions to recognize and identify the root cause (RC) of the existing problem. Then it guides to apply the best containment actions (CA), advance, and then implement the corrective actions and preventive actions (CAPA) that will make the problem solve eternally.

The 8D problem-solving model helps to separate and encompass the most fundamental causes of any objectionable condition. The model is applied as a tool to recognize the factors that cause the problem into effect. And it eradicates and controls the factors that cause the problem. The 8D problem-solving model launches a permanent corrective action based on a statistical analysis of the problem and emphasises the source of the problem by identifying and determining its root causes.

Y. Y. Tesfay, *Developing Structured Procedural and Methodological Engineering Designs*, https://doi.org/10.1007/978-3-030-68402-0_7

7.2   Steps of the 8D is a Problem-Solving Model

The 8D problem-solving model is a problem-solving approach designed to investigate the root cause of a problem. Then it proposes and implements a short-term fix and a long-term solution to prevent the recurring problem. Study the 8D Report in Table 7.1. The following steps are the major components of the 8D.

7.2.1   Step 1: Planning and Formation of Team

This step is all about the formulation of a team that will be involved in the problem-solving process. According to Fig. 7.1, we intended to formulate and establish an efficient and effective operational cross-functional team (CFT) of professionals who

Table 7.1 The 5W2H method

	Is	Is not
Who	Who is affected by the issue/problem? Who first recognized/observed the issue/problem? To who the issue/problem is reported?	Who is not affected by the issue/problem? Who did not observe/disregard the problem?
What	What sort of issue/problem is it? What is occurring/happening? What has the issue/problem? Do we have evidence or indication of the issue/problem in our proprietorship?	What does not have the issue/problem? What possibly will be happening but is not? What possibly will be the issue/problem but is not?
Why	Why it is an issue/a problem? Is the process where the issue/problem happened unchangingly?	Why is it not an issue/a problem?
Where	Where was the issue/problem seen? Where does the issue/problem happen?	Where possibly will the issue/problem be placed but is not? Where else possibly will the issue/problem be placed but is not?
When	When the issue/problem was initially observed? When has it been observed in the meantime?	When the issue/problem possibly will have been observed but was not?
How much/ How Many	How much/many is the quantity of issue/problem? How much is the issue/problem in terms of money, time, and manpower?	How much/many is the quantity of issue/problem? How much are the issues/problems in terms of money, time, and human resources?
How Often	How is the trend of the issue/problem: cyclic, seasonal, random, or continuous? Has the issue/problem happened before? If Yes, you should use the former data and any analysis.	What possibly will the trend be but is not?

Fig. 7.1   Cross-functional team (CFT) task, action, and the required output

have knowledge and experience with the product and process. The Cross-functional team (CFT) is composed up of members from different professional disciplines. So, before we establish the cross-functional team (CFT), we should ask ourselves.

- Is the champion of the cross-functional team (CFT) selected/identified?
- Are the people affected by the problem embodied in the cross-functional team (CFT)?
- Does the cross-functional team (CFT) have the right personnel (technical skills and expertise specialist skills)?
- The cross-functional team's (CFT) goals and membership roles have been segregated, focused, and clarified?
- Should the cross-functional team's (CFT) include customers' and/or supplier's expertise?

We will take the principle one step further by having different levels of CFT:

1. **The core team (CT)** uses either inductive or deductive (convergent) data-driven techniques. The key difference among the inductive (constructive) and inferential (convergent) data-driven research method is that at the same time as a deductive (convergent) method is proposed to test a model, and an inductive (constructive) approach aims at the innovation of a new model from the data. The Core Team (CT) should subject matters on the product, the process, and the data. The team should know profile compiling, scheduling, cyclic terms, manufacturing volume, tools and capacity availability, parametric measures, and so on. The responsibilities of top authorities are:

   (a) A champion is a decision-making sponsor not involved in the team member functions. The champion is ultimately accountable for fixing the issue.
   (b) A team leader is an expert who coordinates and manages the whole 8D project through all its disciplines. Usually, the team leader is a quality engineer. However, sometimes a product engineer or other related professional can also be the team leader. The team leader must create a list defining essential team structure and function to certify and confirm that the team was established for the 8D project. This list is also valuable to determine the role

each team member will play in the 8D project. The team leader essentially upholds the minutes of the 8D meeting and documents the necessary information in the meeting. The meeting minutes include:

- The team functional progress
- Planned against actual accomplishment dates for all activities of the team
- Important and critical decisions agreed by the team, members in the meeting, and
- Decide who will take new tasks? Where? When? How?

(c) The 8D expert is a quality engineer who has substantial knowledge of all the 8D steps. At each step, the 8D Expert must guides the team through the 8D by employing suitable quality tools.

2. **Subject matter experts (SME)** are the team comprised of members who brainstorm, study, and observe (inductive (constructive) and deductive (convergent) techniques). Individuals selected as the subject matter experts (SME) are characteristically leveraging their sole expertise. The SME should use their knowledge and experience to resolve specific problems. Additionally, the subject matter experts (SME) are often fetched to assist with thinking, data collection, and analysis. Examples of subject matter experts (SME):

(a) Manufacturing/industrial engineers are responsible for analyzing current manufacturing processes, producer, tooling, equipment function, and maintenance.

(b) Process engineers are responsible for research projects, design, workflow, and advance processes to optimize output. They are also responsible for investigating and troubleshooting problems.

(c) Product engineers are responsible for selecting materials for manufacturing, creating product design, and testing product prototypes, managing the design of products and their process.

(d) Design engineers are responsible for the research and design of new products, methodologies, and systems. They do several farm duties in the engineering department and draft blueprints, produce test prototypes, and supervise the manufacturing process.

(e) Quality engineers are responsible for reviewing and developing quality specifications and technical design quality documents of the product. They develop thorough, wide-ranging, and well-structured test plans and test prototypes and managing quality test tasks.

(f) Reliability engineers are the components of the engineering department. They are responsible for confirming and validating the new product's reliability, dependability, and maintainability. Similarly, they are responsible for the reliability of machine installations.

(g) The purchasing/procurement specialist is responsible for crafting operative and effective procurement strategies, searching cost-effective agreements with suppliers from preliminary requisition to invoice payment.

(h) Operations managers are responsible for operational management systems, processes, and best practices to ensure all operations and capacity are carried on suitably and cost-effectively.

3. The core functional team (CFT) necessitates proper preparation for the tasks that they are assigned. Here, setting the ground rules is paramount for the core functional team (CFT). Execution and implementation of disciplines like check-lists, procedures, forms, and methods will safeguard stable progress.

4. Two key members are engineering team leader (ETL) and a champion (sponsor). They can affect change by agreeing or disagreeing with the verdicts and can deliver the final approval on the changes.

7.2.2   Step 2: Define and Describe the Problem

Problem-solving should relay on facts, not thoughts. It is always essential to clarify the nature and type of problem. We should ask what is wrong? When did it occur? How immense is the problem degree? And how many times has it happened? etc. According to Fig. 7.2, the explanation should be precise and specific, and easy to comprehend.

1. Employing the inductive tool, lay down the problem in quantifiable, sensible, logical, and understandable terms. The 5W2H technique is one of the most efficient and effective means to recognize the problem. (5W) Who? What? Where? When? and Why (2H) How? and How many?

   Here you should ask the following vital questions

   (a) Has the problem been adequately described and well-defined?
   (b) Are all data included and properly analyzed?
   (c) Has the problem description been established against specifications?
   (d) The specific or related questions that you will need in the problem analysis are summarized in Table 7.1.

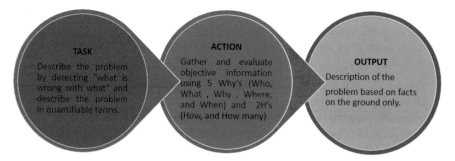

Fig. 7.2 Cross-functional team (CFT) task, action, and the required output for defining the problem

2. Pay attention to historical records of the problem. Here is a memorandum of caution! Do not let these mislead. Past issues are just that—previous problems, and they are not essentially the problem that the Core Team is working on right now. Also, please do not take the recorded or issues reported as a fact. Always use them as information and check. For example, after the problem is reported to you, you should go where the problem occurs. You will then look at signs of mechanical or electrical damage, overheating, unusual sounds, smells, and so on.
3. Most faults make available obvious and apparent clues as to their root cause. Through cautious and careful observation, reasoning, and testing brings the actual problem.
4. When confronted with equipment, which is not operational appropriately, we must be certain that we comprehend "how the equipment is designed to operate?" It makes it much easier to examine the faulty operation.
5. Test the operation of the equipment together with all its components, structures, and features. Please make sure we observe these operations and processes very carefully. In the report, you should record notes on any failures that we should follow in its components, structure, or feature that are not operating properly. That gives a lot of valuable information regarding the problems with the parts of the equipment.

7.2.3   Step 3: Interim Containment Action (ICA)

At this step, we primarily focus on separating the defects and ensuring and safeguarding the customer from receipt of additional parts with the same defect. Interim containment action (ICA) applies different treatability tools to ensure that the defective parts control each operation and warehousing. Interim containment action (ICA) start in Step 2, defining and describing the problem, and it goes until Step 6, which is validating the permanent corrective and preventive actions (CAPA).

Interim containment action (ICA) serves only as a safety measure and habitually accepts no relation to the problem's root cause. Therefore, describing, verifying, and implementing the interim containment action (ICA) to segregate the problem's effects while waiting for corrective and preventive actions (CAPA) are implemented.

Primary data and implementation of interim containment actions (ICAs) are done as fast as possible. Before the permanent corrective and preventive action (CAPA) has been determined and implemented, the interim containment action (ICA) will be done as follows:

1. Prepare cost estimate and schedule for executing the containment actions.
2. Determine the most appropriate containment actions. At this step, we apply traceability analysis using the production, tooling, equipment, and maintenance data. Analyzing the data will help us to identify or the road map of identifying the problem area.

Note that often when equipment faults, certain parts of the equipment will work properly while others not. Furthermore, the part defect may be random or with some pattern. So, the method of traceability for the tool may differ from the defective part.

3. Starting with the entire problem area, take each noted observation and remarks and ask yourself, "what does this tell me about the process, parts, and operation?" If the observation designates that a segment gives the impression to be operating properly, we can eliminate it from the problem area.
4. Use quality specification procedures and apply measurement to narrow the problem area and identify the problem component.
5. Restrict all parts in the process in the stock. The entire suspicious inventory in all locations should reflect accurately isolated.
6. Sorting of parts: Identifying the nonconformance parts from the process or the stock. Use a "divide and eliminate" testing approach until the defective component is identified. Ensure the customer receives only conformance parts (certified material) while examining and determining the problem stay.

To ensure the sorting is done correctly, the project manager of the 8D should examine the following questions:

(a) Are the workers accountable for implementing the containment actions adequately instructed?
(b) Do effective and operational containment actions been executed and verified?
(c) Are the containment actions adequate and functional?
(d) Does the containment action provide serviceable parts to deliver to the customer?
(e) Are defective parts being identified and rejected?

7. Track all our information and record them in the sorting sheet and approve.

7.2.4   Step 4: Determine, Identify, and Verify Root Causes and Escape Points

In engineering, root cause analysis (RCA) is a problem-solving technique, which applies to recognizing and identifying the root causes of failure. Root cause analysis (RCA) is extensively involved in engineering, manufacturing, the service industry, software development, and so on.

Root cause analysis (RCA) is the technique and process of realizing the causes of the failure (problem) to identify appropriate corrective measures and solutions. Root cause analysis (RCA) adopts an effective way to prevent and solve the primary problem rather than systematically fix peripheral issues.

Defining and describing the root causes of the problem is the essential and critical step of the 8D problem-solving process. Determining the actual root causes is usually the hardest and challenging feature of the problem-solving process. Hence, the causes that give the impression to be the problem are often indications, not the actual root causes. Furthermore, the specific causes that allowable the apparent or deceptive symptoms to arise are the actual root causes and every so often hidden deep in the process.

Root cause analysis (RCA) can be achieved with a collection of methods, principles, techniques, and procedures that can altogether be leveraged to recognize and identify the root causes of the issue. Observing outside apparent cause and effect (CAE), root cause analysis (RCA) can show where systems or processes failed in the first place.

There are three major goals of root cause analysis (RCA).

1. Effectively and efficiently recognize, identify, and determine the root cause of a given problem.
2. To comprehend how to fix, compensate, or study any primary issues or issues within the root cause.
3. We will use what we acquire from the root cause analysis (RCA) to avoid future related issues and problems systematically and scientifically.

7.2.4.1 The Fundamental Principles of Root Cause Analysis (RCA)

We apply the root cause analysis (RCA) to also adjust the essential process and system matters to avoid future issues and problems. Dealing with the individual warning sign may seem productive. Nonetheless, if we don't essentially diagnose the actual root cause of a problem, we will probably have several of the same or similar issues in the future. Therefore, the detailed root cause analysis (RCA) will give us a permanent solution to those problems. Thus, the fundamental principles that lead and guide operative and efficient root cause analysis (RCA) are:

- Recognize there can be various root causes for a problem.
- Emphasis on correcting core root causes rather than adjusting sections of the warning signs.
- For short-term solution, never disregard the importance of treating symptoms the warning signs.
- Give attention to how and why the problem happens rather than spending your time on who done it.
- Be scientific (methodological and technical) and try to find actual cause effect evidence (CEE) to back up for the root cause. Provide adequate evidence to support your analysis.
- Give your recommendation to prevent the actual root cause of the problem.

7.2.4.2   Types of Root Causes

It is essential to take a complete method to provide a framework and evidence produced in the operative decision. Therefore, when we are detecting the root cause, the cross-functional team (CFT) must emphasize why the problem occurred and how the defect was not caught? And finally, why the quality system fails? To answer these questions, we need to see the different types of root causes.

1. **Technical Root Causes (TRC)**
   Technical root causes (TRC) is when the root cause is the general for the manifestation of the technical issue. Root cause on the technical level comes from the description and picture of functional cause and effect (CAE) relationships. Examples of technical root causes (TRC) are:

   (a) The physical (shape, color, density, area, volume, strength, strain, brittleness, absorption, temperature and heat transfer, magnetism) and chemical (reactivity, test, pH value, odor, flammability, ductility, toxicity, oxidizing agent, and reducing agent, malleability) function properties of materials.
   (b) In technical, methodological, and producorial process.

2. **Escape Point (EP)**
   An escape point (EP) is a place somewhere in the process the actual root cause possibly will have been noticed, but then again, it perhaps will pass. Therefore, escape point (EP) is considered as the initial control point in the process of succeeding in the actual root cause of a problem that ought to have recognized and detect that problem but unable to do so. The core idea is to know the exact root cause and figure out what went mistaken with the process control system in letting the problem outflow. This root cause denotes to identification and detection process control system. Examples of escape point (EP) are:

   (a) Incorrect calibration cycle, incapable machine, or tool
   (b) Inadequate design or drawing
   (c) Dimension may not include in the statistical process control chart
   (d) Preliminary inspection criteria and procedure

3. **Systemic Root Cause (SRC)**
   Systemic root cause (SRC) root cause denotes the product's process quality and manufacturing procedural systems. The systemic root cause (SRC) could originate from either the problem or the escape point (EP) root causes. Examples of escape point (EP) are:

   (a) Inadequate process failure mode analysis and control plans
   (b) Instructions and procedures for the process may be unclear, inferior, vague, not created, or incomplete
   (c) Wrong tooling, or untrained workforce

To know how we can apply the types of root causes, let us see the following illustration.

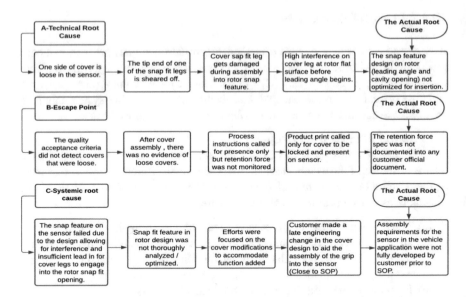

Fig. 7.3 Root cause analysis of the HNXE13 machine with loose sensors problem

Problem Description The customer reported a problem that the HNXE13 machine was wobbly in numerous sensors and spiking to others' clutch. The solution is given in Fig. 7.3.

7.2.4.3 Techniques of Root Cause Analysis (RCA)

The viewpoint behind the root cause analysis (RCA) suggests that it is finest to solve process problems by removing the root causes when doing so. The minute we try to alleviate the root causes of the problem, we should emphasize that the problem does not appear again. There are several methods, techniques, and procedures involved in the root cause analysis (RCA). So, we will review some of the most important methods.

7.2.4.4 Comparative Analysis

This technique is just listing and analyzing differences and changes between "Is" and "Is Not." Arnheiter and Greenland (2008) have been applied the comparative analysis to determine that management decision-making and corporate fundamental strategy principles adept by automotive companies: Honda, Toyota, and Nissan. The research showed that the US automotive industry's problems could be reduced if corporations apply their effort to understand their problems' root causes completely. No matter how big or minor the issues are, the paper suggests that it is beneficial if managers must figure out the root causes of their problems.

Practical Implications

Managers learn that it is beneficial to properly recognize and comprehend the problem's root cause (RC). No matter how severer (or small) the problem may seem, we must start by questioning all prior assumptions and arguments while focusing on simple fundamental truths.

7.2.4.5   Pareto Charts

Root cause analysis (RCA) uses several methods to realize to control the problem and accomplishing the goal. Pareto analysis is one of the utmost essential tools of root cause analysis (RCA).

In the late 1800s, after observing 80% of the land was owned by 20% of the people, an Italian Economist Vilfredo Pareto introduced the Pareto Analysis. He called the phenomena the 80/20 rule. Later after analyzing the logarithmic mathematical model, Joseph Juran recognized that the Pareto analysis is applicable in quality management. The model is named the Pareto Principle, which is the rule of "The Vital Few and Trivial Many." For example, 20% of the time used in each day leads to 80% of the work.

Based on statistical data, the Pareto analysis, also known as the 80/20 rule, provides a vibrant picture of the problem's core causes. Employing Pareto diagrams, we can recognize, identify, characterize, and elucidate issues, many corporations can make the most important jump on the way to better problem-solving practice.

The Pareto method clarifies that 80% of the problem's effects originate from 20% of the root causes. It means time and cost management can be improved if working mostly on 20% of the root causes.

The root cause analysis is reinforced by demonstrating data in the Pareto chart and Pareto diagram where the bar's altitude signifies the amount of a root cause, sorted by the highest impact to the smallest impact (from left to right). That provides a piece of accurate visual information and a clear and letting fast indication of when to invest our thinking exertions and actions in resolving the problem.

The benefits of the Pareto analysis are as follows:

- It aids in recognizing the topmost root causes.
- It helps to line up the topmost subject matter for the problem and attempt to eradicate it at first.
- It gives an awareness of the cumulative effect of the issues.
- The corrective and preventive action (CAPA) can be well organized, planned, and efficiently implemented.
- It provides a specific, simple, and vibrant way to improve problem-solving and the decision-making process.
- Improves and advances the efficiency of quality management systems.
- It helps in planning, time management, change management, analysis, and troubleshooting.

Fig. 7.4 Pareto chart of customer service performance failures of IT helpdesk

Consider the following example: The IT helpdesk's customer service performance was found insufficient and unsatisfactory by most users. The Pareto Chart is given in Fig. 7.4.

Here we can see that almost 80% (79%) of the complaints of the IT department's poor customer service comes from the lack of training and inadequate pay. Therefore, fixing the proper training capability and improving the IT professional's salary will reduce the complaint by setting other factors (communication issue, high turnover, and managerial support).

7.2.4.6 The 5Whys Model

A Japanese industrial engineer Taiichi Ohno invented the 5-Why root cause analysis technique. The 5Whys is very efficient and effective in root cause analysis (RTA). The five whys use a systematic inquiry (the whys) technique to hunt for the problem's root causes. The central idea behind the 5Why method is investigating a provable, demonstrable, and verifiable interrelationship among the problem and processes. Taiichi started the analysis by asking, "Why did the robot stop?" then let us look at how he reached the root cause of the problem in Fig. 7.5.

Steps of Applying the 5Why Method

1. Define and describe the issue/problem to be investigated.
2. Ask "Why" the initial level causes the issue/problem?
3. Develop a diagram and record the causes of taking place in the diagram.

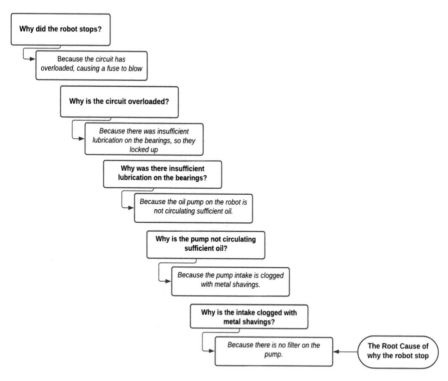

Figure 7.5 Taiichi's root cause analysis using the 5Why method

4. For every cause, ask "Why" and record the responses in the subsequent column, associated with the previous response.
5. Continue asking "Why" up until the root cause discover.

7.2.4.7 The Fishbone Diagram: Cause and Effect (CAE) Diagram

The Fishbone diagram is one of the well-known models of identifying the problem's possible root causes. The Fishbone analysis, also known as the cause and effect (CAE) diagram. The model sorts thoughts, ideas, and data into useful classes. The fishbone analysis is a highly structured approach to analyze the causes and their effects using a graphic way. The issue, problem, or effect is presented at the head of the fish. As shown in Fig. 7.6, under different cause categories (people, methods, machines, materials, measurements, and environment), the bones' possible causes are recorded.

Hence, the systematic categorization of the problem areas is the key to the Fishbone analysis. The approach is useful in identifying and recognizing possible causes for the problem that possibly will not be analyzed by other alternative methods (e.g., like the 5Whys method). As shown in Fig. 7.7, the advantage is that we can use the 5Why model to analyze the root cause in each category.

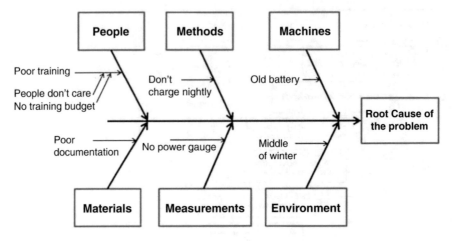

Fig. 7.6 Fishbone analysis

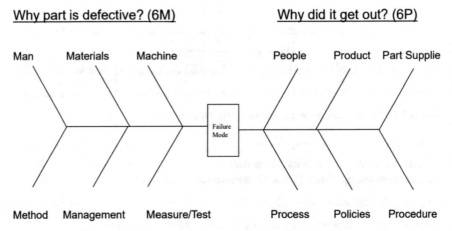

Fig. 7.7 Structured fishbone diagram to apply 5Whys to investigate why a bad part is produced and why it sent to the customer

Man/personnel error: abusing of requirements or procedures, communication problem (both verbal and nonverbal), human error, ignorance, limitation of skills, lack of training provided, insufficient practice.

Material problem: inappropriate material (contamination, misspecification) and bad material handling.

Machine/equipment problem: failed material part, manufacturer error, calibration error, maintenance, electrical noise, contamination.

Procedure problem: inadequate procedure, lack of procedure, lack of systems, inadequate content, inadequate presentation.

Method problem: wrong method, inadequate model, error in equipment, error material selection, Drawing error, specification error.

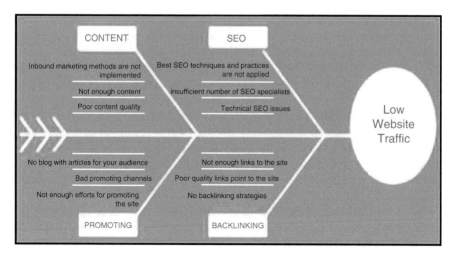

Fig. 7.8 Fishbone diagram analysis of the low-slung website traffic flow problem

Management problem: inadequate managerial system, organization, and planning deficiency, inadequate supervision, improper resource allocation, inadequate policy (definition, dissemination, and enforcement).

Measurement/test error: bad gauge, bad reading due to temperature or environment, random error, absolute error.

Figure 7.8 gives the illustration of root cause analysis (RCA) using the fishbone diagram. The problem is those online marketing professionals working for a firm that experiences low-slung website traffic flow. Using the fishbone diagram, let us analyze the root cause of the problem.

As we have seen from Fig. 7.8, we can see that the root cause of the diminutive website traffic by the users is analyzed based on the content, search engine optimization (SEO), promotion, and backlinking. Poor content quality, poor promotion, bad backlinking, and technical SEO issues are found to be the root causes of the problem.

7.2.4.8 Control Charts

Revise Chaps. 4 and 5 to comprehend the uses of control charts in identifying special cause variations.

7.2.4.9 Design of Experiments (DoE): Analysis of Variance (ANOVA)

Often the Six Sigma practices need data to improve a process, costs, and lead time. Data analysis and interpretation determine the essential decisions that we make.

Therefore, more importantly, the way we analyze the data is the core issue to have the right interpretation and give the right operational and strategic decisions. Researching and experimentation are the basic way to understand the design product and produce the product. Design of experiments (DoE) or analysis of variance (ANOVA) is a well-known and most suitable way to experiment for this purpose.

Design of experiments (DoE) comes to quantify, verify, validate, and confirm the main effects and the interactions of important factors on the product, the process, and engineering performance. The common design of experiments (DoE) application is to identify the factors that have effects on the product or system performance and to define the proper procedure, working window (i.e., specification limits), the process for the tested parameters to get the desired or improved product with adequate performance.

Design of experiments (DoE) is a set of models that statistically analyze the differences between different categories or groups of data. In more naïve terms, the design of experiments (DoE) aids in finding and visualize mean and variance differences in the dataset.

One of the main aims of six-sigma is to find and innovate customs of improving prevailing systems. When doing the root cause analysis (RCA), either DMAIC or DFSS, cannot precisely and accurately show where faults are initiating. However, to assist the Six-Sigma missions, the design of experiments (DoE) has numerous important applications. That helps improve and advance processes and products by Comparing alternatives, reducing variability to increase precision, finding the significant input factors that affect the output, accomplishing an optimum process yield, and extricating the "vital few" from the "trivial many." Precisely, the model pictures where differences are, which factor correlates to the process or production mistakes. Therefore, the design of experiments (DoE) will be done effectively and efficiently than the DMAIC or DFSS process.

Design of experiments (DoE) method is useful for defining an appropriate process, product, working environment, and developing procedure. So, design of experiments (DoE) is not always the technique of root cause analysis (RCA) due to the following reasons:

1. Usually, the DoE result demonstrates and verifies that more than a few factors have a comparable response to the outcome. Therefore, the DoE cannot deliver a strong and clear sign for which of the examined factors is the root cause of a specific problem.
2. The actual root cause may not be among the selected factors in the DoE calculation.
3. Using the results of the DoE, we adjust the process. The change of the process by itself may create a new problem for the product.

7.2.5   *Step 5: Select and Implement Permanent Corrective and Preventive Action (CAPA)*

Once we analyze the root cause of the problem, the next step is selecting and implementing the corrective action to avoid it. Our tasks are planning and applying the carefully chosen permanent disciplinary actions, controlling the plan to eliminate the temporary containment actions, and managing the permanent results.

The corrective action should confirm the effectiveness after executing the corrective actions and ensuring no undesirable, harmful, risky, and negative consequences. Here we need to monitor the corrective actions, any similar significant defect and document the results. Selection and implementation Permanent Corrective and Preventive Action (CAPA) requires the following steps:

1. Ensure that all the company safety procedures have been applied during the permanent corrective and preventive action (CAPA) process.
2. Once we detect the root cause, we should essentially verify it before proceeding with the next steps.
3. Including the mandatory requirements, we need to establish the acceptance criteria.
4. Perform a comprehensive and complete risk assessment (failure mode and effects analysis FMEA) on the corrective and preventive action (CAPA) choices.
5. Based on the risk assessment, perform a well-adjusted choice for corrective and preventive action (CAPA).
6. Choose control point enhancement for the escape point (EP). Here, confirmation of efficiency and effectiveness for both the CAPA and the escape point (EP) is needed.
7. Complete pre-production programs (quantitatively) should confirm that the selected correction will resolve the problem. Here checklists are required.

7.2.5.1   Plan–Do–Check–Act (PDCA)

It is known that Plan–Do–Check–Act (PDCA) a good model for continuous improvement. See the Plan–Do–Check–Act (PDCA) in Fig. 7.9.

The Plan–Do–Check–Act (PDCA) model is a quality management system (QMS) for planning, operation, and continuous improvement tools. By applying the method of the Plan–Do–Check–Act (PDCA) cycle, we can launch and implement effective and operational corrective actions as follows:

1. *The Planning Steps*

      Depending on the corrective action's size, the core functional team (CFT) plans to accomplish the project.

   (a) **Comprehend corrective action systems and requirements**

Fig. 7.9 Plan–Do–Check–
Act (PDCA)

Here the primary step in generating an effective and operational corrective action system. The corrective action systems and requirements should be correctly evaluated, verified, and documented for the next steps.

(b) **Planning and scheduling the Process**

Planning is the design phase regarding the corrective action system's outline and procedure and the requirements and how to integrate the new system into existing operations. Planning should address personnel tasks and responsibilities for the implementation of the corrective action process. Precisely, procedures must be premeditated for the accomplishment of the following vital duties:

- Identifying, review, and evaluate nonconformance or problem causes such as customer complaints, inspection and audit verdicts, process failure, and operational parameters.
- Apprehending issues into the corrective action implementation process and identifying top priorities.
- Check and verify and the root cause of the problem and corrective action implementation mechanisms.

(c) **Develop and Document**

In the development phase, an organized corrective action system is formed based on the developed plan. Precisely, the teams assign the authority and their duty to implement the program.

2. *The Do Steps*

(a) **Conduct Training**

Training must involve collaborative learning events among the team members to accomplish the new system's implementation and how it functions. The training must create the opportunity for each team member to acquire the necessary understanding, skill sets, and knowledge that the project needs.

(b) **Implement**

As soon as the training is completed, the corrective action system's implementation and operation must occur. In this phase, the corrective action measures go alive and need to be fully operational.

3. *The Checking Step: Test the functionality of the corrective action*

After implementation in a full circle, the subsequent step is checking the functionality of the corrective action. Checking the corrective action's functionality is done by taking samples and verifying the specifications of the product design, the adequate functionality of the process, or the issue of interest.

4. *The Action Phase: Adjust and Improve*

At this step, we do adjustments to improve the effectiveness and efficiency of the corrective action. The goal of this step is to make the corrective action consistent with the plan and effectively address the issue, and if not, implement continuous improvement to upgrade the quality of the product, the process, or the system of our interest.

7.2.5.2   Failure Mode and Effects Analysis (FMEA)

We need data to demonstrate and confirm the success and effectiveness of the corrective actions taken. After implementing the corrective actions, we need to recalculate and update the failure mode and effect analysis (FMEA) to perform a risk analysis.

The FMEA is a very powerful qualitative method that is used to recognize, identify, assess, and evaluate the effects of a definite failure mode in the process. Failure mode and effects analysis (FMEA) encompasses the procedure of determining the impact of different faulty operations in the process under certain circumstances.

The FMEA is used:

- To detect the potential failure modes and the related risks introduced to the product or process.
- To prioritize the most and least important action plans to control and eradicate the process's possible failures.
- Record, track, and evaluate the consequences of the corrective action plans.

The steps to complete the FMEA

1. Using the process map, review, and label the process sub-steps.
2. Study the possible failure modes for each process and subprocess and their corresponding purpose (impute–output)
3. Determine the possible failure effects related to every failure mode and analyze all the potential root causes.
4. For each root cause, classify up-to-date process controls.

5. Allot a Severity (S, ranging from 1 to 10, where 10 is the impact is very high and severity leads to a hazard deprived of warning) ranking to every effect that has been known.
6. Allot the occurrence (O, ranging from 1 to 10, where 10 is the impact is very high) ranking.
7. Allot the detection (D, ranging from 1 to 10, where 10 is the effect is very high) rankings.
8. Compute the risk priority number (RPN) = (Severity) $\times$ (Occurrence) $\times$ (Detection)

    (a) RPN = 1 means no risk at all.
    (b) RPN = 1000 means the risk is at the level of considering as a hazard.

9. To reduce RPN by lowering either Severity (S), Occurrence (O), or Detection (D), we need to develop an action plan.
10. Assign responsibility for the action plan and act (take action).
11. Recalculate the RPN and evaluate whether the action is reduced or not. To recalculate the risk priority number (RPN), reassess the Severity (S), Occurrence (O), and Detection (D) rankings for each of the failure modes subsequently, the action plan has been done.

Study an example of FMEA in Table 7.2.

7.2.6   Step 6: Implement and Validate Corrective Actions

The 8D problem-solving system controls and determines to advance and improve capabilities to meet customer expectations and regulatory requirements. A successful 8D problem-solving problem's primary accomplishment needs to be evaluated— the success of the project is assessed by the degree of effectiveness and adequacy of addressing the issues.

In step 5, we have seen that appropriate planning is indispensable to successfully implement corrective and preventive actions (CAPA). A project plan must incorporate steps-to-complete, measurement of project success, lessons gained during the project life span, and proof of improvement. So, we will follow the following steps to validate the corrective and preventive actions (CAPA).

1. Define (or describe), and implement the finest corrective actions done in all the specifications.
2. Once we have properly identified the faulty operation's cause, we can proceed to swap the defective component.
3. After replacing the component, we must test operate all features of the process/operation, be sure we have replaced the proper component, and no other faults in the process/operation.
4. Be sure that the work environment is safe for the personnel.

Table 7.2 FMEA calculation example

Item / Function	Potential Failure Mode(s)	Potential Effect(s) of Failure	S e v	Potential Cause(s)/ Mechanism(s) of Failure	P t o b	Current Design Controls	D e t	R P N	Recommended Action(s)	Responsibility & Target Completion Date	Actions Taken	New Sev	New Occ	New Det	New RPN
Coolant containment. Hose connection. Coolant fill. M	Crack/break. Burst. Side wall flex. Bad seal. Poor hose rete	Leak	8	Over pressure	8	Burst, validation pressure cycle.	1	64	Test included in prototype and production validation testing.	J.P. Aguire 11/1/95 E. Eglin 8/1/96					

Write down each failure mode and potential consequence(s) of that

Severity - On a scale of 1-10, rate the Severity of each failure (10= most severe). See Severity sheet.

Likelihood - Write down the potential cause(s), and on a scale of 1-10, rate the Likelihood of each failure (10= most

Detectability - Examine the current design, then, on a scale of 1-10, rate the Detectability of each failure (10 = least detectable). See Detectability sheet.

Risk Priority Number - The combined weighting of Severity, Likelihood, and Detectability.
RPN = Sev × Occ × Det

Response Plans and Tracking

7.2.7   Step 7: Take Preventive Actions

Preventive measures provide the outlook to preserve and share the knowledge, preventing problems on similar products, processes, or families of processes. Bring up-to-date documents, procedures, and work instructions are anticipated to apply in future use.

Double-check that the problem is resolved in the required degree. These actions include:

1. Check the updated procedures and work instructions for prevention.
2. Check the implementation of standard operational practices.
3. Modify or adjust the operation systems, management systems, and techniques to prevent the recurrence of all similar problems.
4. Check the updated process flow diagram.
5. Check the updated failure mode effect analysis (FMEA) and complete the documentation.
6. Assure action control plan (ACP) is updated.
7. Review and evaluate similar products and processes for problem prevention.

7.2.8　Step 8: Closure and Team Celebration

One of the key secret of the most successful companies globally is their recognition of their personnel. So, it is recommended that the company should acknowledge the team. Congratulate and give praise to individual and collective Core Functional Team efforts to realize the accomplishments of the 8D process. Such activities include:

1. Recording and documentation for future reference.
2. Educations and thought gained on how to make problem-solving realized and how we will make it better.
3. The before and the after operational comparison of the problem (Fig. 7.10).

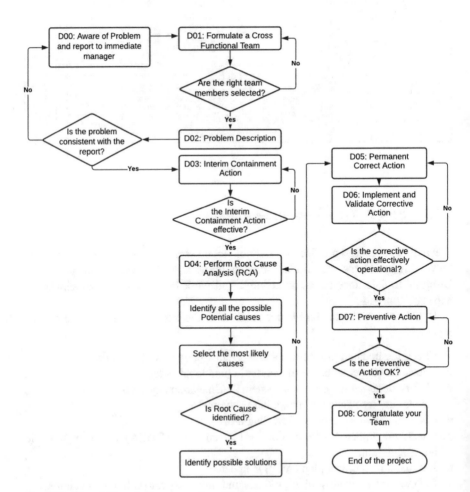

Fig. 7.10 8D problem-solving process flow diagram

7.3   The 8D Report Form

8D Report			CAR#	
Customer Number			Part Name	Part Number
Date				
D01: Formulate a Cross Functional Team				
Role formulate	Name	Position	Responsibility	
Team champion				
8D project leader/responsible				
Other 8D team members				

D02: Problem Description

D3: Interim Containment Action

D04: Root Cause

Root cause analysis—"5 Why"	
Problem description:	
Reason why?	
Reason why?	
Reason why?	
Reason why?	
Reason why?	
Final root cause:	

Root cause analysis—non-detection "5 Why"

Problem description:	
Reason why?	
Reason why?	
Reason why?	
Reason why?	
Reason why?	
Final escape root cause	

D05: Permanent Correct Action

Corrective action detail	Responsible	Completion date	Evidence	Notes

(continued)

Corrective action detail (non-detection)	Responsible	Completion date	Evidence	Notes

D06: Implement Corrective Action

Implementation detail	Responsible	Completion	Evidence	Notes

Implementation detail (non-detection)	Responsible	Completion date	Evidence	Notes

D07: Preventive Action

Defined action	Responsible	Completion date	Evidence	Notes

Current products	Notes

Future products	Notes

D08: Congratulate Our Team

Check

Checked and approved by	Name		Signature	

Bibliography

American Society for Quality (ASQ) Certified Quality Engineer (CQE)., http://prdweb.asq.org/certification/control/quality-engineer/i

E.D. Arnheiter, J.E. Greenland, Looking for root cause: A comparative analysis. TQM J. **20**, 18–30 (2008)

K.C. Chai, K.M. Tay, C.P. Lim, A perceptual computing-based method to prioritize failure modes in failure mode and effect analysis and its application to edible bird nest farming. Appl. Soft Comput. **49**, 734–747 (2016). https://doi.org/10.1016/j.asoc.2016.08.043

G. Gilbert, Quality improvement in a defense organization. Public Product. Manage. Rev. **16**(1), 65–75 (1992)

Goddard Space Flight Center (GSFC), *Performing a Failure Mode and Effects Analysis* (Goddard Space Flight Center, Greenbelt, MD, 1996) Accessed 25 Aug 2013. 431-REF-000370

S. Hill, Why quality circles failed but total quality management might succeed. Br. J. Indus. Relations **29**(4), 541–568 (1991)

A. Hyde, The proverbs of total quality management: Recharting the path to quality improvement in the public sector. Public Product. Manage. Rev. **16**(1), 25–37 (1992)

K. Ishikawa, *What Is Total Quality Control? The Japanese Way* (Prentice-Hall, Englewood Cliffs, NJ, 1985)

ISO, *ISO 9000 Quality Management System - Fundamentals and Vocabulary* (The British Standards Institution, London, 2005)

J.M. Juran, J.A. De Feo, *Juran's Quality Handbook*, 6th edn. (McGraw-Hill, New York, NY, 1999) ISBN 978-0-07-162973-7

Y.W. Kerk, K.M. Tay, C.P. Lim, An analytical interval fuzzy inference system for risk evaluation and prioritization in failure mode and effect analysis. IEEE Syst. J. **11**(3), 1–12. Bibcode:2017ISysJ.11.1589K (2017). https://doi.org/10.1109/JSYST.2015.2478150

L. Martin, Total quality management in the public sector. Natl. Product. Rev. **10**, 195–213 (1993)

B. Poksinska, J.J. Dahlgaard, M. Antoni, The state of ISO 9000 certification: A study of Swedish organizations. TQM Mag. **14**(5), 297 (2002). https://doi.org/10.1108/09544780210439734

W.F. Pruitt, A Disciplined Approach (2019), asq.org. Accessed 25 Oct 2019

T. Pyzdek, *Quality Engineering Handbook* (CRC Press, Boca Raton, FL, 2003) ISBN 0-8247-4614-7

L. Rambaud, *8D Structured Problem Solving: A Guide to Creating High Quality 8D Reports*, 2nd edn. (PHRED Solutions, Breckenridge, CO, 2011) ISBN 978-0979055317

M. Rausand, A. Hoylan, *System Reliability Theory: Models, Statistical Methods, and Applications*, Wiley Series in Probability and Statistics, vol 2004, 2nd edn. (Wiley, Hoboken, NJ, 2004), p. 88

SAE, *Potential Failure Mode and Effects Analysis in Design (Design FMEA) and Potential Failure Mode and Effects Analysis in Manufacturing and Assembly Processes (Process FMEA) and Effects Analysis for Machinery (Machinery FMEA)* (SAE International, Warrendale, PA, 2008)

A.K. Smith, Total quality management in the public sector. Quality Progress **1993**, 45–48 (1993)

J. Swiss, Adapting TQM to government. Public Admin. Rev. **52**, 356–362 (1992)

N. Tichey, *Managing Strategic Change* (John Wiley & Sons, New York, NY, 1983)

C.S.P. Visser, *8D Problem Solving Explained – Turning Operational Failures into Knowledge to Drive Your Strategic and Competitive Advantages.*, ISBN 978-1543000931 (CreateSpace Independent Publishing Platform, Scotts Valley, CA, 2017)

Chapter 8
Models of Continuous Improvement

8.1 Introduction

A continuous improvement process (abbreviated as CIP) is an unending determination and effort to improve or advance products, services, or processes. These continuous efforts seek out "incremental" enhancement over time or "breakthrough" improvement all at once. Delivery (i.e., customer valued) processes are constantly assessed and enhanced in the light of their efficiency, effectiveness, and flexibility.

The continuous improvement process (CIP) means getting better all the time. Some experts comprehend the continuous improvement process as a meta-process for utmost management schemes (e.g., business process management, total quality management, project management, and program management). Edwards Deming, the innovator of the field, observes that the continuous improvement process is a part of the operating system. The feedback from the process (and also the customer) assessed against company goals. The fact that the continuous improvement process can be called a management process does not mean that it needs to be executed by management; rather, it makes verdicts about the execution of the process and the design of the delivery process itself.

A comprehensive characterization is that of the institute of quality assurance (QA) whose well-defined continuous improvement is a never-ending advancement that is attentive to increasing the company's effectiveness and/or efficiency to accomplish its policy and objectives. Continuous improvement is not limited to, quality functions and initiatives. It is also an improvement in business strategy, business results, customer, employee, and supplier relationships subject to continual improvement.

Formally, the continuous improvement process is defined as the on-going efforts within an organization to improve the quality, value, and related activities against the products, processes, or services' functionality. Continuous improvement upsurges the quality and value of the product or service. The degree of the continuous improvement programs of a given company makes it more competitive. It guarantees that an organization possesses to keep up with the varying needs and expectations of the customer. To truthfully be productive with the continuous improvement program, the organization must be premeditated about using the process to accomplish improvements and advancements and sustain the methodologies over time.

Various engineering methodologies have been developed to achieve continuous improvement programs. In this chapter, we will learn how to define and distinguish the following continuous improvement methodologies.

1. Total quality management (TQM)
2. Six Kaizen
3. Plan–Do–Check–Act (PDCA)
4. Six Sigma
5. Theory of Constraint (TOC)

8.2   Systematic Continuous Improvement Categories

There are several methods of continuous improvement programs. Each method may come with an exclusive outlook or approach. However, despite the fact these methods are different somehow, they all share common core principles.

The continuous improvement processes involve:

- Customer focus
- Employee focus
- Process focus
- Leadership

8.2.1   Customer Focus

In simple logic, the company's improvement is only an improvement if and only if the customer agrees with the achieved goals of the continuous improvement programs. It should not be a wonder that each improvement methodology keeps the customer as the primary focus. Through focusing on the customer, we safeguard providing only what the customer expects or wants and nothing else (waste) both now and in the future. Therefore, understanding and trying to respond to customer requirements is one of the company's critical tasks.

8.2.2   Employee Focus

Personnel is the organization's biggest resource; hence, it is not operative without machinery, equipment, intellectual property, and so on. Thus, the success of the organization is a function of proper and full utilization of its personnel. Fully utilizing human resources, the organization's biggest resource, means engaging the personnel expertise in continuous improvement and authorizing and empowering them to make essential changes to the processes that they work within. Therefore, upgrading the personnel's knowledge, skills, and abilities is the key to improving the company's performance.

8.2.3 Process Focus

A process is the set of activities that changes a certain input into an output. A well-defined process is a tool that converts a routine-work environment to an inspired work environment. After the innovative duties have been identified, bounded, and separated, the routine tasks will turn more precise and efficient. Thus, the entire activities within a corporation should be regarded as part of the process, and then sections of operations can be improved. Process improvement is one of the most important tasks of the company.

8.2.4 Leadership Support

Quality is the accountability of everyone in the company. A truly effective and efficient continuous improvement requires a culture focused on the continuous improvement process. As we have studied in Sect. 1.7, quality leadership is a precondition and requirement for executing a quality management system. This step emphasizes "how the structure of organizational leadership and direct the organization and what manner they accomplish within the organization are vigorous elements to the realization and comprehension of the effective and efficient quality management process." Therefore, organizational culture and focus start with leaders. Engaged and supportive leaders are an essential condition for the continuous improvement process.

8.3 Tools of Continuous Improvement

Habitually, some personnel adamant about there being a "correct" or "the best improvement methodology." They can mention some of the methodologies like lean engineering, Six Sigma, kaizen, total quality management, and so on. However, that viewpoint is incorrect and causes to have a suboptimal result.

Continuous improvement materializes when we make the right improvement at the right time, using the right tools to take full advantage of the continuous improvement efforts. The best improvement tool, technology, or methodology be subject to the problem at hand and the customer's needs or operation. To optimize the continuous improvement results, we need to use the best combinations of the tools we have under appropriate knowledge, cost, and time constraints.

Based on the depth of the problem, continuous improvement results can be little. However, every little but positively affecting improvement should be acknowledged and encouraged to keep up what we are doing. Either the small incremental improvements or huge break-through innovations, we are not characterizing or distinguishing here. What matters is that we are involved in the process of continuous improvement.

8.3.1 Total Quality Management (TQM)

Total quality management (TQM) relished mainstream acceptance in the late 1970s, 1980s, and early 1990s due to the teaching of many quality gurus discussed in management and leadership. The philosophy of total quality management (TQM) is based on the leadership and management approach to continuous improvement.

In the 1990s, TQM was outshined by ISO 9000, Six Sigma, and lean engineering and has fallen out of vogue. Although TQM does not enjoy the admiration it once did, we should note that TQM laid the groundwork for future continuous improvement technologies, methodologies, or tools.

Many of the other continuous improvement technologies, methodologies, or tools come with unique characteristics or perspectives. However, the TQM effort is characterized by composing the advantages of various techniques for quality improvement.

Total quality management (TQM) is well-defined by those three terms that make up the appellation. These are total, quality, and management. Let us see the definitions of the terms as follows.

8.3.1.1 Total

Quality specialists who introduce TQM comprehended the necessity of engaging all personnel in the quest for improvement. Thus, the term total in TQM infers that all individuals in the company are responsible for quality via continuous improvement. Tools corresponding improvement squad and the quality spheres are employed within TQM to authorize and involve employees to solve the problems. Everyone in the company needs to accomplish some goals or objectives.

8.3.1.2 Quality

The central objective of the TQM is improving or advancing quality via a continuous improvement process. Customer requirement specifications define quality.

Therefore, TQM's emphasis is on the customer, and the customer establishes quality. And what the customers find valuable, both in the present day and in the future. That is based on the conception that if the company focuses on the customer, they will be successful.

8.3.1.3 Management

TQM applied a process-centered approach. TQM emphasizes that the company's entire activities are part of the process (or system) that needs an appropriate management system. The philosophy of TQM clarifies that the results (which can be good or bad) reflect the effectiveness of process management. Bad results (i.e., poor quality) are signs of inadequate process management and represent an opportunity for improvement.

Thus, the term management emphasizes the degree of involvement and the active support of top leaders within the company. The top leaders are accountable for guiding and supervising the organization to accomplish the organization's overall quality objectives. That includes proper goal definition, creating a positive culture, proper time management, controlling cost functions, focusing on the customer, empowering employees to deliver improvements, providing training and feedback, and managing progress towards the essential organizational objectives.

TQM consumes a lot of statistical data analysis. Thus, TQM accelerates its legacy in the modern-day quality management system, lean manufacturing, and Six Sigma methodologies.

8.3.2 Kaizen

The word Kaizen comes from Japanese. Kaizen consists of two words, namely Kai and Zen. Kai means change, and Zen means for the better. Kaizen's central philosophy is that any improvement (it can be big or small) will have a positive impact. Kaizen does not directly translate into or imply the word "continuous." Nevertheless, Kaizen's essence and its industrial application as a methodology for improvement have taken on a continuous nature.

Similarly, despite the fact the name Kaizen itself does not essentially point toward anything about the change's size, Kaizen habitually infers a smaller scale. According to Kaizen, daily (it could be weekly or monthly) and small improvements involving everyone in the organization.

Most of the successful Kaizen experts emphasize that Kaizen functions better via the engagement of all employees. Specifically, the engagement of those on the shop floor that performs the value-added operation. Kaizen practitioners strongly recommend that it combines the collective talents within a company to create a very powerful engine for improvement via engaging all associates in improvement.

Employees who are engaged in detecting and resolving problems within their specialization area, driving out errors, waste, and process variation. That emphasis on employee engagement is based profoundly on the process-centered approach, where every activity is viewed as a process. Once we view each activity as the process, then we can empower the employees to improve the processes within which they operate.

Since the approach infers the change for the better (i.e., process improvement), specific tools are used within the Kaizen approach. Commonly, standardizing work is one of the essential tools of the Kaizen approach.

The standard work is a living document that represents the "best way" to perform the task to execute an effective process. Standard work is like work instructions that we have been discussing in Sect. 1.8. Quality assurance (QA) detailed how work instructions are applied to describe the explicit actions needed to execute the process in the right manner. Similarly, having standard work for the process generates a groundwork for improvement and process consistency (i.e., reduce variation).

The entire employees perform the task in a matching way and accurate way, and then enhancements can be made to that communal continuous improvement style. The standard work characterizes the "current state" of the process, and employees are stimulated to progress on the contemporary state and generate the future state of the process.

Having the process-centered mindset, mutually, each employee's engagement and perform Kaizen a very prevailing approach and procedure for development and upgrading the process.

The supplementary note on Kaizen value discussing is the robust focus on carrying out tests and consequently taking actions.

Kaizen experts will habitually engage employees via improvement teams (like quality circles or squad) to contest them to experimentation with several ways to solve the problem or portions of the problem. While considerable activity within Kaizen is done continuously, an alternative method within Kaizen has become popularized on a small scale called the Kaizen Blitz. Here, the continuous effort to upgrade the process within the Kaizen makes the Kaizen procedure like the Plan–Do–Check–Act (PDCA) approach.

The Kaizen Blitz is a focused and intense improvement effort. The team of subject matter experts and process participants are assigned to solve the large problem or advance cost and time constraints. Thus, some experts suggest that the Kaizen Blitz is counter to Kaizen's true spirit, which is a continuous activity. Nevertheless, the approach remains one of the popular tools for rapid improvement.

8.3.3   Plan–Do–Check–Act (PDCA)

The Plan–Do–Check–Act (PDCA) cycle is a natural yet influential process for continuous improvement. The PDCA cycle was introduced by Dr. Edwards Deming, who gave acknowledgment for the thought to Walter Shewhart. Consequently, the

Fig. 8.1 Plan–Do–Check–Act (PDCA)

cycle is habitually called the Shewhart cycle or the Deming cycle. Note that although Deming and Shewhart introduced the four-step process, they unquestionably did not discover the methodology. See the Plan–Do–Check–Act (PDCA) in Fig. 8.1.

Later in his professional career, Deming needed to underline the "check phase." He has done a lot of study about the essential activities required for the "Check Phase." For the reason that he called the model Plan–Do–Study–Act. Consequently, the cycle dynamisms by several different names, nevertheless the concept is the same.

The PDCA cycle reflects the scientific method, creating a hypothesis and Design an Experiment (DaE) to test that hypothesis (Plan), implement the experiment (Do), analyze the results (Check), and put that new-fangled knowledge and acquaintance into use for forthcoming experiments (Act). The fundamental characteristic of both the PDCA cycle and the scientific method is that they are iterative, where the term continuous improvement practice originates.

As mentioned earlier, scientists use knowledge from the experimentations to build on their thoughtful of a badly-behaved topic and design future experiments. Consequently, the scientific method is an iterative process. In the same way, the PDCA cycle is an iterative process where the outcomes of the first cycle ought to get-up-and-go the subsequent iterative improvement or cycle. That is where the influence of the PDCA cycle truthfully comes from.

This iterative nature situates continuous improvement and gives the PDCA its tangible power. The PDCA cycle is so influential that it is the suggested improvement procedure within ISO 9001. The PDCA cycle can consist of numerous other tools from other methodologies and the seven quality tools (QC) tools: Fishbone diagram (cause-and-effect diagram, also called the Ishikawa diagram), histogram, control chart, quality check sheet, scatter diagram, Pareto principle, and flow chart. The seven quality management and planning tools:

1. Process decision program chart
2. Prioritization matrix
3. Activity network diagram
4. Matrix diagram or quality table
5. Tree diagram

6. Interrelationship diagram
7. KJ method

Additionally, other statistical methods (i.e., regression methods, design of experiments, categorical data analysis, predictive models, time series models, reliability analysis).

8.3.3.1   The Steps of the PDCA Cycle

Step 1: Plan

The first step in PDCA is to generate a plan on how to realize the goal. Deprived planning can overturn the entire PDCA process. Having a good plan is essential (critical), and we necessarily do our due thoroughness in this phase to maximize our effort's outcomes. As it relates to the design of planning, we should be clear about the achievable goal or objective.

The PDCA cycle is adaptable enough to be used once solving the problem, modifying the process, or accomplishing a new-fangled operation or procedure. Thus, the planning step is like the defining step in DMAIC of the Six Sigma approach, where we define the problem statement. That is, the planning phase is about establishing the goals/objectives of the project.

Example 1 If the goal is to solve a given problem, then the plan should outline the experiment that can determine the root cause. That might include using a design of experiments (DoE) or the establishment of the null hypothesis that we want to test.

Example 2 If the goal is to accomplish an improved operational change (i.e., a breakthrough goal), then the plan should be to design the new processes needed to bring the expected result (i.e., an improved process change). Thus, studying the new approach is the key to deliver the new change in the PDCA cycle.

Note that

- The planning phase should also include discussing the possible risks to accomplishing the goal and any countermeasures that can be applied to mitigate the risk.
- Depending on the complication of the circumstances, the planning step may comprise the project charter, a flow diagram, brainstorming, plus an affinity diagram (or an Activity Network Diagram).
- As we have seen above, the seven-quality control (QC) tools and seven-quality management and planning tools are used in PDCA's planning phase.
- The planning phase includes creating new check sheets to facilitate effective data collection on the current state of the process and future state of the process.

Step 2: Do

Step 2 in the PDCA cycle is to DO what we have already planned in the first phase. That might mean implementing experimentation and gather data from the process being premeditated.

Collecting data and observing the process (where the experiment is performed) are essential throughout this phase. We will use this data (or experimental observations) in the subsequent step to determine if the experiment was successful. That means executing the new process to update (or change) the old method.

Step 3: Check (Study)

Step 3 in the PDCA cycle is the check. At this phase, we need to study and analyze the experiment's outcomes conducted in the DO phase. That should comprise data analysis, which is tested using ANOVA, Pareto Analysis, hypothesis testing, control charts, histograms, process capability analysis, scatter plotting, linear regression, productivity analysis, and so on.

We should also compare the outcomes observed in the experiment against the expectation in the planning phase. That is the time in the cycle where we study to some degree about the new process. That comprises revising what went well and what didn't align with our initial plan and recognition gaps in our acquaintance.

Step 4: Act (or Adjust)

- Step 4 is to take action based on what we learned in step-3 or make adequate adjustments to our original plan. That might comprise implementing a corrective measure to solve the problem or completely implementing a change to the process.
- If the outcomes are adequate, the PDCA cycle stops there! We can start the next goal if we set several goals. Concerning the results, start the cycle all over again, utilizing the acquaintance from the preceding cycle.
- If the experiment's outcomes suggest inappropriate (inadequate) process changes, then we will plan how to make the result to be useful for process improvement. Thus, the PDCA cycle will begin again. That is, we need to adjust the initial plan.

8.3.4 Six Sigma Methodology

The name Six Sigma (6σ) literally comes from six times the standard deviation of the normal probability distribution. The probability of the normally distributed observation on the internal $\mu - 3\sigma$ to $\mu + 3\sigma$ is 99.73%. That is, 99.73% of the normally distributed observation lays within ($\mu - 3\sigma$, $\mu + 3\sigma$). Now, if we equate the length of the interval as $\mu + 3\sigma - (\mu - 3\sigma) = 6\sigma$. Therefore, if our operation operates at the level of 3σ, it means our process is expected to produce a minimum of 99.73% conforming products. The result implied that the process accepts only 2700 nonconforming parts per million (ppm).

The Six Sigma methodology shares a lot of features with total quality management (TQM) methodologies. Hence, the core concepts of total quality management (TQM) are used in the Six Sigma methodology. For instance, a customer-focused strategy is one of the big components of total quality management (TQM) is widely applied in the Six Sigma methodology.

In this chapter will discuss the concept of Six Sigma by breaking it down into four major topics.

- Foundations and focus areas of Six Sigma.
- Define–Measure–Analyze–Improve, and Control (DMAIC) process: Tools in the DIMAIC process.
- The infrastructure of Six Sigma: breakdown of the Six Sigma belts.
- Six Sigma implementation.

8.3.4.1   Foundations and Focus Areas of Six Sigma

In the late 1980s, Motorola senior engineers (namely Dr. Mikel J. Harry and Bill Smith) introduced Six Sigma methods. Initially, the Six Sigma methodology technologically advanced as an iteration. But, since then extensive statistical methods are applied in total quality management (TQM), which gives significant space to popularize the Six Sigma methodology.

In 1988, due to such an important industrial innovation and excellence in manufacturing, Motorola won the first-ever Malcolm Baldridge National Quality Award (MBNQA).

Six Sigma was like a house on fire accepted by GE's then-CEO Jack Welch, which once more powered the admiration of Six Sigma through the 1990s and the twenty-first century. However, other methods characterize a certain mindset (or viewpoint) on continuous improvement. Six Sigma has eased up from not solitary on philosophy on process improvement and the whole tool kit of improvement approaches that have had a massive impact on the industrial quality society since the last 30 years.

Nowadays, it turns out that in the circumstances, the likelihood of manufacturing the product within these specifications is 99.73% (i.e., 2700 nonconforming ppm). That denotes the quality performances of three Sigma (3σ), and it sounds good. Nevertheless, if we operate at 3σ for each component, assume we have an assembled product made up of 50 independent components. Since each component is produced independently, the joint probability of the conforming components. That prevails to 126,444 nonconforming assembly ppm, which is really bad performance. Look at Table 8.1 to comprehend the sigma levels and nonconforming assemblies regarding the number of components produced at the three Sigma (3σ) level.

Table 8.1 Sigma levels and PPM of assembled parts

Sigma level	Probability of each confirming component	Number of components	Probability of conforming assembly	PPM
1	0.6827	50	5.15E-09	1000000
1	0.6827	100	2.65E-17	1000000
1	0.6827	200	7.01E-34	1000000
1	0.6827	300	1.86E-50	1000000
1	0.6827	400	4.92E-67	1000000
1	0.6827	500	1.30E-83	1000000
1	0.6827	1000	1.70E-166	1000000
2	0.9545	50	0.097453319	902547
2	0.9545	100	0.009497149	990503
2	0.9545	200	9.02E-05	999910
2	0.9545	300	8.57E-07	999999
2	0.9545	400	8.14E-09	1000000
2	0.9545	500	7.73E-11	1000000
2	0.9545	1000	5.97E-21	1000000
3	0.9973	50	0.873556404	126444
3	0.9973	100	0.763100792	236899
3	0.9973	200	0.582322818	417677
3	0.9973	300	0.444371003	555629
3	0.9973	400	0.339099864	660900
3	0.9973	500	0.258767375	741233
3	0.9973	1000	0.066960554	933039
4	0.999968329	50	0.998417666	1582
4	0.999968329	100	0.996837836	3162
4	0.999968329	200	0.993685671	6314
4	0.999968329	300	0.990543474	9457
4	0.999968329	400	0.987411213	12589
4	0.999968329	500	0.984288856	15711
4	0.999968329	1000	0.968824553	31175
5	0.999999713	50	0.999985668	14
5	0.999999713	100	0.999971335	29
5	0.999999713	200	0.999942671	57
5	0.999999713	300	0.999914008	86
5	0.999999713	400	0.999885346	115
5	0.999999713	500	0.999856684	143
5	0.999999713	1000	0.999713389	287
6	0.999999999	50	0.999999951	0
6	0.999999999	100	0.999999901	0
6	0.999999999	200	0.999999803	0
6	0.999999999	300	0.999999704	0
6	0.999999999	400	0.999999605	0
6	0.999999999	500	0.999999507	0
6	0.999999999	1000	0.999999013	1

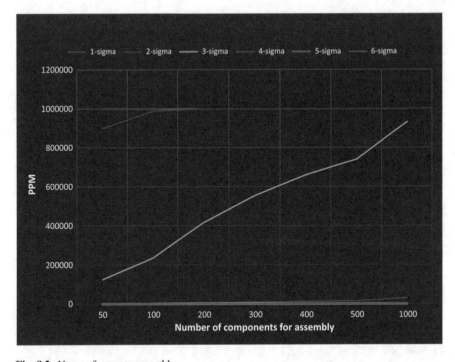

Fig. 8.2 Nonconformance assembly

From Table 8.1, we observe that

- Production at the 1 sigma level followed by the 2 sigma level is the worst-performing operations.
- Fixing the sigma level, as the number of components increases, the nonconforming assembly ppm also increases.
- Production at the 6 sigma level followed by and the 5 sigma level is the best performing operations. For figurative information, look at Fig. 8.2.

If we consider 100 components, the likelihood that any specific unit of the assembled product is nonconforming is 0.237. That brings about 23.7% of the assembled product manufactured under 3 Sigma quality performance will be nonconforming. However, such production is ineffective and, therefore, unacceptable. Now imagine the assembly operations in the real world, the assembled product of many components. A characteristic automobile has about more than 100,000 components. If you go to the aerospace industry, a typical airplane has more than one million components.

For instance, a Jumbo aircraft has 3,000,000–7,000,000 components. If we produce each component at a 3 sigma quality level, then the probability that the given Jumbo aircraft functions properly is less than $0.9973^{3,000,000} \approx 0$. It means none of the Jumbo aircraft will have an acceptable quality. Therefore, production at a higher sigma production level is required for Jumbo aircraft to have satisfactory quality. Production at the higher sigma level produces fewer defects. However, the problem is that it is challenging to have a simple operation with a higher sigma level. Therefore, in the real world, production at a higher sigma level is the best-performing manufacturers' identity matrix.

The Motorola Six Sigma thought is to decrease process variability. They set the specification limits at 6-standard deviations from the mean (i.e., $\mu - 6\sigma$, $\mu + 6\sigma$). Thus, production under Six Sigma quality, the likelihood that each product nonconforming is 0.9999998. That brings only 0.2 ppm, which is an exceptional quality performance.

After the Six Sigma notion was originally developed, a statement was made. When the process achieved the Six Sigma quality performance level, the process mean (i.e., target) is still an issue to instabilities that can cause the process is off by as much as 1.5σ (1.5 standard deviations) from the target. In that situation, the 6 Six Sigma process could produce about 3.4 nonconforming ppm. But, still, the process seems good. However, there is a superficial contradiction in this.

In Chap. 6, we can only forecast the process performance as we learned from the process capability. In the stable (i.e., both in terms of accuracy (mean) and precision (small standard deviation)) process: if the mean is drifting everywhere 1.5σ (1.5 standard deviations) off-target, the estimate prevails that about 3.4 nonconforming ppm possibly will not be reliable. Since the mean may shift by more than the allowable 1.5σ (1.5 standard deviations), in this scenario, the process performance is unpredictable without the process performance is characteristically stable.

- As a starting point, the Six Sigma process emphasizes a single (or can be fewer) improvement focal point. Today, that technique influences many Six Sigma experts. According to the Six Sigma experts, the focus of improvement is synergy about eliminating variation via attaining the target within the processes, which achieve higher quality levels. Then, gradually (i.e., a careful step-by-step), the model encourages to include more improvement areas.
- Some of the new and unique focus areas within the Six Sigma process characterize how the model provides tactical and strategic goals for the company. The focus areas of the Six Sigma process comprise.
- Six Sigma process includes a robust emphasis on a project assortment process. Project assortment process is meant to enhance the bottom line of an organization, which assists to accomplish the organization's strategic goals.
- The Six Sigma projects are established, essentially on business-focused results and the project choice process needs extensive analysis about the impact of projects.
- The other essential focus area for the Six Sigma process is that it heavily applies statistical models for data-based decision making. That brings more evident metrics and tools within Six Sigma.
- Like the Kaizen process, the process-centered approach is one of the Six Sigma process's critical focus areas. The whole activity is considered a process. Each process has a value-added operation that converts a set of inputs to produces an

output(s) based on customer requirements. Thus, the success of the Six-Sigma process is based on the performances of each process.

8.3.4.2   Essential Metrics of Six-Sigma

The notion of the (1) process-centered approach, (2) process capability, and (3) quality engineering tools critical central to Six Sigma methodology. Every process has three common elements: inputs, the process (operation), and outputs. Figure 8.3 gives a simplistic representation of the process.

The given process outputs can be any value-added operation for the vertically integrated process, or they are the final products with the quality features and specifications based on the customers' requirements. Hence, the output (i.e., final product) of the process is measured against the customer-specific requirements, reflecting the quality of the process as we have studied in Chap. 6, process capability (is also called the voice of the process).

A normally distributed process is characterized by its mean (μ, which is the centerline of the process) and the standard deviation (σ) of the process output distribution. Assume we have the normally distributed process output. In Fig. 8.4, we can see how to represent specification limits and nonconforming products. The specification limits represent acceptable process tolerance, whereas nonconforming products are those products whose quality measurements are out of the specification limits.

- To evaluate the process's capability, we need to compare the output of the process against the customer-specific requirements. The process capability index, Cpk (or Cp for a centered process), will give us important feedback about the process performance.

Fig. 8.3 The three-common elements of the process

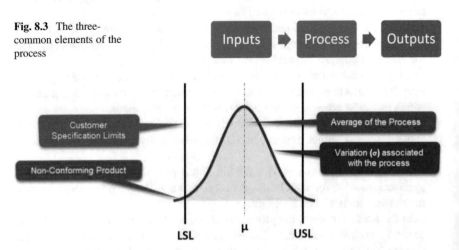

Fig. 8.4 Representation of specification limits and nonconforming products of the process

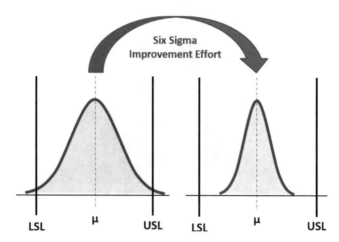

Fig. 8.5 Six Sigma improvement efforts

- If the Cpk is sufficiently large, then the process is characterized as a good performing process. There is no cutting point to say the Cpk is large enough or not. For instance
- In a typical automotive industry, the minimum Cpk required is 1.67.
- In some other industries, the minimum Cpk required can be tighter than the automotive industry, 2. Or, it can be softened in other sectors that they accept Cpk of 1 or 1.33.
- Usually, if the Cpk is positive and lower than 1, the process needs to adjust its variance.
- If the Cpk is negative, it means the process misses its target. So, the process is characterized as an incapable process.

Based on the capability analysis, we will have the percentage of nonconforming product from the process. If the percentage is as small as possible, it is encouraging to proceed with the current process. Otherwise, the poor-quality performance of the process is an opportunity for improvement.

Whenever we have room for improvement, the Six Sigma process effort is all about pushing the nonconforming measurements towards the center of the process. That is done by simultaneously adjusting the process mean towards the target and reducing the process variance. In Fig. 8.5, we observe what happens to the process when we apply Six Sigma tools to eliminate the most important variation sources that are triggering the poor quality.

From Fig. 8.5, we can observe that afterward, we have reduced the variation within the process, the tails of the process output distribution are getting well within the specification limits. That means we have eliminated substantial rejects, which assisted in improving quality. The term Six Sigma shows the number of standard deviations (σ) that fit within the specification limits.

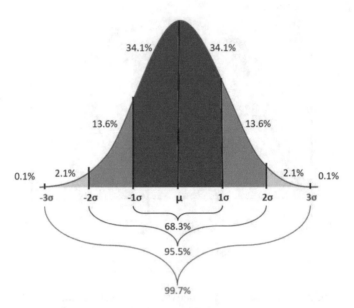

Fig. 8.6 Percentage of the normally distributed population

Figure 8.6 shows the percentage of the normally distributed population that fits within each standard deviation. So, 68.3% of a normally distributed population falls within one standard deviation $(1 - \sigma)$ of the mean, 95.5% falls within two standard deviations $(2 - \sigma)$, and 99.7% falls within three standard deviations $(3 - \sigma)$.

Figure 8.6 shows that a large percentage of the distribution becomes nonconforming if we operate at a 1 sigma quality level within the mean. On the other hand, a large percentage of the distribution becomes conforming if we operate at a 6-sigma quality level within the mean. That is, only a few parts are found nonconforming if we operate at the higher sigma quality levels. Therefore, having the process within 6σ (six standard deviations) from the mean (i.e., the expected target of the process) is the Six Sigma methodology's major aim.

Table 8.2 describes the common Six Sigma metrics. The Table designates the sigma level and the proportion of the population that will be nonconforming (defective) in defects per million opportunities (DPMO) and the process capability. The percentage of nonconforming is higher if the within sigma shifts are observed in the process.

We can perform considerable improvements to the process by removing the sources of variation. Consequently, the percentage of the population falling within the specification limits in that way eradicating rejects (via improving quality). The process that one can observe today may not be the same as after a month. Thus, the short-term process capability performance doesn't predict the long-term process capability performance. Six Sigma experts have suggested that the process's long-term understanding may not have equivalent to the short-term process capability.

Table 8.2 DPMO metrics

Sigma level	Without 1.5 sigma shift			With 1.5 sigma shift		
	DPMO	Yield (%)	Defect rate (%)	DPMO	Yield (%)	Defect rate (%)
1	317,310	68.2690000	31.7310000	697,612	30.23880	69.76120
1.1	271,332	72.8668000	27.1332000	660,082	33.99180	66.00820
1.2	230,139	76.9861000	23.0139000	621,378	37.86220	62.13780
1.3	193,601	80.6399000	19.3601000	581,814	41.81860	58.18140
1.4	161,513	83.8487000	16.1513000	541,693	45.83070	54.16930
1.5	133,614	86.6386000	13.3614000	501,349	49.86510	50.13490
1.6	109,598	89.0402000	10.9598000	461,139	53.88610	46.11390
1.7	89,130	91.0870000	8.9130000	421,427	57.85730	42.14270
1.8	71,860	92.8140000	7.1860000	382,572	61.74280	38.25720
1.9	57,432	94.2568000	5.7432000	344,915	65.50850	34.49150
2	45,500	95.4500000	4.5500000	308,770	69.12300	30.87700
2.1	35,728	96.4272000	3.5728000	274,412	72.55880	27.44120
2.2	27,806	97.2194000	2.7806000	242,071	75.79290	24.20710
2.3	21,448	97.8552000	2.1448000	211,927	78.80730	21.19270
2.4	16,395	98.3605000	1.6395000	184,108	81.58920	18.41080
2.5	12,419	98.7581000	1.2419000	158,686	84.13140	15.86860
2.6	9322	99.0678000	0.9322000	135,686	86.43140	13.56860
2.7	6934	99.3066000	0.6934000	115,083	88.49170	11.50830

In the long term, the process may experience new sources of special cause variation, or current causes of variation may grow in scale. For instance, the machine's performance may decrease with time so that assuming other things are fixed. The processing capability will be declined due to the depreciation of the machine. According to Bill Smith, generally, the long-term process variation is continuously greater than short-term process variation if we consider other factors.

To study the long-term process performance, Bill Smith performed an analysis based on empirical data. He used the long-term shift of 1.5σ (1.5 sigma) to estimate the long-term process performance. We consider that shift and generate new estimations for Defects Per Million Opportunities (DPMO). For instance, consider a 2.7 sigma quality level. If the process is without the 1.5 sigma shift, the DPMO is 6984. However, considering the same sigma quality level, if the process is with the 1.5 sigma shift, then the DPMO will be 115,083. That means, if the current process has 6984 DPMO, then in the long-term, the process will produce 115,083 DPMO.

Table 8.1 shows drastic changes in percent nonconformance (i.e., percentage defectives) and DPMO. Undoubtedly it agrees with the evidence behind the process shift, so it is a suitable approach to long-term process variation to do proper adjustment or modification.

Fig. 8.7 Define–Measure–
Analyze–Improve–Control
(DMAIC)

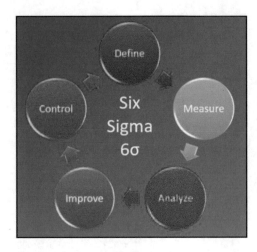

8.3.4.3   Define–Measure–Analyze–Improve–Control (DMAIC)

This section clarifies the DMAIC cyclic process and familiarizes many of the applied tools in each step. Several of the DMAIC tools will be presented. Define–Measure–Analyze–Improve–Control (DMAIC) is a well-structured effective problem-solving model. DMAIC is extensively used in various industries for quality and process improvement. DMAIC is habitually related to the Six Sigma core activities.

After studying this section, we will able to do the following

- Comprehend how the DMAIC cycle fits into the framework of the Six Sigma approach.
- Comprehend the criteria of prioritizing and then selecting projects for improvement programs.
- Clarifying the five steps of DMAIC.
- Explain the purpose of tollgate reviews.
- Comprehend the applications of tollgate review decision-making requirements for each DMAIC step.
- Recognize when and where to use DMAIC.

Projects are an essential aspect of quality and process improvement. Projects are an integral component of Six Sigma. Nevertheless, quality and business improvement via projects traces their backgrounds back to Juran. Juran suggested that the project-by-project method is beneficial and productive to improve quality. He continues; first, we need to define the goals to be achieved (i.e., the problem to be solved or the process that needs improvement), then we should partition the goals into manageable and achievable objects.

Practically the entire implementation of the Six-Sigma methodology applies the DMAIC process to complete a project. Quality and process improvement ensues most efficiently on a project-by-project basis. The Six-Sigma DMAIC cyclic process improvement model is presented in Fig. 8.7.

In Fig. 8.7, we see that the DMAIC is organized into five cyclic steps problem-solving method. The DMAIC cyclic process can be applied to efficaciously complete the given project by implementing the appropriate solution(s) that are planned to resolve the root cause of the problem.

To establish the best DMAIC operation, which ensures the solution(s) is/are gives a permanent change, each of the DMAIC steps should adequately be implemented. The DMAIC cyclic process's essential advantage is that it can be replicated in other applicable business operations.

DMAIC is not essentially tied to the Six Sigma process. DMAIC can be applied regardless of the organization may or may not use the Six Sigma model. DMAIC is a wide-ranging methodology for problem-solving applications. For instance, lean engineering projects focus on production time or cost reduction by making direct process improvement or waste elimination in various industries. Similarly, it can be straightforwardly and efficiently conducted using DMAIC in various industrial problem-solving applications.

8.3.4.4 Tollgate Review Process

A tollgate review process comprises a comprehensive list of queries to be responded at the end of each stage of the DMAIC cyclic process. As the name signposts, the tollgate review is just like checkpoints in the Six Sigma project. The several team associates meet and govern whether the task has been completed as designated in the initial project plan and whether task goals stated in the plan have been accomplished or not.

Between each major step, there are tollgates in the DMAIC process. At the tollgate, each project team presents its task to the project manager or their immediate manager in the Six Sigma process. That is, the tollgate reviews will be conducted after the team completes each step.

The tollgate participants should include the project champion (he/she can be a master Black Belt certified professional, other Black Belt, or experienced Green Belt). In reviewing the project, tollgates are used to confirm that the process is going on track. The tollgates deliver a continuing opportunity to assess whether the team can efficaciously realize and complete the project on time.

Additionally, the tollgates present the opportunity to deliver guidance concerning detailed technical tools and other evidence on the problem. During tollgate reviews, other barriers to success like organizational problems and/or poor strategies are also needed to be identified. Thus, the tollgates are essential to the overall problem-solving process.

8.3.4.5 DMAIC Cyclic Structure

The DMAIC cyclic structure inspires innovative thinking about the problem's characteristic nature and the solution within the specifications of the product.

When the process operates inadequately, it is essential to switch the process and start a new one. For instance, if we know that a new product is required, then the "improve" step of the DMAIC becomes the "design" step.

Design for Six Sigma (DFSS) is a design process method related to the Six Sigma process. In the Six Sigma organization, that probably infers that the DFSS effort is needed. One of the whys and wherefores that DMAIC is so fruitful is that it emphasizes the productive use of available tools.

8.3.4.6   The Five-Steps of DMAIC

The Define–Measure–Analyze–Improve–Control (DMAIC) process is certainly the heart of Six Sigma. In terms of the cycle, the DMAIC has comparable with the PDCA. However, the DMAIC cycle can resolve advanced industrial problems than the PDCA cycle. The DMAIC cycle has a five-step improvement process, which makes essential change or advancement to the process.

Step 1: Define

Having a well-defined problem statement (or project goal) is an essential critical starting point to find the best solutions. Therefore, in the define step, we need to describe and specify the problems details in a clear and structured manner.

- To define the problem, we should do the following activities.
- We should note that we never assume anything (which can be a minor or major issue) without reference or test results.
- Comprehend the clear picture of the voice of the customer. That helps to identify the critical quality requirement of the part to know the customer's needs and distinguish what you what to know. To do that, the revisions of customer-specific requirements and other relevant documents are vital.
- Create the process map (or value streams) of the current-state-process.
- Comprehend the problem statement using data, texts, pictures, videos, interviewing experts, and so on. Note that every piece of information is important, so don't ignore what you observe.
- Based on evidence from the process, clearly state the problem.
- Identify cross-functional team participants.
- Revise the problem and present it to the cross-functional team for approval.
- After generating a unique identification number, document the problem.
- Brainstorm possible solutions to the problem and perform a Cost-Benefit calculation for the proposed solutions.
- Benchmark or brainstorm the future state of the process. Will you be able to call me?

Step 2: Measure

- Once we have well-defined the problem, we should know to measure and collect the relevant data from the current-state-process. The measuring step of the DMAIC can include designing an experiment to recognize key input and output variables. Additionally, this step may require adequate traceability (properly time-framed) of data from the process.
- Some of the essential activities of this step are:

  – Using the current-state process data generate a business case for resolving the problem.
  – Gather data from other relevant sources (such as customer complaints, supplier surveys, etc.).
  – Measure of the current-state process capability and evaluate the result.
  – Measure the accuracy and precision measurement methods that is required to do a gauge R&R study.
  – Brainstorm the root cause of the problem using the cause-and-effect matrix.
  – Set hypotheses to be tested.
  – Based on the measurement results, propose possible process changes, and recommend.

Step 3: Analyze

Once the measurement is done, to identify the root cause of the problem we will do deep data analysis. This step can also comprise the identification of the key variables in the process, which have impacts on the problem and related process outputs. Some of the essential activities of this step can be:

- Paired t-test for the before and after change comparisons.
- Independent t-test for comparing two independent variables.
- Chi-square tests for correlation analysis of categorical data.
- ANOVA for comparing several means of different levels.
- DoE to identify critical inputs to a process and its outputs.
- Linear regression analysis to cause-and-effects tests and to quantify the relationship between inputs and outputs of the process.
- Pareto analysis to categorize influential factors in the root causes.
- Calculation of process capability of future-state-process.
- Perform advanced MSA.
- Fishbone analysis to systematically analyze the root cause.
- 5 Why root cause analysis to determine the root cause and contributing factors associated with the problem.
- Evaluate the impact of an experimental change on the process (future-state-analysis).

Step 4: Improve

The analysis step of the DMAIC helps to identify the root causes of the problem. Once we determine the root cause of the problem, we will make the process improvement, called the improvement step of DMAIC. The improvement step is the implementation of an appropriate corrective and preventive action (CAPA). Sometimes we may face multiple root causes to have various corrective and preventive actions (CAPAs). In this circumstance, we need to prioritize and naturally sequence the corrective and preventive actions (CAPAs), and we need to implement them step-by-step in the proper time frame.

- In the improvement step, we need to consider all the information from the last three sequential phases (i.e., Define–Measure–Analysis) to improve the process.
- Some of the essential activities of this step can be:
  - Make changes to the existing process by a new (or modified one) to remove variation (or eliminate waste).
  - Implement corrective and preventive actions (CAPAs) to the current-state-process and test the future-state-process results.
  - Apply poka-yoke (mistake-proofing) or error-proofing tools to eliminate errors from their origin. These tools are applied in FMEA (Failure Mode and Effects Analysis), FTA (Fault Tree Analysis), and APQP (Advanced Product Quality Planning).
  - Standardize work instructions to define the best way for operating.
  - Smear effective communication among the stakeholders about the proposed changes.
  - Update the operator training matrix, FEMA, control plan, IQC, OC, and so on.
  - Give appropriate training for the affected employees and stakeholders on the new process.
  - Set measures and countermeasures of consequential risks of the CAPA and their mitigation mechanisms.

Step 5: Control

Control is the last step of the DMAIC process improvement. That is habitually the most forgotten or overlooked section of the improvement cycle. This final step is the improvement over the Plan–Do–Check–Act (PDCA) cycle since it forces users to contemplate how they are going to control the modifies recently executed and sustain the improvement over time.

Consideration of the existing problem and the new (or modified) process is essential to assess the effectiveness and adequacy of the corrective and preventive action (CAPA). Therefore, we should spend time concentrating on how we will control the change (new or modified) we made to the process. That ensures that the root cause of the problem is permanently eliminated (objectives achievement) via the CAPA or not.

Some of the essential activities of this step can be:

- Implement statistical process control (such as control charts) techniques to evaluate and control the process.
- Generate the control plan that maps the association between CTQs (critical to quality) and CPIs (critical process inputs).
- Monitor whether the standard work procedures are properly implemented or not.
- Track improvements after the CAPA and check the functionality based on the original estimate.
- Brainstorm additional (or alternative) potential process changes and evaluates the best performances.
- Perform an AAR (After-Action Review) to control what went right or wrong along the way.
- Recommend the alternative cycle through the DMAIC process.

8.3.4.7 Summary of Tools Uses in Each Step of DMAIC Process

In Table 8.3, we summarize the DMAIC cyclic process and the various tools available to each step for the Six Sigma practitioner.

In each step of the DMAIC process, the tools may not be unique to the Six Sigma program. Nevertheless, the Six Sigma experts expected to integrate the tools concisely into the DMAIC cyclic process to optimize the improvement. The also helps to minimize the number of cycles needed to bring the DMAIC process to get the optimal solution.

Table 8.3 Summary of the DMAIC process

Define	Measure	Analyze	Improve	Control
Project Charter (A3)	DOE & Analysis	Pareto Analysis	Poka-Yoke	Control Plan
Project Business Impact	Ishikawa Diagram	Hypothesis Testing	Standard Work	Control Chart (SPC)
Cost/Benefit Analysis	Check Sheet	Histogram	Activity Network Diagram	Standard Work
Stakeholder Analysis	Gauge R&R	Linear Regression Analysis	Communication Strategy	Process Flow Diagram
Team Formation	FMEA/FTA	Scatter Plot	Training Plan	Visual Management
VOC Analysis (CTQs)	Process Capability Analysis	Matrix Diagram	Project Execution	Visual Controls
Benchmarking	Activity Analysis	Prioritization Matrix	Verification of Improvement	After Action Review
Brainstorming	Yield Analysis	Goodness of Fit Testing		
SIPOC Analysis	Interrelationship Digraph	5 Whys		
Process Flow Chart	Waste Walk	Root Cause Analysis		
Affinity Diagram	Cycle Time Analysis			
Activity Network Diagram	Value Stream Map			
Tree Diagram & PDPC	KPIV, and KPOV Analysis			

8.3.4.8   Substructure and Belts Within the Six Sigma

Human power is the main asset of each organization. Educating and authorizing human resources is an essential investment for that organization. Hence, knowledgeable, skillful, and accountable personnel is the key to driving the organization's operation and helps to achieve its goals.

That is one of the whys and wherefores that the Six Sigma has advanced into lean Six Sigma (LSS), consequently allowing the Six Sigma experts to take advantage of the lean and Kaizen manufacturing methods. That results in the full engagement of associates in the organization.

One exclusive feature of the Six Sigma is its special substructure of professionals within the organization who are specialists in the methods and tools of the Six Sigma applications. Six Sigma's substructure can be white belts, yellow belts, green belts, black belts, and master black belts.

One of the most important reasons for distinction within the Six Sigma professionals is that the experts' degree and statistical analysis ability. The other important aspect is the experience of problem-solving. So, based upon that master black belt, Six Sigma professionals are the highest level of experts in the field. Thus, the idea of having "special employees" can be related to their level of knowledge in the application and the practices of Six Sigma methodology (i.e., TQM, Kaizen, PDCA, Lean).

8.3.4.9   Guiding Principle on the Six Sigma Implementation

The first step that leads to operative implementations of the Six Sigma methodology is a substantial emphasis on educational training employees on the various tools and techniques within the Six Sigma. The academic training ought to focus on how to apply the Six Sigma tools and techniques in industrial applications. It is easy to comprehend how specific tools and methods can be used and successfully executed.

It takes a profound sympathetic of all the tools or the techniques to recognize which tool or technique is the right tool to solve the given industrial problem. Essentially, successful implementation of and execution of the Six Sigma methodology requires outstanding supervision of the higher management. Productive time management, critical project execution milestones, troubleshooting, communication, and accountability are the Six Sigma project manager's skills. Without the timely and operational support of the higher management, the Six Sigma project's execution may not be effective and efficient.

8.3.5 Theory of Constraint

Goldratt was made known to and popularized the Theory of Constraints (Abbreviated as ToC). Goldratt, in his book in title, "the Goal," and undertakes that every single process has a constraint that resists or limits it from realizing its preestablished goals.

Most of the continuous improvement methodologies discussed in this chapter can be applied in any problem-solving (i.e., root cause analysis, process improvement, CAPA, etc.). However, the theory of constraints hyper-focused on improving the productivity of the process. To advance the throughput of the process, the theory of constraints focuses on constraints (i.e., bottlenecks). Constrains are any factors, which resist or limit the process not to produce the highest yield.

In ToC, we can switch the terms of constraints and bottleneck. Hence, the comprehensive definition of the constraint gives apostle characteristics of the process bottleneck. The bottleneck is any factor that limits the processing capacity reduces the productivity of the production. The bottleneck can be a machine, process performance, demand, warehouse, supply, and so on. The bottleneck is a point of congestion in the production system, which occurs when capacities arrive too speedily for the production process to accomplish handle. Thus, the bottlenecks can have a weighty effect on the flow of production.

Constraints (bottlenecks) can be exterior to the organization. Conceivably, vendors can affect the quality and the quantity of incoming material. Similarly, customers can affect the organization's effectiveness in terms of switching the demand for the product. Sometimes, the customer's demand can be higher than the actual production, whereas customer demand can be lower than the output.

For instance, in production, the constraint can be the production machine. Here our machine may not be effective to produce the quality based on the customer requirement. Or the machines' capacity that we have may not make the required quantity of the product. Similarly, skilled personnel can also be the constrain for production. The level of technology and the complexity of output are functions of the professional human resources we have in the production. If we do not have such advanced human resources, we cannot produce the product.

- Constraints can also be ineffective:
- Standard operating procedures (SOPs)
- Maintenance cycles
- Calibration cycles
- Buddy checks (buddy checks are systems that need crosschecking of the operation)
- Work in process (WIP)
- Manufacturing execution systems (MES)
- Statistical process control (SPC), and so on (Fig. 8.8)

Precisely, bottlenecks can make the resources whose processing capacity is less than the demand-positioned on it. If the resource (input) has more processing

Fig. 8.8 Visualizing the
bottleneck

Fig. 8.9 The Steps of
process improvement

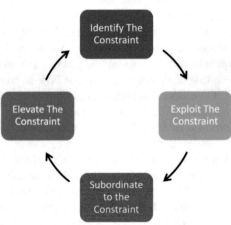

capacity than the demand-positioned on it, we can suspect there is some bottleneck
in the process. Thus, our next task will identify the possible constraints and how to
elevate them from the process.

8.3.5.1   Four-Steps of Operational ToC

As can be seen in Fig. 8.9, operational ToC can have four steps. These are

1. Identify the constrain
2. Exploit the constraint
3. Subordinate the constraint
4. Elevate the constraint

Step 1: Identify the Constraint

The first step in the ToC is to identify and recognize the bottleneck, which limits the processing capability. That step is the essential step. Hence, without knowing the correct bottleneck of the process, we cannot proceed further or solve the problem. There are a lot of techniques to identify bottlenecks in the process. The most efficient one is based on an evaluation of the process input and output. That can be done by measuring the production on the process steps. Then, we can observe the lowest production rate in the process steps.

Thus, the factor that causes the lowest rate of production can be considered as the bottleneck. Another method of identifying the constrain (bottleneck) can be just observing the Work in Process (WIP), which is piling up. Inventory tends to pile up just before the bottleneck due to the slowest step in the process.

For instance, suppose the given process has five steps (i.e., A, B, C, D, and E, which is presented in Fig. 8.10). Furthermore, they assume the average process time of the production is 25 s per unit.

- Operation A consumes 20 s per unit of the product.
- Operation B consumes 15 s per unit of the product.
- Operation C consumes 20 s per unit of the product.
- Operation D consumes 45 s per unit of the product.
- Operation E consumes 25 s per unit of the product.

The slowest operation is D, which consumes 45 s per unit of the product. Therefore, operation D is the bottleneck.

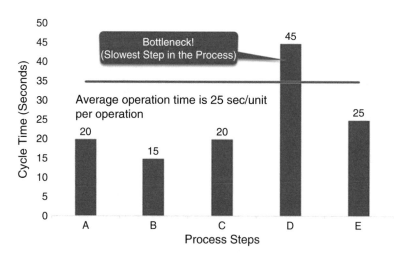

Fig. 8.10 Identification of bottleneck in the process

Step 2: Exploit the Constraint

Once we have identified and recognized the bottleneck (i.e., operation D), we will exploit the bottleneck (see Fig. 8.11). In this instance, let us plan to improve the bottleneck by 10%. That is, we plan to reduce the production time from 45 to 50 s per unit of the product.

Exploiting the bottleneck can be done using:

- The knowledge of Kaizen (or other lean engineering tools) can help make some changes to speed up operation D.
- The Plan–Do–Check–Act (PDCA) or the Define–Measure–Analyze–Improve–Control (DMAIC) cycles can be applied to identify the bottleneck. Both the PDCA and DMAIC can identify the "low hanging fruit" improvements that can speedily improve and exploit the bottleneck.
- Inventory buffer analysis (IBA). Another method for exploiting the constraint comprises an inventory buffer in front of the constraint. The inventory buffer can indicate the bottleneck by analyzing the inventory after each operation—the inventory has the smallest output compared to the other operations.
- The downstream processes can be reconfigured (or reengineered) to ensure any failures in those processes do not avoid the constraint from continuing to produce the product.
- Employee rotation (ER). Here we can alternate the employees to confirm the capacity is fully (100%) utilized at operation D. Here, our objective is to evaluate the employee's productivity.

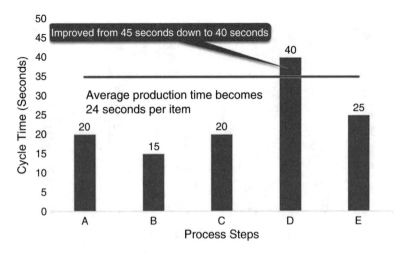

Fig. 8.11 Exploiting the bottleneck

Step 3: Subordinate the Entire Process to the Constraint

Once we have finished exploiting the constraint (bottleneck), we will try to implement the process improvement steps. That step requires subordinating all the process steps to the constraint. From our example, we know that operation B is the most productive one. It only consumes 15 s to produce a unit of the product. Similarly, operations A, C, and E are also moderately productive.

These non-bottleneck operations (i.e., A, B, C, and E) need to be subordinated to the bottleneck (i.e., D). Operating the non-bottleneck operations at their maximum is a waste of time and improper inventory. Therefore, adjusting the productivity of the non-bottleneck operations (i.e., subordinating the process) will be an essential task.

Such a capacity generates waste in the form of extra inventory. That is in the form of overproduction. Therefore, those non-bottleneck processes should be subordinated to the constraint, even if it sometimes means running inefficiently. That means that each operation in the process needs to operate as close as the productivity of the bottleneck.

Step 4: Elevate the Constraint

If step 2 is incapable of eliminating (or break) the bottleneck, then we can elevate the constraint utilizing breakthrough methodologies or capital investments. When the constraint is (a piece of) equipment, conceivably, we purchase a new machine or execute an improvement effort to advance the process and reduce the cycle time. In our manufacturing instance, in Fig. 8.12, let us assume we purchased a machine and consequently the cycle time down in half from 40 to 20 s.

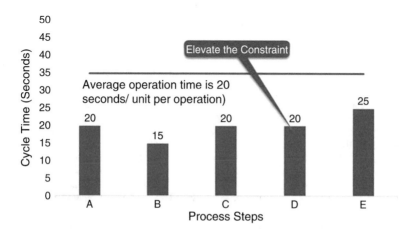

Fig. 8.12 Elevate the constraint

Note that

- If the constraint is human resources, then the solution can be hiring more people after considering the cost of hire and the demand.
- If the constraint is the supplier's quality, then the solution can be changing the supplier or requesting the supplier to improve.
- If the constraint is demand, then the solution can be finding a new market or reducing production.

Bibliography

C.W. Adams, P. Gupta, C.E. Wilson, *Six Sigma Deployment* (Butterworth-Heinemann, Burlington, MA, 2003) ISBN 0-7506-7523-3. OCLC 50693105

T. Bertels, *Rath & Strong's Six Sigma Leadership Handbook* (John Wiley and Sons, New York, NY, 2003), pp. 57–83. ISBN 0-471-25124-0

F.W. Breyfogle III, *Implementing Six Sigma: Smarter Solutions Using Statistical Methods* (John Wiley & Sons, New York, NY, 1999) ISBN 0-471-26572-1. OCLC 50606471

C.E. Cordy, L.R.R. Coryea, *Champion's Practical Six Sigma Summary* (Xlibris Corporation, Bloomington, IN, 2006)

T. Dasgupta, Using the six-sigma metric to measure and improve the performance of a supply chain. Total Qual. Manag. Bus. Excell. **14**(3), 355–366 (2003). https://doi.org/10.1080/1478336032000046652. ISSN 1478-3363

J.A. De Feo, W. Barnard, *JURAN Institute's Six Sigma Breakthrough and Beyond – Quality Performance Breakthrough Methods* (Tata McGraw-Hill Publishing Company Limited, New Delhi, 2005) ISBN 0-07-059881-9

B. El-Haik, N.P. Suh, *Axiomatic Quality* (John Wiley and Sons, New York, NY, 2005), p. 10. ISBN 978-0-471-68273-8

K.J. Fryer, J. Antony, A. Douglas, Critical success factors of continuous improvement in the public sector: A literature review and some key findings. Total Qual. Manag. **19**(5), 497–517 (2007)

C. Gygi, N. DeCarlo, B. Williams, *Six Sigma for Dummies* (Wiley Publishing, Inc., Hoboken, NJ, 2005), p. 23. Front inside cover, ISBN 0-7645-6798-5

G.J. Hahn, W.J. Hill, R.W. Hoerl, S.A. Zinkgraf, The impact of six sigma improvement-a glimpse into the future of statistics. Am. Stat. **53**(3), 208–215 (1999)

M.J. Harry, P.S. Mann, O.C. De Hodgins, R.L. Hulbert, C.J. Lacke, *Practitioner's Guide to Statistics and Lean Six Sigma for Process Improvements* (John Wiley and Sons, New York, NY, 2011), p. 30. ISBN 978-1-118-21021-5

M. Imai, *Gemba Kaizen: A commonsense, low-cost approach to management*, 1st edn. (McGraw-Hill, New York, NY, 1997) ISBN 0-07-031446-2

P.A. Keller, *Six Sigma Deployment: A Guide for Implementing Six Sigma in Your Organization* (Quality Publishing, Tucson, AZ, 2001) ISBN 0-930011-84-8. OCLC 47942384

P.A. Keller, P. Keller, *Six Sigma Demystified* (McGraw-Hill Professional, New York, NY, 2010), p. 40. ISBN 978-0-07-174679-3

Y.H. Kwak, F.T. Anbari, Benefits, obstacles, and future of six sigma approach. Technovation **26** (5–6), 708–715 (2006). https://doi.org/10.1016/j.technovation.2004.10.003

P.S. Pande, R.P. Neuman, R.R. Cavanagh, *The Six Sigma Way: How GE, Motorola, and Other Top Companies are Honing Their Performance* (McGraw-Hill Professional, New York, NY, 2001), p. 229. ISBN 0-07-135806-4

K. Richardson, The 'six sigma' factor for home depot. Wall Street Journal Online (2007)

S.A. Ruffa, *Going Lean: How the Best Companies Apply Lean Manufacturing Principles to Shatter Uncertainty, Drive Innovation, and Maximize Profits* (AMACOM (a division of American Management Association), New York, NY, 2008)

R.D. Snee, R.W. Hoerl, *Leading Six Sigma: A Step-by-Step Guide Based on Experience with GE and Other Six Sigma Companies* (FT Press, Upper Saddle River, NJ, 2002) ISBN 0-13-008457-3. OCLC 51048423

G. Taylor, *Lean Six Sigma Service Excellence: A Guide to Green Belt Certification and Bottom Line Improvement* (J. Ross Publishing, New York, NY, 2008) ISBN 978-1-60427-006-8. OCLC 271773742

G. Tennant, *SIX SIGMA: SPC and TQM in Manufacturing and Services* (Gower Publishing, Ltd, Aldershot, 2001), p. 25. ISBN 0-566-08374-4

D.J. Wheeler, *The Six Sigma Practitioner's Guide to Data Analysis* (SPC Press, Knoxville, TN, 2004), p. 307. ISBN 978-0-945320-62-3

Part IV
Quality Design for New Product Introduction

Chapter 9
Quality Design for New Product Introduction

9.1 Introduction

New product introduction (NPI) is an engineering and business process, where a new product is launched in the market, research center application, or any other business or engineering applications. After the product is developed, it will be introduced to the marketing or sales operation team

Effezctive NPI is a vital keystone of the company's success. There are substantial motivations for companies to make available new products to the market in a highly competitive market environment. The financial payout on or after the efficacious NPIs can assist several companies overwhelmed by the decelerating progress and profitability of prevailing products or services, closing the maturity stages of their life spans

The NPI process is a comprehensive blueprint that comprises seven well-defined stages and outcomes in the finished product being made known into the market. The steps of the NPI process are (1) product idea, (2) product architecture, (3) product design, (4) design validation and test, (5) manufacturing ramp, (6) mass production, and (7) product support. Figure 9.1 gives precise information about the NPI process

An operative cross-functional team is the major driver of the NPI process. The submanagers of the stages control each stage and the NPI process's reported, project manager. It is recommendable that at each stage, the decision should be made by the project manager.

9.2 Design Inputs and Outputs

Definition 1 A design input (DI) refers to both the product's physical and performance requirements, which are the basis for the product design.

Y. Y. Tesfay, *Developing Structured Procedural and Methodological Engineering Designs*, https://doi.org/10.1007/978-3-030-68402-0_9

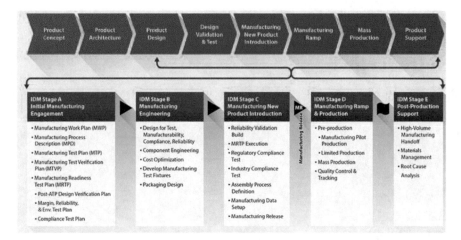

Fig. 9.1 New product introduction (NPI) process

In the design process, the inputs are the customer requirement. These requirements can be features, performance, reliability, quality, and so on. The product that causes the customer is willing to pay to get the values.

Definition 2 A design output (DO) defines all the drawing (specifications of sub-assemblies or assemblies) and all manufacturing procedures.

One of the critical characteristics of design output (DO) is the conception of engineering drawings as a communication tool and a control tool to certify that our product is manufactured and then inspected continuously throughout its lifecycle.

Quality engineers (QEs) are essential experts to support the development of new products. QE uses three critical steps to design the product.

1. The first step in generating raving fans is recognizing all the sources of design inputs (DI) and comprehend customer requirements. These design inputs (DIs) will be the design targets.
2. The next essential step is transforming specific customer requirements into the design concept. Here the QE can use tools like Quality Function Deployment (QFD), Robust Design, Design for X (DFX), Design for Six Sigma (DFSS), and Quality by Design.
3. The third step is utilizing the design review process all over the development process to confirm the development step of the new product.

Proper executions of these three steps, QE will ensure the fulfillment of the specific customer requirement fulfilled by the products.

As we move through those the new design methodologies, the QE is required to generate the documentation of the design output details (DO). The detailed design outputs consist of the engineering drawings, specifications, labeling requirements, bill of material, inspection criteria, and assembly procedures that describe the final product.

The documentation of design outputs (DOs) must be organized to describe all the subassemblies and assemblies of the product. The design outputs (DOs) will be verified during design verification to confirm that the new product reliably brings all the design inputs (DIs) that the QE created. The design outputs (DO) should also result in the final product that meets customer requirements. The requirement also should confirm during the design validation stage of development.

9.2.1   Sources of Design Inputs

No matter the final goal is different, developing the new product is just like the development process of any other product. Thus, every product production has input, process, and output. Designing a new product is no more different. To ensure that the output (final product) is suitable, we must determine design inputs since they will guide and drive the design process. The quality characteristics of the final product tell if there are opportunities for improvement. Using this feedback will help to realize the improvement.

According to Fig. 9.2, to produce the product, there should be a process input, then the process converts the input into the output. Furthermore, we should document the current process's performance, and that becomes feedback to improve the future process.

Customer-specific requirements are design inputs. However, before determining the customers' specific needs, we should be sure that the customer requirements are properly documented and presented. Furthermore, we must be sure that we properly understand the customer-specific requirements.

Many of us have the impression that customers are like only as external customers as immediate or end-users. However, customers are of different types. For instance, operators are customers of engineers and process technicians. So, as we have external customers, we can categorize them as internal customers too.

If we operate in a regulated industry, we will have regulatory agencies as an external customer. For instance, the Federal Drug Administration (FDA) is an external customer in the pharmaceutical industry.

Every company has two distinct customers: Internal Customers (Safety, Quality and Reliability Engineering, Manufacturing, Sales and Marketing, Top Management, etc.) and External Customers (End Users, Regulatory and Environmental Agencies, dealers, distributors, suppliers, etc.). Study Fig. 9.3 for the details of customer input.

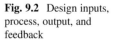

Fig. 9.2 Design inputs, process, output, and feedback

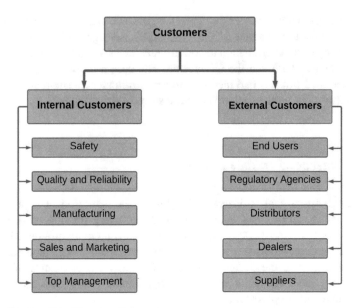

Fig. 9.3 Sources of customer inputs

During designing the new product, proper categorization and identification of customers are essential. For instance, the manufacturability of the design is the function of the engineering and operational team's ability. Additionally, the design's serviceability is a crucial factor for the end user to buy the new product. Hence, if the product is easily serviceable, then the customer is willing to purchase the product. Otherwise, although the product is of high quality, the end users are unwilling to buy the product if the product's serviceability is difficult or costly.

9.2.2   Models to Capture Customer Needs

The three key topics that will help to capture and interpret our customer's needs and expectations are:

1. *Kano Model: The model of analyzing of customer satisfaction*

    In Sect. 1.2, we have stated that "Noriaki Kano (1984) states that a lot of professionals explain quality from the two-dimensional model: (1) must-be quality to describe fitness to use and (2) appealing quality to describe what a customer would like. However, according to Kano, the quality of the product (and its services) meets customer expectations, or it is even exceeding customer expectations.

    The Kano model is suitable for knowing a systematic understanding of the customer's requirements and needs. The Kano model assists in determining the key features, which will maximize customer needs. For instance, product and quality engineers use the model to line up new features by sorting them into groups (i.e., the extent of customer satisfaction).

2. ***Juran Model: Model of Classifying Customer Needs***

According to Juran, quality refers to the product's ability to meet customer needs and the degree of the product satisfying or exceeding the customer expectations. Juran develops a notion called a quality trilogy. The term trilogy describes the three critical sections of quality management. According to Juran, each organization needs to have:

(a) Quality planning
(b) Quality control systems and tools
(c) Quality improvement tools

All three quality processes are extremely interrelated and will positively and negatively affect the quality control and assurance process.

3. ***Customer-Defined Value***

Customer value is the degree of fulfillment of the customer requirements, and consequently, the level of customer satisfaction has the product that the company delivers to the customer.

The company can create customer value in three ways. These are through providing the customer:

(a) High-quality product
(b) The best service
(c) Reasonable or competitive cost

9.2.3   What Are Design Inputs?

Precisely, design inputs are documentations that states the technical requirements of the customer needs. The design inputs reflect the final product in terms of performance, durability, reality, safety, functionality, and other related predefined quality requirements.

Additionally, the design inputs documentations need to present any regulatory requirements. The regulatory requirements can be from internal or external customers' and should be realized by the company. There are many models to integrate design inputs with the process to produce the design output, which is the input to the final product. One of the best models to show such integration is the waterfall model. Study the structure of the waterfall design model in Fig. 9.4.

Definition 3 Verification is the process of cross-checking the design output's accuracy as the exact reflection of the design input.

Verification is done by documentation approval both from the manufacturer and the customer.

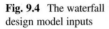

Fig. 9.4 The waterfall
design model inputs

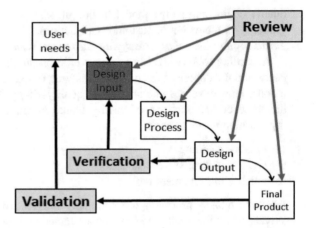

Definition 4 Validation is the process of cross-checking whether the final product is produced based on the customer-specific requirements.

Validation is done by the measurement of the final product against its specification requirements.

Design inputs are the requirements, which must meet the ultimate design concept. In the end, the design inputs serve as the design output verification acceptance criteria. In that way, the design inputs will be the leader of all the decision making in the generation of the design output.

9.3   Translation of Design Input (DI) into Design Output (DO)

Once we get all the documentation and made all the essential communications with the customer about the design input, our next process is creating the design concepts by translating the design input. Here, we must have appropriate decision-making tools and methodologies to confirm the final product meets the customer-specific requirements. Several design concepts start by apprehending the customer-specific needs (voice of the customers) and then making the accurate translation of the requirements into the technical inputs, which can provide the final design concept.

This section will learn the five most essential tools for translating the design input to design concepts. These tools are the most applied tools for engineering applications to deliver a high-quality product. These models are:

1. Quality Function Deployment (QFD)
2. Robust Design (RD)
3. Design for Excellence (Design for X, or DFX)
4. Design for Six Sigma (DFSS)
5. Quality by Design (QD)

9.3.1   Quality Function Deployment (QFD)

Quality Function Deployment (QFD) is the most applied tool to translate the customer-specific requirements into design inputs. Figure 9.5 gives a typical Quality Function Deployment (QFD) diagram.

In Fig. 9.6, which is the House of Quality (HQ), we see that the customer-specific requirements (voice of the customer) are presented on the left side of the house, and at the top of the house is the technical requirements (design inputs).

The relationship matrix is presented in the center of the house. The relationship matrix equates the customer-specific requirements (voice of the customer) to the

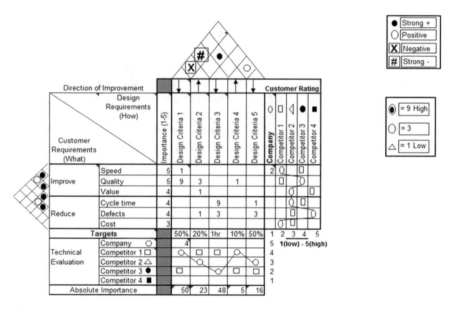

Fig. 9.5   Quality function deployment (QFD) diagram

Fig. 9.6   Framework of VOC–DI relationship matrix–correlation matrix

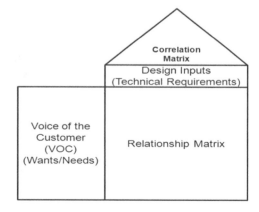

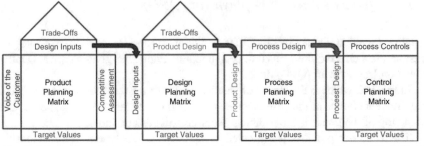

Phase 1 – Performance requirements   Phase 2 – Product Design   Phase 3 – Process Design   Phase 4 – Process Control

Fig. 9.7   Design inputs process phases

design inputs to analyze each design input's association with each customer-specific requirement.

Figure 9.6 shows the correlation matrix is the rooftop of the house, which associates the relationship of the different product specification features. That will make known any contending design inputs. Although most quality engineers are acquainted with the House of Quality (HQ), the complete Quality Function Deployment (QFD) process considers a multiphased method that interprets customer-specific requirements (voice of the customer) down to the process control systems. All of which is intended to consistently bring the design inputs, as shown in Fig. 9.7.

9.3.2   Robust Design

Genichi Taguchi introduced the quality engineering model of Robust Design (RD). So, the robust design (RD) model is also called the Taguchi method. Originally, Robust Design (RD) was introduced to improve the quality of manufactured products. Taguchi was a statistician engineer and applied a statistical approach to quality management systems. As a result, robust design (or Robustification) is one of his most essential contributions for quality system analysis.

The robust design method is introduced to deliver a high-quality product by identifying and controlling all sources of variations. Thus, the goal of robust design is to confirm a design output for the final product to resist any changes. As it is a statistical tool, robust design (RD) is all about process variation and the effect on the final product. Both the final product and the production process will involve or experience variations. According to Taguchi, these variations have two forms. These are:

1. Controllable variation
2. Uncontrollable variation

9.3.2.1 Uncontrollable Variations

The uncontrollable variation is also called noise (or random) variation. Robust Design (RD) is the method of measuring and analyzing both forms of interpretations. Robust Design (RD) aims to optimize the design input to decrease the effect that uncontrollable (random) variations on the final product. Furthermore, the model's prediction will assist us in selecting the key parameters for the controllable variation (i.e., the optimal set point and acceptable range). Study the Robust Design in Fig. 9.8.

For example, environmental changes are one of the best examples of uncontrollable variations; on many occasions, product and quality engineers cannot control the sorts of environmental changes that the final product is exposed to. Usually, the design output and the final product may significantly be affected by the varying environments.

Therefore, initially optimizing the final product's design output the best practice so that the uncontrollable variations (such as environmental changes) do not affect the quality (performance, durability, reliability, functionality, etc.). Applying the robust design methodology, we can optimize the design output by determining the optimal process parameters.

9.3.2.2 Controllable Variations

The robust design applies to analyze controllable variations to set the optimal nominal (target) value for each variable (process parameter, preliminary material dimensions, etc.). Furthermore, the model helps us determine the efficient ranges (tolerance) of variation in the process and the raw material. Thus, the final product's quality will be unaffected by the allowable range of the controllable variation.

Fig. 9.8 Robust design

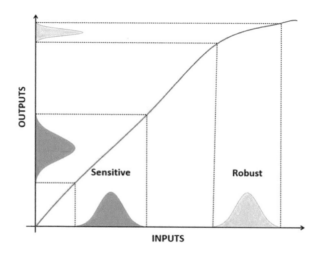

Taguchi's robust design process relies on the applications of the usage of well-organized Designed Experiments (DoE), which will help analyze variations and quantify the impacts of the variations on the quality level of the final product. This is done by analyzing variations of the controllable variables, uncontrollable variables, and the process inputs (materials, etc.). The appropriate model of Design of Experiments (DoE) is a powerful model to estimate the variations' effects. Consequently, the DoE allows us to control the optimal setup point and ranges for the controllable variation and input materials.

9.3.3   Design for Excellence

Design for Excellence (Design for X, or DFX) common method used in the conceptual phase of the new product introduction (NPI). Design for X is initially called Design for Excellence (DFX). It has subsequently progressed into a collection of exclusive procedures where the X stands for the diverse design models, which include:

- Design for assembly and manufacturing
- Design for reliability
- Design for testing
- Design for life cycle cost
- Design for environment
- Design for maintainability and serviceability

One of the hallmarks of DFX is simultaneous engineering, where the product and the production process are technologically advanced in analogous to assimilate the unique procedures above for improved manufacturability, performance, reliability, functionality, life cycle cost, and so on. That simultaneous engineering necessitates cross-functional teams and can significantly condense the time required to advance the product. For each of the methods below, the design team would include subject matter experts from those areas.

9.3.4   Design for Manufacturing and Assembly

As the title implies, the design for manufacturing and assembly (DFMA) is a model that refers to the design for the final product so that it is with no trouble manufactured and assembled. To achieve that, the cross-functional design team needs to comprise the subject matter experts (SME) from manufacturing, engineering, operation, and supply chain. Study the process flow diagram of DFMA in Fig. 9.9.

Vital factors that are significant to study for the design for manufacturing and assembly (DFMA):

Fig. 9.9 DFMA process
flow chart

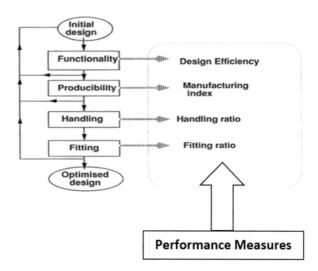

- **Planning**
  Optimizing production costs are easily realizable at the design phase, consult with subject matter experts (SME) on the design.
- **Materials**
  Select an appropriate material management system that determines production style and quality control.
- **Processes**
  Recognize the production to optimize costs.
- **Standardization**
  Standardization of components and parts will significantly optimize costs and improve inventory management.

The six best practices that are applied for designing for manufacturability and assembly are:

1. Decreasing the number of parts to build the product in the manufacturing or assembly of the product.
2. Use of modular design for assembly applications.
3. Use of standardizing procedure in manufacturing.
4. Designing spare parts and components that are easily manipulated.
5. Reducing tight tolerance or needless part finishes.
6. Use error-proofing and mistake-proofing techniques to reduce waste.

9.3.5   Design for Reliability (DFR)

Design for reliability (DFR) is a design approach that collaborates with our company's reliability engineers to certify that the design concept is improved and optimized for reliability (quality over time). The approach normally comprises an in-depth review of the various failure points of the proposed design concept employing hazard analysis (HA) tools like failure mode effect analysis (FMEA), failure tree analysis (FTA), failure risk analysis (FRA) along with model and prototype testing.

Design for reliability (DFR) analysis is expected to reveal the design weaknesses. Consequently, the reliability engineer can then make the necessary design changes and improvements to optimize the DFR.

Advancing DFR will decrease warranty costs, upsurge product safety, performance, and ultimately increase customer satisfaction.

9.3.6   Design for Maintainability and Serviceability (DMS)

When designing large-scale equipment (LSE) requires serviceability (maintenance operation) all over the product lifecycle. Usually, LSMs are expensive and need to work for a long time. So, it is essential to optimize the design for maintainability and serviceability (DMS) for such a product.

MSD significantly decreases the downtime related to repairs and also reduce the lifecycle cost related to maintenance. The DMS eliminates hazards to engineers, technicians, and operators during repair and comprehends customer-specific requirements, and, consequently, enhances customer satisfaction. Essentially, the MSD approach enhances the service team's capability to detect, diagnose, remove damaged parts, replace, or repair the product back to its original condition.

9.3.7   Design for Six-Sigma (DFSS)

Design for Six-Sigma (DFSS) is applying the various Six Sigma tools for the development of improved products. The goal of DFSS is to increase product robustness and, as a result, to increase customer satisfaction. The DFSS involves five stapes Define–Measure–Analyze–Design–Verify (DMADV).

The Six Sigma model and the Design for Six Sigma (DFSS) are similar in the methodological approach. The usual Six Sigma approach is used for continuous process improvement after the production is already started. However, DFSS is applied during the product's lifecycle's design output phase to handle the product efficiency proactively. Study the brief process of the DFSS in Fig. 9.10.

Like the Six Sigma method, the Design for Six Sigma (DFSS) process exploits statistical analysis to develop a data-driven decision. The DFSS is applied during the translation of specific-customer requirements (voice of the customer) into the robust design concept. That aimed to the DFSS to comprise the customer-specific requirements and optimize the process variation for the final product. Principally, quality,

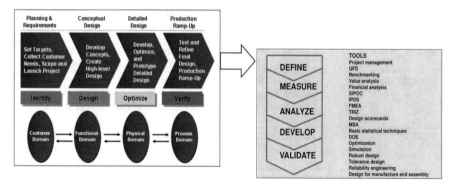

Fig. 9.10 Design for Six Sigma (DFSS) process

Table 9.1 DMADV and IDOV methodologies

Define–Measure–Analyze–Design–Verify (DMADV)	
Define	The scope and objectives (goals) of the project.
Measure	Gather customer-specific requirements (such as critical to quality), translate the requirements into the technical specifications, gather up-to-date performance of products, and prioritize the customer's voice that needs the utmost attention.
Analyze	The various design concepts and their capability of satisfying the critical to quality whereas maintaining robustness.
Design	Describing the design of the final product and production process. Then performing tests and optimize design concepts.
Verify	The design concept meets the customer-specific requirements and the intended use of the final product.
Identify–Design–Optimize–Verify (IDOV)	
Identify	The customer-specific requirements (such as critical to quality), the project scope, objectives (goals), and business strategy.
Design	The final product by generating multiple design concepts.
Optimize	The chosen design concept and optimize the critical attributes (such as quality, reliability, functionality, etc.) and production.
Verify	Confirmation and justification that the design concept the chosen design concept fulfills the customer-specific requirements.

product, and reliability engineers use DFSS to modify the Six Sigma tools for process improvement and the design phase of the product.

The Six Sigma experts apply the define-measure-analyze-improve-control (DMAIC) process for the continuous improvement process. However, DFSS experts use two roadmaps: Define–Measure–Analyze–Design–Verify (DMADV) and Identify–Design–Optimize–Verify (IDOV). Study the DMADV and IDOV methodologies in Table 9.1. The DFSS method, whether it is defined in the DMADV or IDOV, the method initiates every single design process by apprehending the needs of the customer.

Table 9.2 Identify–Design–Optimize–Verify (IDOV) process

Identify	Design	Optimize	Verify
Project generation	QFD Phase 1 and Phase 2	QFD Phase 3 and Phase 4	QFD Phase 4 accomplishment
Strategic planning	Design of experiments (DoE) Fractional–factorial design	Design of experiments (DoE) Full-factorial design	SOP's and work instruction (WI) Preparation
VOC	Critical to quality (CTQs) analysis	CTQ conduit clause to design concepts	CTQ conduit clause to production quality control
PERT and Gantt charts	Design for X (DFX)	Design error proofing	Model adequate Manufacturing lines
Shareholder analysis	System and development engineering	Process failure mode effects analysis (PFMEA)	Process flow diagram
Brainstorming	Measurement system analysis (MSA)	Robust design (RD) methods	Process error proofing
Benchmarking	Design failure mode and effect analysis (DFMEA)	Detection of critical process parameters	Design verificationplan and report (DVPR)
SIPOC (COPIS) analysis	Input–output process maps	Markov chain Monte Carlo (MCMC)	Lean methods
Cross-functional team (CFT) development	Lifecycle analysis	Reliability studies	SPC control plan and control charts for
Kano model (KM) analysis	Statistical tolerancing	Production capacity planning (PCP)	Process capability analysis (PCA)
Strength, weakness, opportunity, and threat analysis (SWOT)	Axiomatic design	Fault tree analysis (FTA)	Process validation

Both approaches, that is, the DMADV and IDOV, comprise an assortment of tools and techniques destined to bring the optimal design concept based on the customer-specific requirements.

The DFSS calls the customer-specific requirements critical to quality (CTQ). The critical to quality (CTQ) is applied through the design concept process to direct the design and drive the key decisions to confirm that the final product should satisfy the customer's needs.

Once the CTQ's are recognized, DFSS applies statistical tools and techniques to translate those requirements into various potential design concepts. That results in determining the optimal design concept.

Once a final design concept has been chosen, the DFSS models (DMADV and IDOV) will move into the optimization process to refine the design concept, producing a good-quality product.

Table 9.2 presents the IDOV process with a list of the well-defined tools, which are possibly used for analysis in the design process.

The last phase of any DFSS is the Verification and Validation phase. Here final design concept is tested in various ways to confirm that it fulfills the customer-specific requirements.

9.3.7.1 Why DFSS?

- The major benefit of the DFSS method is that it allows the manufacturer to avoid many challenges of implementing and executing the DMAIC process of Six Sigma practice.
- Furthermore, the DFSS has comprised both design improvement and the Six Sigma approach. Once the product is in production, DFSS helps eradicate process variation problems, waste reduction, customer relationship, and customer complaint management, optimize warranty issues, and so on.
- Essentially, DFSS is a prevention activity of production problems. That means the model assists in eliminating issues before they begin through improved quality design.
- As a lean manufacturing method, DFSS can result in a shorter development cycle time of the product and increasing speed to market.

9.3.8 Quality by Design

The development of quality planning using a set of quality tools and techniques known as Quality by Design. Quality by Design aims to improve process quality through understanding the process based on referring to the existing process documentation. The application and implementation of quality by design fall into a three-step process approach.

Step 1—Critical Quality Attributes (CQAs)

Identifying and quantifying the critical quality attributes (CQAs) of the product is the initial step of the quality by design. The QBD methodology essentially needs to identify the product quality attributes to comprehend the final product's functionality, performance, safety, reliability, etc.

Thus, the critical quality attributes (CQAs) of the product must be controlled within a prespecified tolerance (range) to make certain about the product's quality.

Step 2—Identify the Process Parameters

Identifying the process parameters and material inputs are crucial to have a data-driven process control system. Here the main concern is that the parameter variability has a significant impact on the CQA's. These can be verified through the application of adequate DoE models to measure the variability.

The second step is all about the analysis of process variations. Specifically, we will analyze the measured process dataset. That will help us to predict the process variation per input material or parameter. Then in step 3, based on the result of DOE, we will develop a system to control the variations. Or we will reduce the impact of the variations on the final products.

Step 3—**Develop a control scheme**

Here the engineering team is expected to generate a system to control the variations. That helps to make sure that the material inputs and process parameters are adequately controlled. Once we have developed the system, we need to check all the critical quality attributes (CQAs) of the final product, whether they meet the customer-specific requirement or not.

The control scheme (or control plan) should include the input materials and process parameters. Furthermore, we should implement appropriate measurement systems and sampling plans to test and confirm the CQAs. Here, you should also note that we can use CQA's as the acceptance criteria during the process execution, process validation, or supplier qualification.

It is always essential to calculate the process capability indexes (Cp or Cpk) for these critical quality attributes to assure whether the process is stable or not.

These three-steps must be conducted during the final product and process development. At last, we will be ready for full-scale production.

9.3.8.1   The Difference Between Quality by Design and QbD

Quality by Design is a method based on process knowledge. In contrast, QbD is a quality system based on assessing a quality document to forecast the future.

The QbD used risk-based tools such as robust process design, design of experiments (DoE), real-time process control, process analytical technology (PAT), and quality management systems (QMS).

In recent times, especially in the pharmaceutical industry, the FDA has placed a strong emphasis on QbD. This initiated the acknowledgment that quality cannot check into the product. Quality should be the requirement to build into the product. That means the QbD approach is a preventive approach. It avoids or reduced problems before they happen.

The QbD method is like DFSS. Hence the QbD method identifies what is critical to the customers and confirming that the process plan dependably delivers a good-quality product.

Fig. 9.11 The waterfall model of design review process

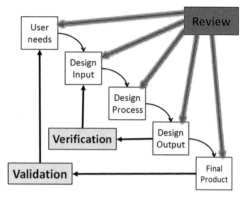

9.4   Design Review Process

The design concept advances through the various stages of the design process. During the development of the design concept, at each step, we need to have both a systematic and unintentional design review process to confirm the procedure is going on the right track. The periodic design review is based on the prescheduled time frame of process evaluation. In contrast, the unintentional design review is required to apply when some unexpected problem occurs in the design process when the core functional team (CFT) detects the problem. The obligatory design review aims to troubleshoot the issue at the right phase, on time, and in the right department.

There are a lot of design review models. One of the well-known models is the waterfall model. Figure 9.11 presents a simple instance of how the waterfall model applies to review each design phase in the design process.

The design review processes are critical to meet the overall goal of the design concept. The review processes allow the cross-functional team (CFT) or independent milestone reviewers (such as third parties or project managers) to understand the project's progress, make timely corrective actions, and find better alternatives. Thus, the review processes are aimed to meet effective and efficient project closure.

Bibliography

T. Bertels, *Rath & Strong's Six Sigma Leadership Handbook* (John Wiley and Sons, New York, NY, 2003), pp. 57–83. ISBN 0-471-25124-0

D. Braha, O. Maimon, *A Mathematical Theory of Design: Foundations, Algorithms, and Applications* (Springer, New York, NY, 1998)

G. Brue, R.G. Launsby, *Design for Six Sigma* (McGraw-Hill, New York, NY, 2003) ISBN 9780071413763. OCLC 51235576

T.A. Carbone, Critical Success Factors in the Front-End of High Technology Industry New Product Development, Doctoral Dissertation, University of Alabama in Huntsville, 2011

R.R. Cavanagh, R.P. Neuman, P.S. Pande, *What Is Design for Six Sigma?* (McGraw-Hill, New York, NY, 2005)

S. Chowdhury, *Design for Six Sigma* (Dearborn Trade Publishing, Chicago, IL, 2002)

C.L. Dym, P. Little, *Engineering Design*, 3rd edn. (John Wiley & Sons, Inc., New York, NY, 2009)

R.J. Eggert, *Engineering Design*, 2nd edn. (High Peak Press, Meridian, ID, 2010)

T. Hasenkamp, Engineering design for six sigma. Qual. Reliab. Eng. Int. **26**(4), 317–324 (2010)

K.B. Kahn, *The PDMA Handbook of New Product Development*, 3rd edn. (John Wiley & Sons, Hoboken, NJ, 2012) ISBN 978-0-470-64820-9

K. Kenneth, *The PDMA Handbook of New Product Development*, 3rd edn. (John Wiley & Sons, Hoboken, NJ, 2013), p. 34. ISBN 978-0-470-64820-9

P.A. Koen, *The Fuzzy Front-End for Incremental, Breakthrough and Platform Products and Services* (Consortium for Corporate Entrepreneurship)

P. Ralph, Y.A. Wand, *Proposal for a Formal Definition of the Design Concept* (Springer, New York, NY, 2009)

P.R. Smith, P.S. Eppinger, Identifying controlling features of engineering design iteration. Manag. Sci. **43**(3), 276–293 (1997)

D.G. Ullman, *2009, The Mechanical Design Process*, 4th edn. (McGraw Hill, New York, NY)

K. Yang, B. El-Haik, *Design for Six Sigma: A Roadmap for Product Development* (McGraw-Hill, New York, NY, 2003) ISBN 9780071412087

Part V
Some Innovative Engineering Quantitative Models and Their Applications

Chapter 10
Analyzing the Impact of the Bullwhip Effect

10.1 Introduction

Supply chain management was familiarized in the late 1980s and came into extensive use in the 1990s. Supply chains hold companies and business activities needed to plan, source, produce, distribute, and use products or services. The supply chain is the alignment and structure of firms that deliver products or services to the market that encompasses all stages and phases involved in satisfying a customer need (Ganeshan and Harrison 1995).

Supply chain management comprises marketing, finance, new product development, customer service, and other interconnected and sophisticated activities. According to Christopher (1998), a comprehensive clarification of supply chain philosophy takes along a schemes approach to comprehend and accomplish the various activities compulsory to organize the flow of products or services to the best work for the final customer. That is, hypothetically, the whole chain is managed as a single object.

Supply chain management's fundamental view is that a single company cannot stand by itself in a chain as a supplier, producer, distributor, or retailer for the end customer. Even we theoretically accept there is a single company that acts like the whole chain. This company is most probably inefficient and ineffective (cost, market coverage, customer satisfaction, etc.). The company cannot be autonomous. Instead, proper coordination with other independent firms is needed to become effective and efficient (Milgrom and Roberts 1992; Williamson 1981).

Many organizational coordination scholars suggest that one of the indicators of effective and efficient supply chain management is the degree of organizational coordination (synchronization) of the business partners in the chain and its capability to be flexible and responsive according to the pattern of the end customer. That shows that the supply chain's performance is measured by its ability to go with the end customer's changing aspects. The pattern of the end customer is dynamic because of many recognizable convincing factors. For example, information technology (IT) created excellent choices regarding price, quality, or time for the end customers. That makes some randomness in most supply chains (Boon-itt and Pongpanarat 2011).

Usually, the supply chain management strategy is theoretically presented as a planning and operation system through the entire supply chain. Conferring the DuPont analysis, supply chain management emphasizes strategy while simultaneously reducing both inventory and operating expense. Likewise, the formulation and configuration of supply chain management strategy integrate the coordination and the control of production, inventory, location, transportation, and information. Consequently, supply chain management encompasses a rigorous and deep analysis of the chain under uncertain conditions (Chopra and Meindl 2001).

This paper's main inspiration is the uncertainty of the end customer demand pattern and its effects through the supply chain. The theory of supply chain management proposes a successful flow of material must be reinforced by the flow of suitable, on time, reliable, consistent information. Proper information flow across the supply chain is a function of the degree of organizational coordination. However, knowledgeably, or unintentionally, the chain companies have shown poor and inappropriate organizational coordination (Kopczak and Johnson 2003). That may be one of the potential factors that resist formulating and implementing making effective and efficient supply chain.

In literature, we can find the Bullwhip effect concept, which is one of the effects of lack of coordination through the chain. A very small change in the demand pattern from the end customer can amplify the unnecessary accumulation of inventories upstream of the supply chain. Therefore, this paper tried to show how the Bullwhip effect attacks the supply chain using the bootstrapping estimation technique on ANOVA models and Signal processing. Then, we try to assess the potential causes of the Bullwhip effect using State-Space models or recessive autoregression estimation methods.

The bootstrapping estimation technique on ANOVA models will help us estimate and compare the means of end inventory levels of the different supply chain actors under stochastic conditions (Cochran and Cox 1992). Furthermore, the model can give us information about the bias of the classical estimator. Spectral density analysis allows us to see the nature of the autocorrelation function on the observed time series data in the Fourier space (Boashash et al. 2003). The Fourier transformation contains the sum of an infinite series of sine and cosine waves of different amplitude. Therefore, we plot the density and periodogram over the series's frequency (Engelberg 2008). That creates a good ground to analyze the scope and nature of the autocorrelation structure of the observed end inventory level of the supply chain actors. That will help us choose and interpret the appropriate econometric model, which can eliminate autocorrelation during the causality using State-Space models or recessive autoregression estimation methods of the Bullwhip effect in the supply chain.

10.2 Literature Review

When we evaluate the efficiency and effectiveness of managing the individual business firm or organization, we have tremendous limitations. In the real world, the main reason is that businesses or organizations can never stand by itself as a manufacturer, distributor, retailer, and customer. Therefore, a new way of managing the whole system (the chain) is another very important task (Larson and Halldarsson 2002).

A new version of the business or organizational management as a chain is required to run safely efficiently and effectively the activities done in each business or organization introduced in the late 1980s. This stream of management is called supply chain management (Burt et al. 2003). The council of supply chain management professionals (Blanchard 2010) given the following definition:

Supply chain management encompasses the planning and managing of all activities involved in sourcing and procurement, conversion, and all logistics management activities. Importantly, it also includes coordination and collaboration with channel partners, which can be suppliers, intermediaries, third-party service providers, and customers. Supply chain management integrates supply and demand management within and across companies.

Supply chain management is an important but complicated topic in organizational coordination. Today many firms are implementing supply chain management to improve profits and increase customer satisfaction. The more detailed understanding of what supply chain management will achieve to the chain's companies is infantile. Supply chain professionals are working to make the management system efficient, effective, and beneficial to all the chain companies (Dess et al. 2005; Bhote 1987).

The study conducted by Ganeshan and Harrison (1995) has given more clues about the decisions involved in supply chain management and the models used to aid in making these decisions. The study conducted by Lee et al. (1998) has given how to study information sharing in the specific context of global supply chain management. The survey conducted by Lummus and Vokurka (1999) has shown how supply chain management got started and how to effectively manage the supply chain.

10.2.1 The Bullwhip Effect

In supply chain management literature, we find three types of flows:

1. The product flow consists of the product's movement from the supplier to the customer and any from the customer returns to the supplier.
2. The information flow involving transmitting orders and updating the status of delivery.
3. The financial flow is consisting of credit terms, payment timetables, and delivery and title ownership.

Order Requested

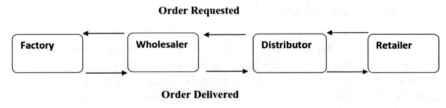

Order Delivered

Fig. 10.1 Simple representation of the product flow across the supply chain

Supply chain management must address problems of distribution network configuration and distribution strategy both in principle and practice. The distribution
network configuration is responsible for the number, the location, and the network
tasks of suppliers, production services, delivery centers, warehouses, cross-docks,
and customers. The delivery strategy, including questions of operating control,
delivery scheme, means of transportation, replacement strategy, and transportation
control (Ellram 1990).

Let us give special attention to the product flow from the upstream. According to
the supply chain downstream under the assumption, demand from customers is
uncertain, and the coordination (both intra-coordination and inter-coordination)
effort of the partners across the chain is low. The simple representation of the
product flow can be seen in Fig. 10.1.

Figure 10.1 introduces possible problems (can be a delay, insufficient transport, a
change in transport cost, a mismatch between orders requested and order delivered,
etc.) that affect the product flow. More importantly, there are uncertainties about the
demand pattern for the end customers over time. Collectively, such uncertainties
create problems in the entire supply chain (Altomonte et al. 2012). This will lead us
to recognize and study the Bullwhip effect. The Bullwhip effect is an important and
major research field for supply chain management scholars.

Forrester (1961) had formulated the equations describing the association between
inventory and orders. In his paper, it is revealed that the central discrepancy delay
equations describing an inventory responding to a flow in consumer demand can be
solved precisely. Forrester founded the simulation method and recognized the
importance of assimilating information and material flow.

Lee et al. (1997) stated that the Bullwhip effect tends to larger variance than the
consumer demand that activated the orders. In other words, the Bullwhip effect is an
extreme change in the supply position upstream in a supply chain generated by a
small change in demand downstream in the supply chain.

Lee also suggested that the Bullwhip effect causes larger unnecessary inventory
to the upstream than the supply chain's downstream. This is caused by the serial
nature of communicating orders upstream the chain with the various inherent reasons
for moving product downstream the chain.

10.2.2   Empirical Studies on the Bullwhip Effect

Lee et al. (2000) analyzed the Bullwhip effect typically based on the principle of inventory cost minimization or profit maximization. The analysis is established on the reduction in the Bullwhip effect against competing processes, such as permanent inventory deficits or inventory excesses. Conferring to this article in several supply chains, the Bullwhip effect can get up 13–25% of operating costs. The study put forward the managerial impression that the Bullwhip effect can have the main power on firm costs. Besides, knowing these costs have a great prominence for supply chain managers.

Croson and Donohue (2005) tried to study the Bullwhip effect's behavioral perspective using descriptive statistics, time series plot, comparison of variance, and nonparametric sign test. The study was conducted from a simple, serial supply chain constrained to information lags and stochastic demand. This study's data was obtained from a Beer distribution game conducting two experiments over two different sets of participants. The first game was conducted by controlling standard operational causes (e.g., batching, demand estimation, price fluctuations, etc.). The second experiment was conducted by sharing information with the actors in the supply chain. The first experiment confirmed that only by controlling the standard operational causes did the Bullwhip effect exist. The second experiment demonstrated that the Bullwhip effect is underweighted when information on inventory levels is shared. The important recommendation from the study to the supply chain management is that the inventory information helps alleviate the Bullwhip effect by helping upstream chain members better anticipate and prepare for fluctuations in inventory needs downstream.

Dejonckheere et al. (2003) explained that the Bullwhip effect is multiplicative in an old-fashioned supply chain. There is enough evidence in many business backgrounds to verify the use of mathematical models. One of the main reasons used to verify investments in inventories is its character as a barrier. Frequently, it is believed that inventories have a calm down influence on material replenishment. However, a change in inventory management policies can strike at the foundations of the Bullwhip effect by increasing the unpredictability of demand in the supply chain.

Warburton (2004) started inquiries on the Bullwhip effect by setting a strong statement on the Bullwhip effect. The paper claimed that the Bullwhip effect's consequence is problematic for all the companies in the supply chain. The paper used the definition of the Bullwhip effect, set appropriate differential equations from the downstream of the supply chain, and solved these equations' systems to quantify their contributions to inventory amplifications upstream of the supply chain. This study suggested that analytical solutions of differential equations' methods agree with numerical integrations and previous control theory results. The study put forward organizationally important ordering strategies. The differential equation solution on the replenishment rate for the retailer's orders to the manufacturer and the manufacturer's inventory, reacting to the retailer's orders, showed that the

Bullwhip effect emerged naturally. The paper accentuated that the customer demand information to the suppliers has a crucial role in managing the entire supply chain.

10.3   Hypotheses and Their Theoretical Foundations

The main objective of supply chain management or logistics is to minimize inventory and availability of the product on the shelf. According to the DuPont analysis (Groppelli and Nikbakht 2000), inventory minimization has two advantages in improving the company's return on investment (ROI). The first one is decreasing the total cost, and the other is reducing the working capital. Giving credit to the importance of minimizing the actors' inventory across the supply chain, we must look at factors that cause the Bullwhip effect (unnecessary accumulation of inventory).

The theory of supply chain management suggested that the Bullwhip effect's causes are twofold, the Operational effect and the Behavioral effect (Lee et al. 1997, 2000; Bhattacharya and Bandyopadhyay 2011). The operational effect contains factors like demand prediction, order batching, regulating and shortage gambling, price volatility, lead time, inventory strategy, replenishment strategy, inappropriate control structure, lack of transparency, number of echelons, multiplier effect, lack of synchronization, misunderstanding of feedback, local optimization deprived of global visualization, company progression, and capacity bounds. Moreover they argued that the strategic contact of two rational supply chain members could also be a cause of Bullwhip Effect. The behavioral causes contain factors like disregarding the time delays in making ordering pronouncements, lack of learning/ training, and fear of empty stock. More importantly, the authors generalize that the root of all the causes of the Bullwhip effect is the lack of coordination among the supply chain partners (Croson and Donohue 2005).

In supply chain management literature, we found that coordination is divided into two major groups. The first one is the intraorganizational coordination that is further categorized into six groups. Mintzberg (1990) identified the six groups as direct supervision, mutual adjustment, standardization of work process, standardization of output, standardization of skill, and standardization norm. The second one is interorganizational coordination. It is identified that there are three types of organizational coordination (Jain et al. 2006). These are market (pricing theory is the major enabler of coordination), hybrid (vertical coordination through relational contract), and hierarchy (vertical integration). Most of the coordination of business-to-business partnerships is accomplished by hybrid (Williamson 1981).

In the real world, it is observed that companies showed a lack of interorganizational coordination. In literature, we found that some of the reasons for the lack of interorganizational coordination are companies usually want power over the supply chain partner, negative implications of inward bound into partnership, opportunism, lack of trust, promise, and trustworthiness. More prominently, scholars suggest that organizational coordination is difficult upstream of the supply chain. However, the Bullwhip effect is investigated from the downstream of the supply chain (Tempelmeier 2006).

Many studies on the Bullwhip effect give much emphasis on interorganizational coordination. However, this study starts by showing how the Bullwhip effect attacks the supply chain. Then, it focuses on the causality analysis of the Bullwhip effect by considering both intraorganizational coordination and interorganizational coordination. This study's hypotheses are building referencing to the simultaneous depreciation of inventory and availability of the product on the shelf and its managerial implications under the lack of organizational coordination. The Beer Game is played by snubbing organizational coordination (both inter and intra) from corner to corner of the supply chain. Therefore, the null hypotheses of the study are:

Hypotheses 1 Regardless of organizational coordination, the supply chain actors can control and regulate their inventory.

This hypothesis evaluates each business partner's potential to know their inventory's overall status in any organizational (both inter and intra) coordination. If the null hypothesis is accepted, then we induce that the impact of organizational (both inter and intra) coordination on inventory management is insignificant.

Hypotheses 2 Regardless of organizational coordination, the actors across the supply chain can have a similar mean level of inventory.

This hypothesis evaluates the status of the product flow in the supply chain. One of the features of a proper supply chain configuration is the adequate flow of product. If we accept our null hypothesis, then the impact of organizational (both inter and intra) coordination is insignificant for the chain's proper product flow.

Hypotheses 3 Regardless of organizational coordination, the actors across the supply chain can have efficient inventory costs.

Hypotheses 4 Regardless of organizational coordination, the supply chain actors can have efficient stock-out costs.

Hypothesis 3 and hypothesis 4 evaluate each business partners' potential to make effective and efficient inventory management in any organizational (both inter and intra) coordination. If the null hypothesis is accepted, then we induce that organizational (both inter and intra) coordination on effective and efficient inventory management is insignificant.

Hypotheses 5 Regardless of organizational coordination, the actors across the supply chain effectively achieve all the competitive advantage dimensions.

This hypothesis evaluates the success of the supply chain in any organizational (both inter and intra) coordination of the business partners. If the null hypothesis is accepted, then we induce that the impact of organizational (both inter and intra) coordination on the supply chain's success is insignificant.

Hypotheses 6 Primary causes and reasons for the inefficient and ineffective management of inventory are not from the organizational coordination of the supply chain actors.

This hypothesis is the reflection of the potential causes of the Bullwhip effect. Many of the previous studies of the Bullwhip effect address that the cause of Bullwhip effect is lack of proper interorganizational coordination. In this analysis, we include the impact of intraorganizational coordination of the business on the Bullwhip effect. If the business's intraorganizational coordination causes the Bullwhip effect, then the existing theories of organizational coordination will be modified.

10.4   Design of Data and Methodology

10.4.1   The Data

In these modern-day theoretical hypotheses using experimental data becomes remarkably important and growing fast in managerial science. Simulation is an important method of the generation of empirical data to test some theoretical hypothesis. The Systems Dynamics Group at the Massachusetts Institute of Technology (MIT) developed the Beer game in the 1960s. Many intellectuals of supply chain management mentioned that the Beer game's simulation process could demonstrate the Bullwhip effect (Boute et al. 2007).

The Beer Game has a rich history and is the best common simulation and the most extensively used games in various business schools, supply chain electives, and decision-making seminars. Because of its prominence, Simchi-Levi, with other Academicians, developed a computerized version of the Beer Game. Several versions of the Beer Game are currently available, protracted from guidebooks to computerized (high technology) and even web-based descriptions (Simchi-Levi et al. 2000).

This study uses a simple and linearly networked supply chain that contains factory, distributor, wholesaler, and retailer. The game is initiated when the end customer places an order at the retailer of the supply chain. The end customer's demand is given, so the pattern showed a small change, but anonymous to the members. The game players have no prior information about supply chain management, and they play the game randomly. Lack of knowledge of the players' supply chain management makes the observations that come from the absence of interorganizational coordination. The random structure of the game makes the observations that come from the absence of intraorganizational coordination. The game's goal is to minimize the overall logistics costs of the simulated supply chain (Senge 1990). See the data in Annex.

10.4.1.1   Exogenous Variables of the Study

- *Indicial inventory* (X_1): the inventory before order delivered and incoming order.
- *Order delivered* (X_2): the amount ordered by the partner in the supply chain.

- *Incoming order* (X_3): the amount of article revived from the partners in the supply chain.

10.4.1.2   Endogenous Variables of the Study

- *End inventory* (Y): the inventory after order delivered and incoming order.

10.4.2   Methodology

10.4.2.1   One-Way Analysis of Variance (ANOVA)

One-way analysis of variance (ANOVA) is used to see the existences of a certain random variable's main differences with a single treatment over its levels. The linear statistical model for ANOVA is given as (Cochran and Cox 1992; Sanford 1985):

$$y_{ij} = \mu + \alpha_i + \varepsilon_{ij}, i = 1, 2, 3, \ldots, a \text{ and } j = 1, 2, 3, \ldots, n \qquad (10.1)$$

where μ the grand mean of y_{ij}, α_i the ith treatment effect on y_{ij}, and $\varepsilon_{ij} \sim iidN(0, \sigma^2)$.

The bootstrapping estimation method is applied to estimate the model parameters. Usually, the estimation of the model parameters is either using ordinal least square (OLS) or generalized least square (GLS) estimators according to the parameters are fixed or random, respectively. Nevertheless, modern econometric methods used bootstrapping to acquire thorough information about the estimated parameters. In this case, we apply the Bias-Estimation Bootstrap technique. The estimation method gives information about the estimates' bias due to resampling in addition to the estimates of OLS or GLS (Davison and Hinkley 1997). Therefore, the following are the point estimators of the parameters of the model:

$$\text{The estimator of the grand mean} : \widehat{\mu} = \frac{\sum_{i=1}^{a}\sum_{j=1}^{n} y_{ij}}{an} + \text{bias} \qquad (10.2)$$

The estimator of treatment effect with the grand mean : $\widehat{\mu} + \widehat{\alpha}_i$

$$= \frac{\sum_{j=1}^{n} y_{ij}}{n} + \text{bias} \qquad (10.3)$$

The estimator of the difference in the treatment effects : $\widehat{\alpha}_i - \widehat{\alpha}_{i'}$

$$= \frac{\sum_{j=1}^{n} y_{ij}}{n} - \frac{\sum_{j=1}^{n} y_{i'j}}{n} + \text{bias}, \quad i \neq i' \qquad (10.4)$$

10.4.2.2   Signal Processing

Signal processing is a stochastic process of time series data formulated as a series of harmonic functions (Hamilton 1994). Signal processing helps to identify the auto-correlation structure of the time series data. The signal processing stochastic model for a discrete variable is given as (David 2000; Priestley 1991):

$$y_t = \mu_t^* + \sum_k [a_k \cos(2\pi v_k t) + b_k \sin(2\pi v_k t)] \tag{10.5}$$

where μ_t^* is the mean of the series at time t, a_k, b_k (Fourier transformation coefficients of *cosine* and *sine* waves) are independent normal random variables, v_k are distinct frequencies, and k is the index of the summation.

The mean, variance, and covariance of the spectrum of the time series data are derived as follows:

$$E[y_t] = E[\mu_t^*] + E\left\{ \sum_k [a_k \cos(2\pi v_k t) + b_k \sin(2\pi v_k t)] \right\}$$

$$E[y_t] = \mu_t^* + \sum_k E[a_k \cos(2\pi v_k t) + b_k \sin(2\pi v_k t)] \tag{10.6}$$

$$\therefore E[y_t] = \mu_t^*$$

$$\mathrm{Var}[y_t] = E\{[y_t] - E[y_t]\}^2$$

$$\mathrm{Var}[y_t] = E\left\{ \mu_t^* + \sum_k [a_k \cos(2\pi v_k t) + b_k \sin(2\pi v_k t)] - \mu_t^* \right\}^2 \tag{10.7}$$

$$\therefore \mathrm{Var}[y_t] = E\left\{ \sum_k [a_k \cos(2\pi v_k t) + b_k \sin(2\pi v_k t)] \right\}^2$$

$$\mathrm{Cov}[y_t, y_{t-\tau}] = E\{(y_t - E[y_t])(y_{t-\tau} - E[y_{t-\tau}])\} \tag{10.8}$$

where $\quad y_t - E[y_t] = \sum_k [a_k \cos(2\pi v_k t) + b_k \sin(2\pi v_k t)] \quad$ and $\quad y_{t-\tau} - E[y_{t-\tau}] = \sum_k [a_k \cos(2\pi v_k t - \tau) + b_k \sin(2\pi v_k t - \tau)].$

Therefore, Eq. (10.5) can be expressed as:

$$\therefore \mathrm{Cov}[y_t, y_{t-\tau}] = E\left\{ \left(\sum_k [a_k \cos(2\pi v_k t) + b_k \sin(2\pi v_k t)] \right) \right.$$

$$\left. \left(\sum_k [a_k \cos(2\pi v_k t - \tau) + b_k \sin(2\pi v_k t - \tau)] \right) \right\}. \tag{10.9}$$

For simplicity of computation, we assume time as a continuous variable. Furthermore, we assume that the spectrum extends infinitely [$T \in (-\infty, \infty)$] in time in both
directions. Then the covariance of the series is given as:

$$\text{Co-var}[y_t, y_{t-\tau}] = \frac{1}{2T} \int_{-T}^{T} \left(y_t - \mu_t^*\right)\left(y_{t-\tau} - \mu_{t-\tau}^*\right) dt$$

$$= \frac{1}{2T} \int_{-T}^{T} [a_k \cos\left(2\pi v_k t\right) + b_k \sin\left(2\pi v_k t\right)][a_k \cos\left(2\pi v_k t - \tau\right) + b_k \sin\left(2\pi v_k t - \tau\right)] dt$$

$$(10.10)$$

where τ is the order of the lag.

In order to compute the variance, we set $\tau = 0$. Therefore, the variance of the
series is given as follows:

$$\text{Var}[y_t] = \frac{1}{2T} \int_{-T}^{T} [a_k \cos\left(2\pi v_k t\right) + b_k \sin\left(2\pi v_k t\right)]^2 dt \qquad (10.11)$$

Therefore, the autocorrelation function (Auto) is expressed as:

$$\text{Auto}[y_t, y_{t-\tau},] = \frac{\int_{-T}^{T} [a_k \cos\left(2\pi v_k t\right) + b_k \sin\left(2\pi v_k t\right)][a_k \cos\left(2\pi v_k t - \tau\right) + b_k \sin\left(2\pi v_k t - \tau\right)] dt}{\int_{-T}^{T} [a_k \cos\left(2\pi v_k t\right) + b_k \sin\left(2\pi v_k t\right)]^2 dt}$$

$$(10.12)$$

Estimation techniques of spectral density can involve parametric or nonparametric approaches based on time domain or frequency domain analysis. For example, a
common parametric technique consists of fitting the observations to an
autoregressive model. A common nonparametric technique is a periodogram (Stoica
and Moses 2005).

10.4.2.3 Recursive Autoregression: Cochrane–Orcutt Autoregression

Whenever we use time series data, we encounter the problem of autocorrelation. This
problem causes to underestimate the variance of the random error term and increase
the coefficient of determination. As a result, the model information gives false
confidence to the researcher. We specify a reasonable model to avoid such a problem
as:

$$y_t = \beta_0 + \beta_1 x_{2t} + \beta_2 x_{3t} + \varepsilon_t \quad \text{where } \varepsilon_t = \rho \varepsilon_{t-1} + v_t, \ |\rho|$$
$$< 1 \ \text{and} \ v_t ii\tilde{d}N(0, \sigma_v^2) \tag{10.13}$$

where β_0 is the constant of the model, β_1, β_2 is the coefficient of the exogenous imputes x_{2t}, x_{3t}, respectively, ε_t, v_t are random error terms, ρ is the coefficient of autocorrelation, and σ_v^2 is the variance of v_t.

Let us apply the Cochrane–Orcutt transformation (Cochrane and Orcutt 1949) to eliminate the model's autocorrelation.

$$y_t = \beta_0 + \beta_1 x_{2t} + \beta_2 x_{3t} + \varepsilon_t \tag{10.14}$$
$$\rho y_{t-1} = \rho \beta_0 + \rho \beta_1 x_{1t-1} + \beta_2 x_{3t} + \rho \varepsilon_{t-1} \tag{10.15}$$

Subtract Eq. (10.13) from Eq. (10.14) we have:

$$y_t - \rho y_{t-1} = (\beta_0 - \rho \beta_0) + (\beta_1 x_t - \rho \beta_1 x_{t-1}) + (\beta_2 x_{3t} - \rho \beta_3 x_{2t-1}) + (\varepsilon_t - \rho \varepsilon_{t-1})$$
$$y_t^* = \beta_t^* + \beta_1 x_{2t}^* + \beta_2 x_{3t}^* + v_t \tag{10.16}$$

where $y_t^* = y_t - \rho y_{t-1}$, $\beta_t^* = (\beta_0 - \rho \beta_0)$, $x_{2t}^* = (x_t - \rho x_{2t-1})$ and $x_{3t}^* = (x_t - \rho x_{3t-1})$
.

Since $v_t ii\tilde{d}N(0, \sigma_v^2)$ regression equation (10.16) has controlled the autocorrelation of the series. So, we can apply the Ordinary Least Square Estimator (OLS) recursively on regression equation (10.16).

10.5 Results and Discussions

Hypotheses 1 Regardless of organizational coordination, the actors across the supply chain can control and regulate their inventory. To test this hypothesis, we use Eq. (10.3) (see Eq. (10.3)). The result of testing hypothesis 1 is given in Table 10.1.

Table 10.1, we discern that at a 5% level of significance, each of the actors' average end inventories across the supply chain is statistically poles apart from zero. The point estimator's bias of the mean inventory for each actor in the supply chain is negligent. The variability of the end inventories of each actor across the supply chain is remarkably high. Furthermore, the point estimator's bias of the end inventory variance for each actor in the supply chain is considerably increased. That confirms that the Bullwhip effect is much observable in variance than the mean of the ending inventory. That means that in the absence of proper organizational coordination, business partners cannot know the variability of their end inventory. That proves the fundamental character tics of the Bullwhip effect statement claimed by Lee et al. (1999).

Table 10.1 Estimates of the mean and the variance of end inventory of the actors

Actors in the supply chain		Estimate	Bootstrap[a]					95% Confidence interval	
			Bias	Std. error	t-cal	DF	Sig.	Lower	Upper
Factory	Mean	21.0278	−0.0230	4.4025	4.78	35	0.0000*	13.1219	29.9452
	Variance	730.713	−28.271	154.168		35		404.801	1013.508
Distributor	Mean	38.8333	−0.0715	3.2941	11.80	35	0.0000*	32.0029	45.1998
	Variance	385.571	−13.675	79.094		35		217.893	532.545
Wholesaler	Mean	11.6667	−0.0362	2.0170	5.78	35	0.0000*	7.7224	15.6313
	Variance	145.086	−4.270	28.414		35		85.682	199.585
Retailer	Mean	−8.0556	−0.0537	1.7095	−4.71	35	0.0000*	−11.0909	−4.2829
	Variance	97.197	−2.727	27.910		35		43.103	151.422

*Significant at 5% level of significance
[a]Unless otherwise noted, bootstrap results are based on 1000 bootstrap samples

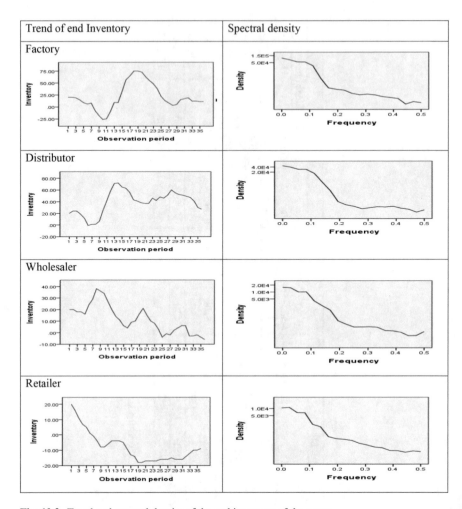

Fig. 10.2 Trend and spectral density of the end inventory of the actors

Since the observation is a time series data, it is not enough to conclude the control and the regulation of the inventory, only from the mean and variance. Still, it is also important to analyze the characteristics of the inventory concerning trends. Figure 10.2 will give us the trend of the ending inventory of the actors across the supply chain.

In addition to the result from Table 10.1, from Fig. 10.2 (left column), we notice that each of the actors' end inventory levels across the supply chain are also different. The inventory management policy of each of the supply chain actors can be evaluated for their ability to have a stable and consistent end inventory. However, from Fig. 10.2 (left column), all the actors have far from having stable and consistent end inventory.

Furthermore, each of the supply chain actors' inventory management policy can be evaluated using spectral density analysis (see Eq. (10.12)). Here strong autocorrelation is one of the indicators of good inventory management policy of the actors. However, from Fig. 10.2 (right column), we observe that the spectral density of the ending inventory of each of the actors of the supply chain is rapidly diminishing with the frequency. That confirms that the ending inventory of the supply chain actors is random rather than well controlled. The overall evidence of the analysis from Fig. 10.2 showed that the inventory management of the supply chain actors is poor.

Hypotheses 2 Regardless of organizational coordination, the actors across the supply chain can have a similar mean level of inventory. To test this hypothesis, we use Eq. (10.4) (see Eq. (10.4)). The hypothesis gives a better insight into quantifying the magnitude and impact of the Bullwhip effect on the different supply chain actors concerning the unnecessary accumulation of inventory. The result of hypothesis 2 is given in Table 10.2.

Table 10.2 shows that at a 5% level of significance, each of the actors' mean inventory across the supply chain is significantly different. Therefore, we reject our null hypothesis. That suggests that without proper organizational (intra and inter) coordination, the supply chain's inventory does not have adequate product flow. More importantly, conjoining the result we found from Table 10.1, we found that the supply chain's mean inventory generally becomes higher in the upstream than the downstream. That proves the Bullwhip effect's basic hypotheses (Lee et al. 1999; Dejonckheere et al. 2003).

In supply chain management literature, it is found that interorganizational coordination difficult in the upstream than the downstream. However, this estimation result confirms that the Bullwhip effect, which is initiated from the downstream highly, attacks the upstream than the downstream of the supply chain regarding the accumulation of unnecessary inventory.

Hypotheses 3 Regardless of organizational coordination, the actors across the supply chain can have an efficient cost of inventory. To test this hypothesis, after we assign zero for when there is stock out conditions, we apply Eq. (10.4) (see Eq. (10.4)). The result of hypothesis 3 is given in Table 10.3.

Table 10.4 shows that at a 5% level of significance, each of the supply chain actors has a significant amount of unnecessary inventory holding cost due to the excessive inventory. Therefore, we reject our null hypotheses. This result endorses that without proper organizational (both intra and inter) coordination, each supply chain actor cannot have an efficiency of inventory cost.

Hypotheses 4 Regardless of organizational coordination, the actors across the supply chain can have efficient stock-out costs. To test this hypothesis, after we assign zero for when there is none-stock out conditions, we apply Eq. (10.4) (see Eq. (10.4)). The result of hypothesis 4 is given in Table 10.4.

Table 10.4 shows that at a 5% level of significance, except for the distributor, each supply chain actor has a significant amount of unnecessary stock-out cost.

Table 10.2 Estimated mean differences of the end inventory of actors

(I) Mean difference of end inventory of actors in the supply chain		Estimate of mean difference $(I - J)$	Bootstrap[a]				95% Confidence interval	
			Bias	Std. error	t-cal	Sig.	Lower	Upper
Factory	Distributor	−17.81	−0.07932	5.6511	−3.15	0.0012*	−28.73	−6.61
	Wholesaler	9.36	−0.12725	4.9540	1.89	0.0315*	−0.68	19.02
	Retailer	29.08	−0.25040	4.8564	5.99	0.0000*	19.43	38.73
Distributor	Wholesaler	27.17	−0.04792	3.7848	7.18	0.0000*	19.82	34.28
	Retailer	46.89	−0.17107	3.5833	13.09	0.0000*	39.41	53.36
Wholesaler	Retailer	19.72	−0.12315	2.6654	7.40	0.0000*	14.44	24.63

*Significant at 5% level of significance
[a]Unless otherwise noted, bootstrap results are based on 1000 bootstrap samples

Table 10.3 Estimated unnecessary holding cost of the actors

Actors in the supply chain		Estimate of inventory holding cost	Bootstrap[a]					95% Confidence interval	
			Bias	Std. error	t-cal	DF	Sig.	Lower	Upper
Factory	Mean	$23.44 \times UC$	0.1028	4.0298	5.82	35	0.0000*	15.72	32.03
	Variance	569.05	−12.470	127.422	4.47	35		304.07	797.58
Distributor	Mean	$38.86 \times UC$	−0.2384	3.2259	12.05	35	0.0000*	32.03	44.59
	Variance	383.32	−6.867	71.507	5.36	35		236.15	516.05
Wholesaler	Mean	$12.36 \times UC$	0.0017	1.8665	6.62	35	0.0000*	8.80	15.98
	Variance	125.21	−3.439	25.353	4.94	35		72.57	173.79
Retailer	Mean	$1.67 \times UC$	−0.0064	0.7503	2.22	35	0.0165*	0.39	3.27
	Variance	21.49	−0.483	10.722	2.00	35		2.52	44.54

UC unit inventory holding cost
*Significant at 5% level of significance
[a]Unless otherwise noted, bootstrap results are based on 1000 bootstrap samples

Table 10.4 Estimated unnecessary stock-out cost of the actors

| Actors in the supply chain | | Estimator | Bootstrap[a] | | | | | 95% Confidence interval | |
			Bias	Std. error	t-cal	DF	Sig.	Lower	Upper
Factory	Mean	2.42	0.0376	1.1723	2.06	35	0.0234*	0.37	5.00
	Variance	45.11	−0.785	22.062	2.04	35		3.46	90.45
Distributor	Mean	0.03	0.0000	0.0270	1.03	35	0.1554	0.00	0.09
	Variance	0.03	−0.001	0.026	1.09	35		0.00	0.09
Wholesaler	Mean	0.69	0.0029	0.2427	2.86	35	0.0035*	0.26	1.21
	Variance	2.22	−0.060	0.839	2.64	35		0.68	3.86
Retailer	Mean	9.72	0.0666	1.0459	9.30	35	0.0000*	7.61	11.89
	Variance	42.38	−1.213	5.651	7.50	35		29.68	51.96

*Significant at 5% level of significance
[a]Unless otherwise noted, bootstrap results are based on 1000 bootstrap samples

Therefore, we reject our null hypotheses. These results confirm that each of the supply chain actors cannot have efficient stock-out costs without coordination. That demonstrates that demand management and customer relationship management of the actors in the supply chain is poor.

Table 10.3 shows that the Bullwhip effect attacks the upstream than the supply chain's downstream concerning inventory holding cost. That resembles the Bullwhip effect attacks more the upstream of the supply chain. However, the stock-out case is higher downstream than upstream of the supply chain. Therefore, all the supply chain actors will suffer from the Bullwhip effect to achieve their goals.

Hypotheses 5 Regardless of organizational coordination, the actors across the supply chain effectively achieve all the competitive advantage dimensions. The competitive advantage of the supply chain is measured from the dimensions of quality (of service or product), cost efficiency (reduction of overall production and logistics cost), and time (availability of the service or product during the order of the customer).

Table 10.1 shows that the accumulations of the inventory of the actors downstream are significant. Such accumulations of inventories will increase the probability of damage, defect, or out-date the product and/or backorder. The occurrence of damage, defect, or outdated effects on the product will decrease the competitive advantage's quality dimension.

Conjoining the results of Tables 10.3 and 10.4, we see that the supply chain has unnecessary holding cost + stock-out cost + backorder cost + other indirect costs. The existence of such unnecessary costs has an important implication of the theory of supply chain management. The supply chain is cost-efficient; this will increase the probability of increasing the product's price to the final customer. Consequently, the price dimension of the competitive advantage is most likely reduced.

From Table 10.4, we see that the stock-out case of the retailer is statistically significant. That implies that such a supply chain cannot meet the time dimension of the competitive advantage. Therefore, we reject our null hypothesis. That shows that the actors across the supply chain cannot achieve all the competitive advantage dimensions without organizational coordination.

Hypotheses 6 Primary causes and reasons for the inefficient and ineffective management of inventory are not from the organizational coordination of the supply chain actors. To test this hypothesis, we apply Eq. (10.14) (see Eq. (10.16)).

Here we try to explain how the supply chain showed poor customer relationship management, demand management, inventory management, and so on. Our hypotheses reflect the identification of the causes of each actor of the supply chain's inventory. We can think that there are primarily two different reasons. These are lack of intraorganizational coordination or interorganizational coordination of the actors of the supply chain. If the cause is the lack of intraorganizational coordination, then the actors' inventory is only explained only by their mean and variance. That means we cannot find any exogenous variable from the other actors of the supply chain that cause the inventory accumulation. Otherwise, if the cause is because of a

Table 10.5 Estimators of causality analysis of the accumulation of end inventory of the actors

Exogenous variable of ending inventory of the actors in the supply chain	Unstandardized coefficients		Standardized coefficients			Coefficient of determination after recursion estimation
	B	Std. error	Beta	T	Sig	
Factory order delivered	−0.901	0.299	−0.467	−3.017	0.005*	0.507
Factory incoming order	0.284	0.225	0.195	1.263	0.216	
(Constant)	19.156	28.413		0.674	0.505	
Distributor order delivered	−0.425	0.238	−0.284	−1.785	0.084	0.464
Distributor incoming order	0.588	0.245	0.382	2.397	0.023*	
(Constant)	40.978	14.443		2.837	0.008*	
Wholesaler order delivered	0.064	0.366	0.025	0.175	0.862	0.604
Wholesaler incoming order	0.549	0.132	0.600	4.148	0.000*	
(Constant)	−16.012	19.843		−0.807	0.426	
Retailor order delivered	−0.098	0.286	−0.054	−0.342	0.735	0.489
Retailor incoming order	0.501	0.161	0.489	3.117	0.004*	
(Constant)	−17.157	3.234		−5.306	0.000*	

*Significant at 5% level of significance

lack of interorganizational coordination of the actors across the supply chain, we will find at least one exogenous variable that can affect each (some) of the actors' inventory in the supply chain. Therefore, the test of this hypothesis has important implications in the theory of supply chain management.

The result of the causality analysis using the Cochrane–Orcutt recursive autoregression is given in Table 10.5.

Interpretations of Table 10.5

From Table 10.5, we observe that the constant of the distributor and the retailer is statistically significant. That shows the intraorganizational coordination of these two actors is poor. That is one of the causes of the distributor's highest inventory accumulation and causes a loss of the time dimension of the competitive advantage of the retailer's supply chain. The exogenous variable order delivered significantly affects the factory's inventory, and the exogenous variable incoming order affects the distributor, wholesaler, and retailer stocks. That shows the inter-coordination of all the actors in the supply chain is poor.

Therefore, we reject our hull hypotheses. We generalize the finding of the causality analysis that the Bullwhip effect exists because of the lack of intraorganizational and interorganizational coordination of the chain actors.

10.5.1 Implications to Theory of Organizational Coordination

The finding of this study confirmed that the intra-organization coordination of the business partner in the supply chain is one of the potential causes of the Bullwhip effect. That result showed that coordination with business partners with poor intraorganizational coordination could cost the other firms good intraorganizational coordination. That has several implications on the drawbacks of the existing theories of organizational coordination.

One of the most influential coordination theories is Williamson's (1981) transaction cost analysis (TCA). The TCA hypothesis tried to give the solution for making efficient coordinating of firms among market (pricing theory is the major enabler of coordination), hybrid (vertical coordination through relational contract), and hierarchy (vertical integration). The TCA used a certain dimension to analyze making efficient coordination. These dimensions are asset specificity, uncertainty, and transaction frequency. For example, the TCA solution suggests that as asset specificity, uncertainty, and transaction frequency increase the making vertical integration (hierarchy) more efficient than hybrid governance.

10.5.2 Critics Toward Transaction Cost Analysis (TCA)

The TCA assumed that transactors (firms) are "opportunistic" and "bounded rationally." These two behavioral assumptions are a tool to develop the TCA. However, TCA ignores the dynamic aspect of these two behavioral factors. Transactors may be opportunist or bounded rationally. However, another factor is directly interacting with these two important factors, that is, "experience." Experience is a dynamic factor and can make firms increase or decrease their "opportunistic" or "rational bounded" behavior according to their benefit from their previous organizational coordination.

Suppose let us consider ideal hybrid coordination that controls TCA aspects, that is, asset specificity, uncertainty, and transaction frequency. At a given opportunist and rational bounded behavior firms, assume that they made efficient coordination; hence their coordination controlled all TCA dimensions. Now let us ask ourselves a question that "do this coordination is the most efficient coordination for the transactors?" The TCA argues that such coordination is the most efficient coordination for the transactors. However, the analysis of the Bullwhip effect (see the result of hypotheses 6) prevails that even under the fulfillment of the TCA's dimensions, the most efficient coordination is not achievable. Hence, if at least one of the transactors has poor intraorganizational coordination, all the transactors will become cost-efficient due to the Bullwhip effect.

Let us consider the following case. To illustrate how "experience" affects organizational coordination, firms may not realize coordinating with another firm having bad intraorganizational coordination in a single transaction. They may recognize that through time (experience). In this aspect, firms prefer to coordinate with another firm having good intraorganizational coordination. Otherwise, if there is supplier risk, firms attack their opportunistic behavior, and they recognize the benefits of their business partner have a positive impact on making efficient coordination. They become willing to assist their business partner in modifying its intraorganizational coordination.

Suppose by considering the dimensions of (i.e., asset specificity, uncertainty, and transaction frequency). The hierarchy is the solution from TCA for a particular organizational coordination problem. However, there are well-known pieces of evidence and reasons that hierarchy may not be realized. Therefore, this leads the firms to coordinate in the relational contract (hybrid governance). However, we have identified that coordinating with firms with poor intraorganizational coordination will cause an ineffective and/or inefficient supply chain. Therefore, another type but important coordination is not discussed by the TCA.

10.5.3   The Hyper-Hybrid Coordination

The causality analysis of the Bullwhip effect introduces another type of organizational coordination called hyper-hybrid governance. This governance has the following characteristics.

1. It is a relational contract.
2. It recognizes that one firm's benefit in the supply chain will benefit all the actors in the supply chain. That is, coordinating with firms having good intraorganizational coordination is preferable to coordinating with firms having poor intraorganizational coordination.
3. Appropriate when supplier risk is high.
4. The coordination extends to recommend, consult, and help firms establish appropriate means of intraorganizational coordination.

10.5.4   The Kraljic–Tesfay Portfolio Matrix

Kraljic (1983) purchasing portfolio model is the dominant solution for firms to coordinate with their business partners effectively and efficiently. The model contains all the advantages of the transaction cost economics, resource dependence theory, pricing theory of markets, and relational contract theory. The two important dimensions of the Kraljic (1983) purchasing portfolio models are the purchasing impact on the firm's bottom line (profitability) and the supplier risk. However, the Kraljic matrix was incomplete in addressing coordination inefficiency and ineffectiveness due to the supplier's intraorganizational coordination. Therefore, to make efficient and effective coordination by considering the supplier's intraorganizational

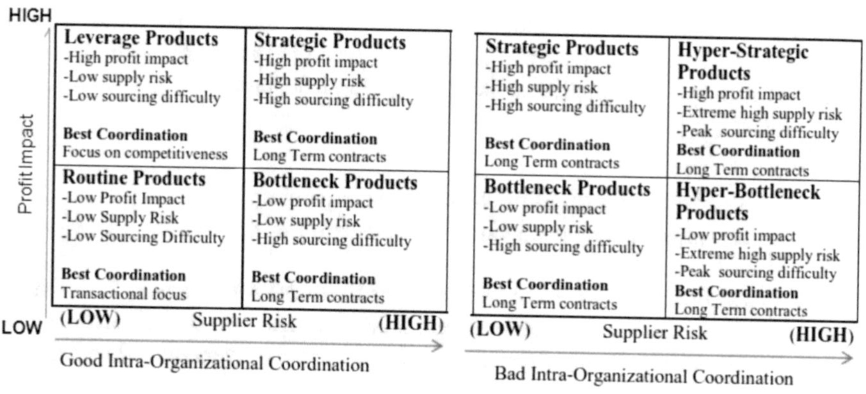

HIGH

Leverage Products	Strategic Products
-High profit impact	-High profit impact
-Low supply risk	-High supply risk
-Low sourcing difficulty	-High sourcing difficulty
Best Coordination	**Best Coordination**
Focus on competitiveness	Long Term contracts
Routine Products	**Bottleneck Products**
-Low Profit Impact	-Low profit impact
-Low Supply Risk	-Low supply risk
-Low Sourcing Difficulty	-High sourcing difficulty
Best Coordination	**Best Coordination**
Transactional focus	Long Term contracts

(LOW)   Supplier Risk   (HIGH)

Good Intra-Organizational Coordination

Strategic Products	Hyper-Strategic Products
-High profit impact	-High profit impact
-High supply risk	-Extreme high supply risk
-High sourcing difficulty	-Peak sourcing difficulty
Best Coordination	**Best Coordination**
Long Term contracts	Long Term contracts
Bottleneck Products	**Hyper-Bottleneck Products**
-Low profit impact	-Low profit impact
-Low supply risk	-Extreme high supply risk
-High sourcing difficulty	-Peak sourcing difficulty
Best Coordination	**Best Coordination**
Long Term contracts	Long Term contracts

(LOW)   Supplier Risk   (HIGH)

Bad Intra-Organizational Coordination

Profit Impact (vertical axis label, LOW to HIGH)

Fig. 10.3 The Kraljic–Tesfay Portfolio Matrix

coordination and hyper-hybrid governance, we will modify the existing Kraljic purchasing portfolio model.

Many dimensions can analyze intraorganizational coordination. However, in this portfolio modeling, we categorized the supplier's intraorganizational coordination into two dimensions, good intra-organizational coordination, and bad intraorganizational coordination. The modified model of the Kraljic portfolio matrix by the inclusion of two intraorganizational coordination classes is given in Fig. 10.3.

10.5.4.1   The Best Coordination with Supplier Having Adequate-Intraorganizational Coordination

If all the suppliers have good intraorganizational coordination, the existing Kraljic-Portfolio model is still the best way of making effective and efficient coordination. Therefore, based on the current model, we have the following four approach classifications:

Routine Product: these purchasing approaches are appropriate when neither the supplier risk nor purchasing impact on its bottom line is high. In this approach, the company worries about minimizing the logistical cost and make e-sourcing with the supplier the most appropriate strategy.

Leverage Product: these purchasing approaches are appropriate when there is no supplier risk but purchasing impact on the company's bottom line is high. In this approach, making competitive bidding of the supplier is the most appropriate strategy.

Bottleneck Product: these purchasing approaches are appropriate when supplier risk is high and purchasing impact on the company's bottom line is low. In this approach, making a partnership with the supplier is the most appropriate strategy.

Strategic Product: these purchasing approaches are appropriate when both supplier risk and purchasing impact to its bottom line are high. In this approach, making a partnership with the supplier is the most appropriate strategy.

10.5.4.2   The Best Coordination with Supplier Having Inadequate-Intraorganizational Coordination

If some (or all) the suppliers have bad intraorganizational coordination, the existing Kraljic-Portfolio model is no more the guideline for effective and efficient coordination. Therefore, we modify the existing Kraljic-Portfolio by the model as follows.

Bottleneck Product: these purchasing approaches are appropriate when the purchasing impact on the company's bottom line is low, and the supplier risk is absent. However, there is heterogeneity in the intraorganizational coordination of the supplier. In this aspect, to make effective and efficient coordination, the buyer must set criteria about intraorganizational coordination for the suppliers. This criterion causes to maximizes the supplier risk. Therefore, the products categorized as "Routine Products" in the Kraljic-Portfolio model are transformed into "Bottleneck Products." Consequently, making a partnership with the supplier is the most appropriate strategy.

Strategic Product: these purchasing approaches are appropriate when the purchasing impact on the company's bottom line is high, and the supplier risk is absent. However, there is heterogeneity in the intraorganizational coordination of the supplier. In this aspect, to make effective and efficient coordination, the buyer has to set criteria for the suppliers' intraorganization coordination. This criterion causes to maximizes the supplier risk. Therefore, the products categorized as "Leverage Products" in the Kraljic-Portfolio model are transformed into "Strategic Product." Consequently, making a partnership with the supplier is the most appropriate strategy.

Hyper-Bottleneck Product: these purchasing approaches are appropriate when supplier risk is high and purchasing impact on the company's bottom line is low. Furthermore, the intraorganizational coordination of the suppliers is bad. To differentiate this product type from the Kraljic-Portfolio model of bottleneck products has given a new name as "hyper-bottleneck products." In this aspect, making only a partnership with the supplier is not an effective and efficient coordination method. This coordination is more efficient and effective if the intraorganizational coordination of the supplier has modified through the business partners' assistance. Therefore, this coordination is more than a partnership. Hence the business partners are needed to support each other to adjust their interorganizational coordination for their mutual benefit.

Hyper-Strategic Product: these purchasing approaches are appropriate when supplier risk is high and purchasing impact on the company's bottom line is high. Furthermore, the intraorganizational coordination of the suppliers is bad. To differentiate this product type from the Kraljic-Portfolio model of Strategic

Products has given a new name as "Hyper-Strategic Products." In this aspect, making only a partnership with the supplier is not an effective and efficient coordination method. This coordination is more efficient and effective if the intraorganizational coordination of the supplier is modified through the business partners' assistance. Therefore, this coordination is more than a partnership. Hence the business partners need to support each other to adjust their interorganizational coordination for their mutual benefit.

10.6 Conclusions and Recommendations

10.6.1 Conclusions

This paper proposed whether actors' coordination across the supply chain makes the supply chain achieve its goals. Based upon applying different stochastic models on the data obtained from the Beer distribution game, we found that without coordination, the supply chain could not have effective and efficient inventory management, demand management, and customer relationship management.

The statistical analyses confirm that the Bullwhip effect has different cost implications for various supply chain actors concerning inventory. The inventory holding cost is higher upstream than downstream of the supply chain. On the other hand, the analyses also confirm that the Bullwhip effect has different stock-out cases for the supply chain actors. The stock-out point is higher in the downstream than the upper stream of the supply chain. These important findings suggest that all the actors of the supply chain will suffer from the Bullwhip effect.

When we evaluate the supply chain under the conditions of the lack of coordination, we found that such supply chains could not achieve all the dimensions of competitive advantages. The study identified that both the lack of intraorganizational and interorganizational coordination could cause the Bullwhip effect.

This study suggests that the scope of the supply chain management is structured from establishing proper interorganizational coordination of the business partners. Then extended to adjusting the intraorganizational coordination of each of the business partners proactively responsive according to the end customer's stochastic demand.

10.6.2 Recommendations

One of the important contributions of supply chain management is profitability by reducing the overall cost of the chain and satisfying the end customer's need. Each company has to identify its internal problem and establish appropriate, flexible, dynamic, and responsive intraorganizational coordination.

Furthermore, by understanding the negative impact of the lack of coordination, companies in the real world must identify their partners and properly coordinate for the joint profit.

There are important theories that are helping companies on how to coordinate with each other. The resource dependence theory deals how making effective coordination (Davis and Cobb 2010). Transaction cost economics deals with how making cost-efficient coordination (Williamson 1981). Relational contract theory deals with attacking opportunism and creating trust among the business collaborates (Uchida 1993).

However, supply chain management is a young and growing science that tried to link the companies from the upstream (from raw material) and the downstream (up to the end customer). Connecting all the chain participants into one direction for the same common objective is much more difficult than property managing a single organization. The difficulty of coordination is becoming complicated, as the network of the supply chain is large. Therefore, it is time for different professionals (management, economics, finance, information technology, psychology, mathematics, statistics, and others) to provide helpful theories of coordination in supply chain management.

When we consider several supply chains, the resource, which is lost due to the lack of coordination at a country level, is significant. That shows the lack of proper coordination of each company within their chain has negative implications for the given country's welfare. Therefore, governments must focus on the antitrust policy of coordination.

Annex: Data from the Beer Distribution Game

Observations	Factory				Distributor				Wholesaler				Retailer			
Week	X_1	X_2	X_3	Y	X_1	X_2	X_3	Y	X_1	X_2	X_3	Y	X_1	X_2	X_3	Y
1	20	4	4	20	20	4	4	20	20	4	4	20	20	4	4	20
2	20	4	4	20	20	4	8	24	20	4	4	20	20	8	4	16
3	20	5	4	19	24	4	4	24	20	6	4	18	16	9	4	11
4	19	8	4	15	24	9	4	19	18	4	4	18	11	8	4	7
5	15	10	4	9	19	12	5	12	18	6	4	16	7	8	4	5
6	9	8	5	6	12	17	4	−1	16	4	12	24	5	8	6	1
7	6	6	8	8	−1	8	10	1	24	4	9	29	1	8	4	−1
8	8	20	4	−8	1	8	8	1	29	8	17	38	−1f	7	4	−4
9	−8	18	8	−18	1	1	6	6	38	10	8	36	−4	8	4	−8
10	−18	18	10	−26	6	1	20	25	36	10	8	34	−8	8	8	−8
11	−26	15	16	−25	25	2	18	41	34	8	1	27	−8	8	10	−6
12	−25	5	20	−10	41	1	18	58	27	8	1	20	−6	8	10	−4
13	−10	2	21	9	58	2	15	71	20	8	2	14	−4	8	8	−4
14	9	7	7	9	71	4	5	72	14	4	1	11	−4	8	8	−4
15	9	7	30	32	72	9	2	65	11	7	2	6	−4	9	8	−5
16	32	0	25	57	65	9	7	63	6	6	4	4	−5	8	4	−9
17	57	0	8	65	63	14	7	56	4	4	9	9	−9	8	4	−13
18	65	0	10	75	56	13	0	43	9	8	9	10	−13	7	6	−14
19	75	1	1	75	43	2	0	41	10	8	14	16	−14	8	4	−18
20	75	3	1	73	41	3	0	38	16	8	13	21	−18	8	8	−18
21	73	10	0	63	38	8	7	37	21	8	2	15	−18	7	8	−17
22	63	8	0	55	37	3	3	37	15	8	3	10	−17	8	8	−17
23	55	7	0	48	37	1	10	46	10	10	8	8	−17	8	8	−17

(continued)

Observations	Factory				Distributor				Wholesaler				Retailer			
Week	X_1	X_2	X_3	Y	X_1	X_2	X_3	Y	X_1	X_2	X_3	Y	X_1	X_2	X_3	Y
24	48	10	0	38	46	12	8	42	8	8	3	3	−17	8	8	−17
25	37	15	0	22	42	1	7	48	3	8	1	−4	−17	9	10	−16
26	23	20	10	13	48	12	10	46	−4	9	12	−1	−16	8	8	−16
27	13	5	0	8	46	10	15	51	−1	8	7	−2	−16	8	8	−16
28	8	5	0	3	51	11	20	60	−2	8	12	2	−16	8	9	−15
29	5	0	0	5	60	10	5	55	2	8	10	4	−15	9	8	−16
30	5	0	10	15	55	8	5	52	4	9	11	6	−16	8	8	−16
31	15	2	4	17	52	1	0	51	6	10	10	6	−16	8	8	−16
32	17	0	2	19	51	2	0	49	6	10	1	−3	−16	7	9	−14
33	19	7	0	12	49	4	2	47	−3	8	8	−3	−14	8	10	−12
34	12	0	0	12	47	6	0	41	−3	8	9	−2	−12	8	10	−10
35	12	1	0	11	41	12	1	30	−2	6	4	−4	−10	8	8	−10
36	11	0	0	11	30	3	0	27	−4	8	6	−6	−10	7	8	−9

Bibliography

C. Altomonte, F.D. Mauro, G. Ottaviano, A. Rungi, V. Vicard, *Global Value Chains During the Great Trade Collapse: A Bullwhip Effect?* Working Paper Series 1412 (European Central Bank, Frankfurt, 2012)

R. Bhattacharya, S. Bandyopadhyay, A review of the causes of Bullwhip effect in a supply chain. Int. J. Adv. Manuf. Technol. **54**(9–12), 1245–1261 (2011)

R.K. Bhote, *Supply Management: How to Make US Suppliers Competitive* (American Management Association Membership Publications Division, New York, NY, 1987)

D. Blanchard, *Supply Chain Management Best Practices*, 2nd edn. (John Wiley and Sons, Chichester, 2010)

S. Boon-itt, C. Pongpanarat, Measuring service supply chain management processes: The application of the Q-sort technique. Int. J. Innov. Manag. Technol. **2**(3), 373–376 (2011)

R.N. Boute, S.M. Disney, M.R. Lambrecht, B. Van Houdt, An integrated production and inventory model to dampen upstream demand variability in the supply chain. Eur. J. Oper. Res. **178**(1), 121–142 (2007). https://doi.org/10.1016/j.ejor.2006.01.023

B. Boashash, M. Mesbah, P. Colitz, Time-frequency detection of EEG abnormalities. Time-frequency signal analysis and processing: A comprehensive reference, 663–670 (2003)

D.N. Burt, D.W. Dobler, S.L. Starling, *World Class Supply Management: The Key to Supply Chain Management*, 7th edn. (McGraw-Hill Irwin, New York, NY, 2003)

S. Chopra, P. Meindl, *Supply Chain Management: Strategy, Planning, and Operations, Chapter 1* (Prentice-Hall, Inc., Upper Saddle River, NJ, 2001)

M. Christopher, *Logistics and Supply Chain Management; Strategies for Reducing Cost and Improving Service*, 2nd edn. (Prentice-Hall, Inc., Upper Saddle River, NJ, 1998)

W.G. Cochran, G.M. Cox, *Experimental Designs*, 2nd edn. (Wiley, New York, NY, 1992)

D. Cochrane, G.H. Orcutt, Application of least squares regression to relationships containing autocorrelated error terms. Journal of the American Statistical Association **44**, 32–61 (1949)

R. Croson, K. Donohue, Behavioural causes of the Bullwhip effect and the observed value of inventory information. Manag. Sci. **52**(3), 323–336 (2005)

G.F. Davis, J.A. Cobb, Resource dependence theory: Past and future, in *Stanford's Organization Theory Renaissance*, (Emerald Group, Bingley, NY, 2010)

A.C. Davison, D.V. Hinkley, *Bootstrap Methods and Their Applications* (Cambridge University Press, Cambridge, UK, 1997)

J. Dejonckheere, S.M. Disney, M.R. Lambrecht, D.R. Towill, Measuring the Bullwhip effect: A control theoretic approach to analyse forecasting induced bullwhip in order-up-to policies. Eur. J. Oper. Res. (2003). https://doi.org/10.1016/S0377-2217(02)00369-7

G.G. Dess, G.T. Lumpkin, M.L. Taylor, *Strategic Management*, 2nd edn. (McGraw-Hill Irwin, New York, NY, 2005)

P. Drucker, *The Practice of Management* (Harper and Row Publishers, Inc, New York, NY, 1954)

L. Ellram, The supplier selection decision in strategic partnerships. J. Purch. Mater. Manag. **26**, 8–14 (1990)

S. Engelberg, *Digital Signal Processing: An Experimental Approach* (Springer, Berlin, 2008)

H. Escaith, N. Lindenberg, S. Miroudot, *International Supply Chains and Trade Elasticity in Times of Global Crisis, MPRA Paper* (University Library of Munich, Munich, Germany, 2010)

R.C. Feenstra, H.H. Gordon, Aggregation bias in the factor content of trade: Evidence from U.S. manufacturing. Am. Econ. Rev. **90**(2), 155–160 (2000)

J.W. Forrester, *Industrial Dynamics* (MIT Press, Cambridge, MA, 1961)

J. Frieden, D. Lake, *International Political Economy: Perspectives on Global Power and Wealth* (Routledge, London, 1995)

R. Ganeshan, T.P. Harrison, *An Introduction to Supply Chain Management* (Department of Management Sciences and Information Systems, Beam Business Building, Penn State University, University Park, PA, 1995)

A.A. Groppelli, E. Nikbakht, *Finance*, 4th edn. (Barron's Educational Series, Inc, New York, NY, 2000)

J.M. Ivancevich, J.H. Donnelly Jr., J.L. Gibson, *Management: Principles and Functions*, 4th edn. (BPI Irwin, Boston, MA, 1989)

K. Jain, L. Nagar, V. Srivastava, Benefit sharing in inter-organizational coordination. Supply Chain Manag. **11**(5), 400–406 (2006)

L.R. Kopczak, M.E. Johnson, The supply-chain management effect. MIT Sloan Manag. Rev. **44**(3), 27 (2003)

P. Kraljic, Purchasing must become supply management. Harv. Bus. Rev. **61**(5), 109–111 (1983)

P.D. Larson, A. Halldarsson, What is SCM? and where is it? J. Supply Chain Manag. **38**(4), 36–44 (2002). https://doi.org/10.1111/j.1745-493X.2002.tb00141.x

H.L. Lee, V. Padmanabhan, S. Whang, The Bullwhip effect in supply chains. Sloan Manag. Rev. **38** (3), 93–102 (1997)

H.L. Lee, K.C. So, C.S. Tang, The value of information sharing in a two-level supply chain. Manag. Sci. **46**(5), 626–643 (2000). https://doi.org/10.1287/mnsc.46.5.626.12047

R.R. Lummus, R.J. Vokurka, Defining supply chain management: A historical perspective and practical guidelines. Indus. Manag. Data Syst. **99**(1), 11–17 (1999)

K. Matsuyama, A Ricardian model with a continuum of goods under nonhomothetic preferences: Demand complementarities, income distribution, and north-south TRADE. J. Polit. Econ. **108** (6), 1093–1120 (2000). https://doi.org/10.1086/317684

P. Milgrom, J. Roberts, *Economics, Organization and Management* (Prentice-Hall, Inc., Englewood Cliffs, NJ, 1992) ISBN 978-0-13-224650-7

H. Mintzberg, The Design School: Reconsidering the basic premises of strategy formation. Strateg. Manag. J. **11**(3), 171–195 (1990). https://doi.org/10.1002/smj.4250110302

M.B. Priestley, *Spectral Analysis and Time Series* (Academic Press, London, 1991)

P.M. Senge, *The Fifth Discipline: The Art and Practice of the Learning Organization* (Currency Doubleday, New York, NY, 1990) 423 pp

Y. Shiozawa, A new construction of Ricardian trade theory: A many-country, many-commodity with intermediate goods and choice of techniques. Evolut. Inst. Econ. Rev. **3**(2), 141–187 (2007). https://doi.org/10.14441/eier.3.141

D. Simchi-Levi, P. Kaminsky, E. Simchi-Levi, *Designing and Managing the Supply Chain. Concepts, Strategies, and Case Studies*, vol 2000 (Irwin McGraw-Hill, New York, NY, 2000)

P. Stoica, R. Moses, *Spectral Analysis of Signals* (Prentice Hall, Inc., Upper Saddle River, NJ, 2005) 07458

H. Tempelmeier, *Inventory Management in Supply Networks Problems, Models, Solutions* (Books on Demand, Norderstedt, 2006)

T. Uchida, The New Development of Contract Law and General Clauses – A Japanese Perspective in the Organizing Committee (Eds), Proceedings of the Symposium: Japanese and Dutch Laws Compared (1993)

UN, Globalization for Development: The International Trade Perspective (2008), http://unctad.org/en/Docs/ditc20071_en.pdf

R.D.H. Warburton, An analytical investigation of the Bullwhip effect. Prod. Oper. Manag. **13**(2), 150–160 (2004)

O.E. Williamson, The economics of organization: The transaction cost approach. Am. J. Sociol. **87** (3), 548 (1981). https://doi.org/10.1086/227496

Chapter 11
Industrial Applications of Tesfay Process and Tesfay Coordination

11.1 Introduction

Contemplate multiple businesses collaborate with companies operating as a serial chain for product development for the specific end customer. Many scholars suggested that product development and end customer satisfaction optimized if the multiple companies managed as a single chain. That leads scholars to introduce supply chain management (SCM) in the late 1980s (Tesfay 2014).

The SCM is designed to fill the gaps in business performance after considering two core ideas. (1) Practically, products that deliver an end customer characterize various organizations' aggregate work, which brings up a supply chain. (2) Notwithstanding supply chains have been real, most organizations have only paid attention to their intraorganizational coordination (David 2010; Tesfay 2015a, b).

SCM is vigorous systematic operational and strategic administration of supply chain activities to take full advantage of the end customer and realize a maintainable economic and modest benefit. SCM is a complicated function of several business factors, which the supply chain accomplishments involve in each stage of product development, entire logistic operations, and information systems (Bhote 1987; Burt et al. 2003; Dess et al. 2005; Ganeshan and Harrison 1995; Larson and Halldarsson 2002).

SCM scholars remark that one of the characteristics of an effective and efficient supply chain is the degree of organizational harmonization of the companies in the chain and its competence according to the end customer's pattern. However, in real business operations, the end customer's design is dynamic and random due to voluminous convincing factors. That randomness of the customer pattern is a fundamental challenge of modern SCM (Boonitt and Pongpanarat 2011; Tesfay 2014).

Referring to DuPont analysis, the SCM strategy design to concurrently decreasing both inventory and operating costs. Correspondingly, SCM methods' formulation and structure integrate production, inventory, location, transportation, and

Y. Y. Tesfay, *Developing Structured Procedural and Methodological Engineering Designs*, https://doi.org/10.1007/978-3-030-68402-0_11

information about the end customer's random pattern. Subsequently, SCM comprehends challenging and in-depth exploration of the chain under circumstances of uncertainty (Tesfay 2015a, b; Chopra and Meindl 2001).

11.2   The Hypotheses

The SCM theory predicts effective and efficient material low can instantaneously reduce inventory and accessibility of the product or service to end customer. Scholars of SCM suggest that applicable information flow through the supply chain is achieved via establishing proper interorganizational coordination. However, the degree of demand fluctuation challenges the entire supply chain (Ellram 1990; Kopczak and Johnson 2003; Tesfay 2015a, b).

DuPont analysis predicts optimization of inventory has two benefits to improve the company's return on investment (ROI), reducing (1) the overall cost and (2) the working capital (Groppelli and Ehsan 2000). Giving tribute to reducing the inventory of business partners across the supply chain, we emphasize factors that root the Bullwhip effects, which take along superfluous accumulation of stock.

According to Bhattacharya and Bandyopadhyay (2011), Croson and Donohue (2005), and Lee et al. (1997, 2000), figured out that the foundations of the Bullwhip effects are lack of interorganizational coordination of the supply chain business partners.

In SCM literature, we found that organizational coordination is divided into two major groups, intraorganizational (Mintzberg 1990) and interorganizational (Jain et al. 2006) coordination. This paper is an extension of the Tesfay (2014, 2015a, b) papers on Bullwhip effects and the in-depth analysis of the sixth hypothesis of the Tesfay (2015a, b) paper. In detail, we try to give the clarifications of the following hypothesis.

Hypothesis 1 Many SCM scholars suggested that the Bullwhip's primary cause is the deficiency of inter-organizational coordination of the chain's business partners. Nevertheless, suppose let us assume an ideal supply chain that all the business partners in the chain have proper interorganizational coordination. Thus, this hypothesis is concerned with the prospective bases of the Bullwhip effects. Here are central interrogations about "do having proper inter-organizational coordination imply all the business partners in the supply chain control the Bullwhip effects?" In this honor, we propose to analyze whether the Bullwhip effects are caused through the intraorganizational coordination at least one of the business collaborates in the supply chain or not (Hypothesis 1 leads us to a new way of understanding the challenges of modern supply chain analysis).

Hypothesis 1 echoes the role of Mintzberg's (1990) classification and practices of apposite intraorganizational coordination of the business partners on the productivity of the supply chain.

Hypothesis 2 If hypothesis 1 is rejected, that is, if the Bullwhip effect can cause intraorganizational coordination at least one of the business collaborates in the chain, then what are its implications on the prevailing theory of the transaction cost analysis (TCA). That is, our examination will give solutions about the organizational coordination types:

Market (rule of the demand-supply theory is the best means coordination)
Hybrid (coordination over and done with relational contract)
Hierarchy (vertical integration or merger) an adequate amount of more productive coordination? (Williamson 1981)

If not, we intend to diagnose the TCA by bringing together new organizational coordination (since working from 1981, Williamson worked with TCA and won the Nobel Prize in Economics in 2009).

Hypothesis 2 leads us to search and innovative method/s to solve complex challenges in forming an effective and efficient supply chain.

Hypothesis 3 If we set up additional organizational coordination, which can diagnose the TCA, does the Kraljic (1983) portfolio model be effective and efficient organizational coordination? If not, how the existing Kraljic (1983) portfolio model is going to be adapted to make more effective and efficient coordination?

Hypothesis 3 is concerned with analyzing the sufficiency of the Kraljic portfolio model in modern organizational coordination.

The paper evaluates the potential causes of the Bullwhip effects using various panel data regression models and structural equations of seemingly unrelated regression (SUR) models on the experimentally simulated data from the Beer distribution game.

11.3 Literature Review

In 1961, Forrester had introduced mathematical differential equations to analyze the relationships between supply chain overall inventory and customer order. The analysis precisely solved the fundamental discrepancy delay equations enlightening an inventory responding to a consumer order flow. Forrester set up the simulation method and accredited the significance of coordinating information and material flow (Forrester 1961).

The Bullwhip effect is a propensity to a more considerable variance than the actual consumer demand that triggered the order. Thus, the Bullwhip effects lead to enormous superfluous inventory to the upstream compared to the downstream supply chain. Such amplification of stock in the upstream supply chain via communicating orders from the downstream supply chain is due to several business operations of moving product to the end-user (Lee et al. 1997).

Lee et al. (2000) examined the Bullwhip effects typically based on profit maximization via inventory cost minimization. These include stable inventory insufficiencies and inventory extremes. The article suggested that the Bullwhip effects can cover 13–25% of operating costs. Recent papers predicted that controlling the Bullwhip effect could improve product profitability by 10–30%. Croson and Donohue (2005) analyze the behavioral perspective of the Bullwhip effects by applying various statistical models. The analysis is conducted in two different experiments on two groups of participants. The significant result from the experimentation to the SCM theory is that "modifications in inventory management strategies should recognize unpredictability of demand side of the supply chain to slow down the Bullwhip effects" (Dejonckheere et al. 2003).

Warburton (2004) analyzed by defining appropriate differential equations of downstream supply chain and solving them to quantify their contributions to inventory amplifications in the upstream supply chain. The outcome of this study revealed that sharing customer demand information with the suppliers has a crucial role in reducing the impact of Bullwhip effects on the entire supply chain.

Tesfay (2014) employed a stochastic analysis to examine six hypotheses on the Bullwhip effects. The first five hypothesis and their test results are summarized as follows. The oversimplifications of the test results of hypothesis 1 to hypothesis 5 indorse that the way of inventory and stakeout cost or another form of attacking competitive advantage of the Bullwhip effects can attack all the business in the chain. The paper endorses that the firm must acknowledge the damaging impact of the Bullwhip effects and enhance them to formulate dynamic, flexible, and responsive organizational coordination to do a successful business. The paper also calls for different professionals to conduct research and provide helpful theories of coordination in SCM. Consequently, this paper design gives much prominence to the Tesfay (2014) paper, has six hypotheses, and tries to show its applications in the airline industry.

11.4   Design of Dataset and Methodology

11.4.1   The Design of Dataset

In this analysis, we use experimental data of Tesfay (2014), which consider supply chain factory, distributor, wholesaler, and retailer (Boute et al. 2007; Simchi-Levi et al. 2000; Tesfay 2014). The game has the following essential characteristics (1) business transaction triggered from the end-customer; (2) demand information is anonymous to business partners of the upstream supply chain; (3) players (business partners) do not know as mentioned above about SCM; and (4) objective of the games to reduce the overall costs of the supply chain (Senge 1990). See the dataset presented in the Appendix.

Lack of knowledge of SCM of the players (business partners) generates observations from the nonappearance of interorganizational coordination. The random transactions of the game generate the observations that come from the nonappearance of intraorganizational coordination.

The list of exogenous variables is indicial inventory (X), order delivered (X_1), and incoming order (X_2). The endogenous variable of this analysis is the ending inventory (Y).

11.4.2   The Methodology

11.4.2.1   Panel Data Regression Models

Panel data (also called longitudinal data) are observed across different cross-sectional (spatial) units over repeated time intervals. Therefore, such data contain both spatial and time effects of the response variable (Davidson and MacKinnon 1993; Davies and Lahiri 1995; Granger 1991; Hsiao 2003; Pearl 2000; Zellner and Ando 2010).

Panel data regression model which contains two erogenous variables (X_1 and X_2) and endogenous variable (Y) are given as follows (Baltagi 2008; Barreto and Howland 2005; Greene 2012; Tesfay 2016a, b).

$$Y_{it} = \beta_i + \beta_1 X_{1it} + \beta_2 X_{2it} + \varepsilon_{it} \tag{11.1}$$

where $i = 1, 2, 3, \ldots, n$, $t = 1, 2, 3, \ldots, T$, β_i is the ith spatial effects, β_1 and β_2 are the common coefficients of X_1 and X_2, respectively, and $\varepsilon_{it} \sim iidN\,(0, \sigma^2)$ is the random error terms.

The structural stability breaks are a statistical test to become aware of the spatial effects panel data regression model's advantage over the pooled panel data regression model. The test statistic of the Chow test expressed as follows (Dougherty 2007).

$$F = \frac{\left(R_{\text{fem}}^2 - R_{\text{pooled}}^2\right)/(a-1)}{\left(1 - R_{\text{fem}}^2\right)/(aT - a - p)} \tag{11.2}$$

where a is the number of cross sections, T is the total number of time observations, p is the number of regressors in the model, R_{fem}^2 is the square of the coefficient of determination of the fit of spatial effects panel data regression model, R_{pooled}^2 is the square of the coefficient of determination of the fir of pooled panel data regression model (Dougherty 2007).

11.4.2.2   Structural Equations of a Seemingly Unrelated Regression Model

Structural equations of seemingly unrelated regression endogenous vector of $(y_{1t}, y_{2t}, y_{3t}, \ldots, y_{at})$ and the exogenous/independent vectors of $(x_{11t}, x_{21t}, x_{31t}, \ldots, x_{a1t})$ and $(x_{12t}, x_{22t}, x_{32t}, \ldots, x_{a2t})$ are given as follows (Davidson and MacKinnon 1993; Tesfay and Solibakke 2015, 2016).

$$
\begin{cases}
y_{1t} = \beta_{10} + \beta_{11}x_{i1t} + \beta_{12}x_{12t} + \varepsilon_{1t} \\
y_{2t} = \beta_{20} + \beta_{21}x_{i1t} + \beta_{22}x_{12t} + \varepsilon_{2t} \\
\quad \vdots \\
y_{at} = \beta_{a0} + \beta_{a1}x_{a1t} + \beta_{a2}x_{a2t} + \varepsilon_{at}
\end{cases}
\tag{11.3}
$$

where $i = 1, 2, 3, \ldots, a$, $t = 1, 2, 3, \ldots, T$, β_{i0} is the constant of the ith regression model, β_{i1} and β_{i2} are the common coefficients of X_{i1} and X_{i2}, $\varepsilon_{it} iidN\left(0, \sigma_i^2\right)$ is the random error terms.

We can express the structural equations of seemingly unrelated regression (SUR) models in Eq. (11.4) with the above specification.

$$
y_{it} = \beta_{i0} + \beta_{i21}x_{i1t} + \beta_{i2}x_{i2t} + \varepsilon_{it}
\tag{11.4}
$$

11.4.2.3   Recursive Autoregressive-SUR Model

Classical estimation in the presence of autocorrelation leads to an inefficient parameter estimate. Thus, to control autocorrelation, we can apply the recursive Cochrane-Orcutt autoregression estimation on each component of the SUR model equations as follows (Ayele et al. 2017; Cochrane and Orcutt 1949; Tesfay 2015a, b).

First, we specify the model as

$$
y_{it} = \beta_{i0} + \beta_{11}x_{i1t} + \beta_{i2}x_{i2t} + \varepsilon_{it}
\tag{11.5}
$$

where $\varepsilon_{it} = \rho_i\varepsilon_{it-1} + v_{it}$, $|\rho_i| < 1$, ρ is the measure of autocorrelation and $v_{it} iidN\left(0, \sigma_{iv}^2\right)$.

Cochrane-Orcutt autoregression procedure is given as

$$
y_{it} = \beta_{i0} + \beta_{i1}x_{i1t} + \beta_{i2}x_{i2t} + \varepsilon_{it}
\tag{11.6}
$$

$$
\rho_i y_{it-1} = \rho_i\beta_{i0} + \rho_i\beta_{i1}x_{i1t-1} + \beta_{i2}x_{i2t-1} + \rho_i\varepsilon_{it-1}
\tag{11.7}
$$

Subtract Eq. (11.5) from Eq. (11.6), we have

$$y_{it} - \rho_i y_{it-1} = (\beta_{i0} - \rho_i \beta_{i0}) + (\beta_{11} x_{i1t} - \rho_i \beta_{i1} x_{i1t-1}) +$$
$$(\beta_{i2} x_{i2t} - \rho_i \beta_{i2} x_{i2t-1}) + (\varepsilon_{it} - \rho_i \varepsilon_{it-1}) \tag{11.8}$$

$$y_{it}^* = \beta_{i0}^* + \beta_{i1}^* x_{i1t}^* + \beta_{i2}^* x_{i2t}^* + v_{it} \tag{11.9}$$

where $y_{it}^* = y_{it} - \rho_i y_{it-1}$, $\beta_{i0}^* = (\beta_{i0} - \rho_i \beta_{i0})$, $x_{i1t}^* = (x_{i1t} - \rho_i x_{i1t-1})$ and $x_{i2t}^* = (x_{i2t} - \rho x_{i2t-1})$.

Since $v_{it} iidN \left(0, \sigma_{iv}^2\right)$ regression equation (11.8) has removed the autocorrelation of the series. Thus, we apply the OLS recursively on regression equation (11.8).

11.4.2.4   Adjusting Cross-sectional Variability: Bartlett's Test of Heteroscedasticity

To take the best fit of data from the panel data regression model and structural equations of seemingly unrelated regression (SUR) model, we apply Bartlett's test of heteroscedasticity. Here, the null states that all the endogenous variables have equal variance. The test statistic is given as follows.

$$\chi_{cal}^2 = \frac{(N - k) \ln \left(S_p^2\right) - \sum_{i=1}^k (n_i - 1) \ln \left(S_i^2\right)}{1 + \frac{1}{3(k-1)} \left[\sum_{i=1}^k \left(\frac{1}{n_i - 1}\right) - \left(\frac{1}{N-k}\right)\right]} \tag{11.10}$$

where $N = \sum_{i=1}^k n_i$, $S_p^2 = \frac{1}{N-k} \sum_{i=1}^k (n_i - 1) S_i^2$, S_i^2 for $i = 1, 2, 3, \cdots, a$, is the sample variance.

According to Snedecor and Cochran (1989), the null hypothesis is rejected if $\chi_{cal}^2 > \chi_{a-1,\alpha}^2$.

11.5   Results and Discussions

11.5.1   Examination of Hypothesis 1: Causes of the Bullwhip Effects

To test hypothesis 1, we must fit the pooled panel data regression model, spatial effects panel data regression model, and the SUR model. The pooled panel data regression model's fit, spatial effects panel data regression model, and the SUR model are given in Tables 11.1, 11.2, 11.3, 11.4, 11.5, and 11.6.

Table 11.1 shows that the F-statistics (p-value $= 0$) prevailing the pooled regression model is adequate. However, the pooled regression model controls only 19.8% of the whole supply chain's ending inventory variation. The model predicts that both the order delivered and incoming order of the supply chain cause

Table 11.1 Model adequacy statistic of pooled panel data regression model

Actor	SS	Df	MS	F-cal	p-value	R	R^2	Adjusted R^2	Std. error
Regression	1373.83	2	686.92	17.30	0	0.45	0.20	0.18	6.30
Residual	5560.33	140	39.72						

Table 11.2 Parameter estimating of pooled panel data regression model

Parameter		Estimate	Std. error	t-cal	p value
Regression coefficient	Order delivered	−0.64	0.14	−4.52	0*
	Incoming order	0.43	0.11	3.96	0*
Constant		14.49	12.63	1.15	0.2530

Note: *Significant is at 5% level of significance

Table 11.3 Model adequacy statistic of spatial effects panel data regression model

Actor	SS	Df	MS	F-cal	p value	R	R^2	Adjusted R^2	Std. error
Regression	2090.57	6	348.40	9.79	0	0.548	0.30	0.26	5.97
Residual	4875.77	137	35.60						

Table 11.4 Parameter estimating of spatial effects panel data regression model

Parameter		Estimate	Std. error	t-cal	p value
Regression coefficient	Order delivered	−0.61	0.13	−4.5	0*
	Incoming order	0.45	0.10	4.33	0*
Spatial effects	Factory	3.30	19.89	0.17	0.869
	Distributor	12.33	19.67	0.63	0.532
	Wholesaler	3.13	19.70	0.10	0.874
	Retailer	26.41	19.91	1.33	0.187

Note: *Significant is at 5% level of significance

unnecessary end inventory. However, the constant of the model is insignificant (Table 11.2).

Table 11.3 shows that the F-statistics (p value $= 0$) prevailing the spatial effects panel data regression model is adequate. However, the pooled regression model controls about 30.00% of the whole supply chain's ending inventory variation. In Table 11.4, the model predicts that both the order delivered and incoming order of the supply chain cause unnecessary end inventory. However, all the spatial effects of the constant of the model are insignificant.

The Chow test prevails F-value of 6.702 (p value < 0.00036). That suggests that at the 5% level of significance, the spatial effects panel data regression model has more predictive power than the pooled panel data regression model for the causality of the ending inventory of the supply chain actors.

Table 11.5 shows that at a 5% level of significance, the F-statistics prevails that the SUR model is adequate. The SUR model controls 25.7%, 21.5%, 36.5%, and

Table 11.5 Model adequacy statistic of structural equations of a seemingly unrelated regression model

Actor		SS	Df	MS	F-cal	p value	R	R^2	Adjusted R^2	Std. error
Factory	Regression	810.96	2	405.500	5.55	0.01	0.507	0.257	0.187	8.567
	Residual	2348.50	32	73.390						
Distributor	Regression	430.62	2	215.300	4.25	0.02	0.464	0.215	0.140	7.113
	Residual	1568.30	31	50.590						
Wholesaler	Regression	247.75	2	123.900	9.00	0	0.604	0.365	0.303	3.732
	Residual	431.69	31	13.930						
Retailor	Regression	21.16	2	10.580	4.87	0.01	0.489	0.239	0.165	1.474
	Residual	67.33	31	2.172						

Table 11.6 Parameter estimating of structural equations of a seemingly unrelated regression model

Parameter		Unstandardized coefficient		Standardized coefficient (beta)	t-cal	p value
		B	Std. error			
Factory	Incoming order	−0.901	0.230	−0.47	−3.02	0.005*
	Order delivered	0.284	0.230	0.20	1.26	0.216
	Constant	19.160	28.410		0.67	0.505
Distributor	Incoming order	−0.430	0.240	−0.28	−1.79	0.084
	Order delivered	0.590	0.250	0.38	2.40	0.023*
	Constant	40.980	14.440		2.84	0.008*
Wholesaler	Incoming order	0.060	0.370	0.03	0.18	0.862
	Order delivered	0.550	0.132	0.60	4.148	0*
	Constant	−15.990	19.822		−0.81	0.426
Retailer	Incoming order	−0.100	0.286	−0.05	−0.34	0.736
	Order delivered	0.500	0.161	0.49	3.12	0.004*
	Constant	−17.160	3.233		−5.31	0*

Note: *Significant is at 5% level of significance

23.9% of the variations of the ending inventory of the factory, distributor, wholesaler, and retailer. The SUR model fit suggests that the causes of the ending inventory for the different actors are different.

The Bartlett test of heteroscedasticity prevails that a Chi-square value is 626.32 (p-value $= 0$). That suggests at the 5% level of significance, the variance of the ending inventory of each business partners of the supply chain is statistically different. That endorses fitting of the spatial effects panel data regression model hides some essential variations of the supply chain business partners. Precisely, the interpretation of the spatial effects panels data regression model. Therefore, we will interpret the SUR model's structural equations for the causality of the ending inventory of the supply chain business partners.

From Table 11.6, we see that the constant of distributor and retailer are found statistically significant. That demonstrates the intra-organizational coordination of these business partners found inappropriate to respond to customer demand. That is one of the primary foundations of the distributor peak unnecessary-accumulation inventory (upstream supply chain) and leads to stock out at the retailer (downstream supply chain). The predictor variable order delivered significantly disturbs all the upstream supply chains (i.e., distributor, wholesaler, and retailer). That shows the inter-coordination of all the supply chain business partners found it inappropriately to respond to customer demand.

When we make a sweeping statement, we found that the causality analysis results about the Bullwhip effects exist without the intraorganizational and interorganizational coordination of the business partners in the supply chain.

11.5.2   Examination of Hypothesis 2: Implications of the Causality Analysis on Transaction Cost Analysis (TCA)

The finding of this analysis confirmed that intra-organization coordination of the supply chain business partner is one of the potential foundations of the Bullwhip effects. The result shows that coordinating with business partners having poor intraorganizational coordination can maximize overall cost or cause potential business risk than conforming with business partners having. That has several implications on the drawbacks of the existing theories of organizational coordination.

One of the most influential coordination theories is Williamson's (1981) transaction cost analysis (TCA). The TCA hypothesis tried to give the solution for making efficient coordinating of firms among market (rule of the demand-supply theory is the best means coordination), hybrid (coordination through relational contract), and hierarchy (vertical integration). The TCA used a certain dimension to analyze making efficient coordination. These dimensions are uncertainty, asset specificity, and transaction frequency. For example, the TCA solution suggests that as uncertainty, asset specificity, and transaction frequency increase the making vertical integration (hierarchy) more efficient than hybrid governance (Tesfay 2015a, b).

The TCA assumed that transactors (firms) are "bounded rationally" and "opportunistic." These two behavioral assumptions are a tool to develop the TCA. However, TCA ignores the dynamic aspect of these two behavioral factors. Transactors may be opportunist or bounded rationally. However, another factor directly interacting with these two important factors is "experience." Experience is a dynamic factor and can make firms increase or decrease their opportunistic or rational bounded behavior according to their benefit from their previous organizational coordination (Tesfay 2015a, b).

11.5.3   Tesfay Process

The Tesfay process illustrated the impact of experience with the firm's opportunism and bounded rational behavior are analyzed as follows. According to the Tesfay-process, the simplest form of opportunism's dynamic aspect and bounded rationally, and the following formula represents interorganizational coordination behavior.

$$\sum_{k}^{t-1} (\text{Value-Creation}_k) \Rightarrow (\text{OP}_t, \text{BR}_t) \Rightarrow (\text{Inter-organizational coordination}_t)$$

$$\Rightarrow (\text{Value-Creation}_t)$$

where OP_t and BR_t are, respectively, the degree of opportunism and bounded rationally behavior of the firm at a time to perform intraorganizational coordination, respectively, $t \in \{1, 2, 3, \ldots\}$ is the index that representing the time of the business interaction event and the previous business interaction periods.

From the formula, we can observe that a given firm has its degree of opportunism (OP_t) and bounded rationally (BR_t) behavior. The firm uses the previous benefit $\sum_{k}^{t-1} (\text{Value-Creation}_k)$ from the previous knowledge of practicing intraorganizational coordination. That knowledge and information assist in new intraorganizational coordination. Then finally, the firm will evaluate the benefit (Value Creation$_t$) of the new intraorganizational coordination.

Let us consider the following case to illustrate how "experience" affects organizational coordination firms, which may not realize coordinating with another firm having inadequate intraorganizational coordination in a single transaction. They may recognize that through time (experience). In this aspect, firms prefer to coordinate with another firm having good intraorganizational coordination. Otherwise, if there is a supplier risk, firms attack their opportunistic behavior, and they recognize the benefits of their business partner, which have a positive impact on making efficient coordination. They become willing to assist their business partner in modifying its intraorganizational coordination.

The coordination type's hybrid and hierarchy may not be an efficient means of interorganizational coordination.

Example 1 Suppose that let us consider ideal hybrid coordination that controls TCA aspects, that is, uncertainty, asset specificity, and transaction frequency. A given opportunist and rational bounded behavior firms assume that they made efficient coordination; hence, their coordination controlled all TCA dimensions. Now let us ask ourselves that our question is that "do such type of coordination is the most efficient coordination for the transactors?" The TCA argues that such coordination is the most efficient coordination for the transactors. However, the analysis of the Bullwhip effects (see the result of hypotheses 1) prevails that even under the fulfillment of the TCA's dimensions, the most efficient coordination is not achievable. Hence, if at least one of the transactors has inadequate intraorganizational coordination, all the transactors will become cost-efficient due to the Bullwhip effects.

Example 2 Suppose that considering the dimensions of uncertainty, asset specificity, transaction frequency, and so on, the hierarchy is a TCA solution for an organizational coordination problem.

However, there are well-known shreds of evidence and reasons that hierarchy may not be realized. Therefore, this leads the firms to coordinate in the relational contract (hybrid governance). However, we have identified that coordinating with firms with poor intraorganizational coordination will cause an ineffective and inefficient supply chain. Therefore, another necessary type of organizational coordination, which the TCA did not discuss, should be defined to synthesize a more effective and efficient supply chain.

11.5.4   The New Discovered Organizational Coordination: Tesfay Coordination

The causality analysis of the Bullwhip effects leads us to introduce a new type of organizational coordination known as the hyper-hybrid (Tesfay coordination) governance. This governance has the following four characteristics.

1. It is the relational contract.
2. It recognizes that one firm's benefit in the supply chain will benefit all the business partners in the supply chain. Coordinating with firms having good intraorganizational coordination is preferable to coordinating with firms having poor intraorganizational coordination.
3. In a buyer-supplier relationship, it is useful when supplier risk is high.
4. The governance extends to recommend, consult, and help firms to establish appropriate means of intraorganizational coordination (Tesfay 2015a, b).

11.5.5   Implications of Examination of Hypothesis 3

Kraljic (1983) purchasing portfolio model is the prevailing solution for firms to form effective coordination with their business partners. The portfolio model encompasses all the advantages of the transaction cost economics, resource dependence theory, pricing theory of markets, and relational contract theory. The Kraljic (1983) purchasing portfolio model's two essential dimensions are the purchasing impact on the firm's profitability and the supplier risk. However, the Kraljic portfolio model was incomplete in addressing coordination inefficiency and ineffectiveness due to the supplier's intraorganizational coordination. Therefore, to make effective coordination by considering the supplier's intraorganizational coordination and using the hyper-hybrid governance, we will modify the existing Kraljic purchasing portfolio model.

Intraorganizational coordination can be analyzed in many dimensions. However, in this portfolio modeling, we categorized the supplier's intraorganizational coordination into two dimensions good intraorganizational coordination and inadequate intraorganizational coordination. The modified model of the Kraljic portfolio matrix by the inclusion of two intraorganizational coordination classes is given in Fig. 11.1 (Tesfay 2015a, b).

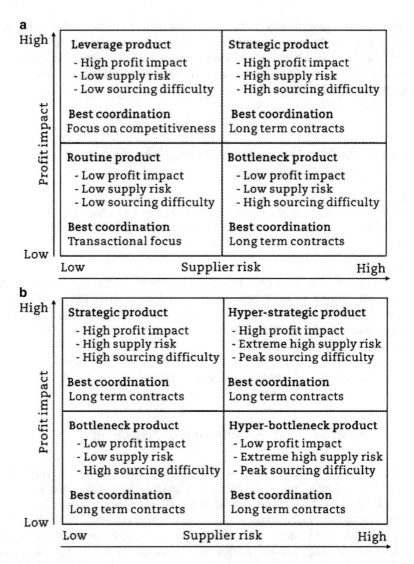

Fig. 11.1 The Kraljic–Tesfay portfolio matrix. (**a**) Good intraorganizational coordination. (**b**) Bad intraorganizational coordination

11.5.5.1   The Best Coordination with the Supplier Having Good Intraorganizational Coordination

If all the suppliers have good intraorganizational coordination, the existing Kraljic portfolio model is still the best way of making effective and efficient coordination. Therefore, based on the current model, we have the following four methodology classifications (Tesfay 2015a, b).

1. *Routine product.* These purchasing techniques are suitable when neither the supplier's risk nor profitability to its bottom line is high. In this method, the company's suspicions about reducing the logistical cost and establish sourcing through the supplier are found the most suitable strategy.
2. *Leverage product.* These purchasing techniques are suitable when there is no supplier risk. Nevertheless, profitability in the company's bottom line is high. In this method, making competitive bidding among suppliers is the most apposite strategy.
3. *Bottleneck product.* These purchasing techniques are suitable when supplier risk is high and profitability to its bottom line is low. In this method, establishing a partnership with the supplier is the most apposite strategy.
4. *Strategic product.* These purchasing techniques are suitable when both supplier risk and profitability to the company's bottom line are high. In this method, establishing a partnership with the supplier is the most apposite strategy.

11.5.5.2 The Best Coordination with the Supplier Having Inadequate Intraorganizational Coordination

If some (or all) suppliers have inadequate intraorganizational coordination, the existing Kraljic portfolio model is no more the guideline to make effective and efficient coordination. Therefore, we modify the current Kraljic portfolio by the model (Tesfay 2015a, b).

1. *Bottleneck product.* These purchasing techniques are appropriate when the profitability to the company's bottom line is low and supplier risk is absent. However, there is heterogeneity in the intraorganizational coordination of the supplier. In this aspect, to make effective and efficient coordination, the buyer must set criteria about intraorganizational coordination for the suppliers. This criterion causes to maximizes supplier risk. Therefore, the products categorized as "routine products" in the Kraljic portfolio model transformed into "bottleneck products." Consequently, establishing a partnership with the supplier is the most suitable strategy.
2. *Strategic product.* These purchasing techniques are appropriate when the profitability to the company's bottom line is high, and supplier risk is absent. However, there is heterogeneity in the intraorganizational coordination of the supplier. In this aspect, the buyer has to set the suppliers' intraorganization coordination criteria to make effective and efficient coordination. This criterion causes to maximizes supplier risk. Therefore, the products categorized as "leverage product" in the Kraljic portfolio model transformed into "strategic product." Consequently, establishing a partnership with the supplier is the most appropriate strategy.
3. *Hyper-bottleneck product.* These purchasing techniques are appropriate when supplier risk is high, and profitability to its bottom line is low. Furthermore, the intraorganizational coordination of the suppliers is inadequate. We call such a product a "hyper-bottleneck product." In this aspect, only establishing a

partnership with the supplier is not the most productive coordination method. This coordination is more efficient and effective if the supplier's intraorganizational coordination is modified through the business partners' assistance. Therefore, this coordination is more than a partnership. Hence the business partners are needed to support each other to adjust their interorganizational coordination for their mutual benefit.

4. *Hyper-strategic product.* These purchasing techniques are appropriate when supplier risk is high, and profitability to its profitability is high. Furthermore, the intraorganizational coordination of the suppliers is inadequate. We call such a product a "hyper-strategic product." In this aspect, only establishing a partnership with the supplier is not the most productive coordination method. This coordination is more efficient and effective if the supplier's intraorganizational coordination is modified through the business partners' assistance. Therefore, this coordination is more than a partnership. Hence the business partners are needed to support each other to adjust their interorganizational coordination for their mutual benefit.

11.6   Applications of Tesfay Coordination in the Airline Industry

The airline industry is responsible for passengers and cargo transportation, which operates in every country globally. The industry has played essential roles in establishing today's global economy (Belobaba et al. 2009). The airline industry is regarded as a complex and dynamic industry. The airline industry is characterized by a capital-intensive, labor-intensive, and high-cost sector. Furthermore, competition (price-war) is a typical feature of the airline industry. Therefore, it is evident that the airline industry is marginally profitable (Borenstein 1992; Chua et al. 2005; Doganis 2010).

The airline industry's primary revenue comes from passengers' transportation, and the minor income comes from the carrier of cargo and postal services. Recently, the expansion of information technology had given the customer power, so the airlines' price war makes the industry less profitable. Therefore, instead of competition (like the price-war), the airlines acknowledge that coordination and collaboration will support their profitability. Due to this, many aviation associations and strategic airline alliances (e.g., Star Alliance, Sky Team, and Oneworld) established to maximize the profit of the airlines (Babikian et al. 2002; Belobaba et al. 2009; Fedorco and Hospodka 2013; Keynes 2009; Oumand Yu 1998).

Several other scholars described the resonant characteristics of the airline industry. Cu and Li (2017) tried to measure airline dynamic efficiency by applying a "dynamic epsilon-based measure (DEBM) model." The paper showed efficiency change in the industry is positively associated with factors of the financial crisis. Thus, the article confirmed that the airline industry is sensitive to financial sectors.

The finding of the Tesfay (hyper-hybrid) coordination helps the airlines to make successful coordination. First, the result informs airlines to acknowledge that coordinating (even collaboration) with the airline characterized by low intraorganization coordination costs the other airlines to have good intraorganization coordination. Therefore, airlines should give much emphasis to intraorganization coordination or their collaborative airlines before agreeing. Second, suppose the intraorganization coordination, the collaborative airline is low, and coordination with that airline is essential (e.g., to operate in wider geographical regions, capacity limitation, etc.). In that case, the airlines having good intraorganization coordination should share their best practices and modify the intraorganization coordination of the poorly performing collaborative airline.

In general, the following are the contributions of hyper-hybrid coordination in the airline industry.

1. It avoids risky price-war competition and leads to beneficiary competition among airlines.
2. It helps to quickly identify, mitigate business risk, and control financial flow.
3. It helps operational cost reduction and quickly maximizes yield.
4. It helps to share business information for the air transport supply chain.
5. It maximizes capacity (ASK) by improving operational distance.
6. It maximizes the operational geographical network.
7. It helps to disseminate technological changes among airlines.
8. It enables the airlines to align operational capacity decisions on passengers and cargo's demand.

11.7 Conclusions and Recommendations

11.7.1 Conclusions

This chapter proposes to test three hypotheses of modern SCM challenges. The first hypothesis was intended to ascertain possible potential variables/factors, which cause the Bullwhip effects. We fitted the pooled panel data regression model, spatial effects panel data regression model, and structural equations seemingly unrelated regression (SUR) model on experimentally generated data from the Beer distribution game to test this hypothesis.

The fit of all the proposed models suggests that interorganizational coordination found the primary cause of Bullwhip effects. This result proved the hypothesizes of Bhattacharya and Bandyopadhyay (2011), Dejonckheere et al. (2003), Lee et al. (1997, 2000), and Forrester (1961).

More rigorously, to select the best-fitted model of the causal analysis of Bullwhip effects, we applied several structural changes and heteroscedasticity tests. The overall test result suggests that the fit of the SUR model structural equations is the best to analyze the causes of Bullwhip effects. The SUR model's fit predicts that the

Bullwhip effect happens because of the lack of proper intra- and interorganizational coordination of each business collaborates in the supply chain (Tesfay 2015a, b).

The outcome of the causality analysis on Bullwhip effects using the SUR model will have several implications on the existing organizational coordination theory. Accordingly, we have set the second hypothesis to analyze the drawbacks of transaction cost analysis (TCA). TCA presents how firms make effective organizational coordination. The model anticipates that business transistors are "bounded rationally" and "opportunistic." On the other hand, the TCA pays no attention to these two behavioral factors' dynamic characteristics. Transactors may be opportunist or bounded rationally.

Conversely, an additional factor directly interacts with these two critical factors, that is, "experience." Experience in dynamic and vigorous characteristics could make firms modify (increase/decrease) their opportunistic or rational bounded behavior. Such behavioral dynamics occur because firms learn about their benefit (value creation) from their previous organizational coordination. Through "experience," firms prefer to coordinate with firms having good intra-organizational coordination. Otherwise, if there is supplier risk, firms attacking their opportunistic behavior and distinguishing their business partner's remunerations positively influence efficient coordination and develop keenness to assist their business partner in transforming its intraorganizational coordination. In this case, another type of coordination, the Tesfay coordination, can make more effective coordination by creating proper interorganizational and then extended to modify the intraorganizational coordination of the business partners in the chain.

Once we have shown the TCA's wickedness and introduced the hyper-hybrid organizational governance, we inspired to modify the Kraljic (1983) purchasing portfolio model in our third hypothesis. The Kraljic purchasing portfolio model encompasses all the TCA pluses, resource dependence theory, pricing theory of markets, and relational contract theory. The two critical dimensions of the model are profitability and supplier risk. Nevertheless, in the buyer-supplier relationship, the Kraljic matrix was incomplete in addressing coordination inefficiency and ineffectiveness due to its intraorganizational coordination. Therefore, to make an efficient and effective buyer-supplier relationship by considering the intraorganizational coordination of the supplier and using the hyper-hybrid governance, we have modified the existing Kraljic purchasing portfolio model by the Kraljic-Tesfay purchasing portfolio model.

11.7.2   Recommendations

The finding of this analysis confirmed that intraorganization coordination of the business partner in the chain is one of the potential foundations of Bullwhip effects. This result showed that coordination with business partners with poor intraorganizational coordination could cost the other firms good intraorganizational coordination. Therefore, firms must use intraorganizational coordination as interorganizational coordination criteria with their business partners.

Tesfay coordination can play several roles to remove superfluous competition and develop better operational performance in the airline industry. In the buyer-supplier relationship, if the supplier risk is higher and the intraorganizational coordination of the supplier is poor, the potential solution that makes the coordination more efficient and effective is hyper-hybrid governance. The hyper-hybrid governance practice can improve many agricultural products (from the developing world to the developed world). Such coordination has an essential role in attacking the factors that resist the flow of goods in the global supply chain. Furthermore, hyper-hybrid governance practice encourages nations to collaborate and formation of regional trade agreements.

The SUR model structural equations were selected as the best model to predict the Bullwhip effects. As we have seen in the model fit, the regression models' coefficient of determination is still low, which implies several variables/factors omitted to explain our endogenous variable's variability. Thus, this analysis encourages professionals to work on theories coordination in the modern challenges of SCM.

Appendix: Experimental Data from the Beer Distribution Game

	Observation															
	Factory				Distributor				Wholesaler				Retailer			
Week	X	X_1	X_2	Y	X	X_1	X_2	Y	X	X_1	X_2	Y	X	X_1	X_2	Y
1	20	4	4	20	20	4	4	20	20	4	4	20	20	4	4	20
2	20	4	4	20	20	4	8	24	20	4	4	20	20	8	4	16
3	20	5	4	19	24	4	4	24	20	6	4	18	16	9	4	11
4	19	8	4	15	24	9	4	19	18	4	4	18	11	8	4	7
5	15	10	4	9	19	12	5	12	18	6	4	16	7	8	6	5
6	9	8	5	6	12	17	4	−1	16	4	12	24	5	8	4	1
7	6	6	8	8	−1	8	10	1	24	4	9	29	1	8	6	−1
8	8	20	4	−8	1	8	8	1	29	8	17	38	−1	7	4	−4
9	−8	18	8	−18	1	1	6	6	38	10	8	36	−4	8	4	−8
10	−18	18	10	−26	6	1	20	25	36	10	8	34	−8	8	8	−8
11	−26	15	16	−25	25	2	18	41	34	8	1	27	−8	8	10	−6
12	−25	5	20	−10	41	1	18	58	27	8	1	20	−6	8	10	−4
13	−10	2	21	9	58	2	15	71	20	8	2	14	−4	8	8	−4
14	9	7	7	9	71	4	5	72	14	4	1	11	−4	8	8	−4
15	9	7	30	32	72	9	2	65	11	7	2	6	−4	9	8	−5
16	32	0	25	57	65	9	7	63	6	6	4	4	−5	8	4	−9
17	57	0	8	65	63	14	7	56	4	4	9	9	−9	8	4	−13
18	65	0	10	75	56	13	0	43	9	8	9	10	−13	7	6	−14

(continued)

| | Observation | | | | | | | | | | | | | | | |
| | Factory | | | | Distributor | | | | Wholesaler | | | | Retailer | | | |
Week	X	X_1	X_2	Y	X	X_1	X_2	Y	X	X_1	X_2	Y	X	X_1	X_2	Y
19	75	1	1	75	43	2	0	41	10	8	14	16	−14	8	4	−1
20	75	3	1	73	41	3	0	38	16	8	13	21	−18	8	8	−1
21	73	10	0	63	38	8	7	37	21	8	2	15	−18	7	8	−1
22	63	8	0	55	37	3	3	37	15	8	3	10	−17	8	8	−1
23	55	7	0	48	37	1	10	46	10	10	8	8	−17	8	8	−1
24	48	10	0	38	46	12	8	42	8	8	3	3	−17	8	8	−1
25	37	15	0	22	42	1	7	48	3	8	1	−4	−17	9	10	−1
26	23	20	10	13	48	12	10	46	−4	9	12	−1	−16	8	8	−1
27	13	5	0	8	46	10	15	51	−1	8	7	−2	−16	8	8	−1
28	8	5	0	3	51	11	20	60	−2	8	12	2	−16	8	9	−1
29	5	0	0	5	60	10	5	55	2	8	10	4	−15	9	8	−1
30	5	0	10	15	55	8	5	52	4	9	11	6	−16	8	8	−1
31	15	2	4	17	52	1	0	51	6	10	10	6	−16	8	8	−1
32	17	0	2	19	51	2	0	49	6	10	1	−3	−16	7	9	−1
33	19	7	0	12	49	4	2	47	−3	8	8	−3	−14	8	10	−1
34	12	0	0	12	47	6	0	41	−3	8	9	−2	−12	8	10	−1
35	12	1	0	11	41	12	1	30	−2	6	4	−4	−10	8	8	−1
36	11	0	0	11	30	3	0	27	−4	8	6	−6	−10	7	8	−9

Source: Tesfay 2014, See Chap. 10

Bibliography

C. Altomonte, F.D. Mauro, G. Ottaviano, et al., *Global Value Chains During the Great Trade Collapse: A Bullwhip Effects?* (European Central Bank, Frankfurt, 2012)

A.W. Ayele, E. Gabreyohannes, Y.Y. Tesfay, Macroeconomic determinants of volatility for the gold price in Ethiopia: The application of GARCH and EWMA volatility models. Glob. Bus. Rev. **18**(2), 308–326 (2017)

R. Babikian, S.P. Lukacho, I.A. Waitz, The historical fuel efficiency characteristics of regional aircraft from technological, operational, and cost perspectives. J. Air Transp. Manag. **8**, 389–400 (2002)

B.H. Baltagi, *Stochastic Analysis of Panel Data*, 4th edn. (John Wiley and Sons, Chichester, 2008)

H. Barreto, F. Howland, *Dummy Dependent Variable Models. Introductory Stochastics: Using Monte Carlo Simulation with Microsoft Excel* (Cambridge University Press, Cambridge, 2005)

P. Belobaba, A. Odoni, C. Barnhart, *The Global Airline Industry* (John Wiley and Sons, Chichester, 2009)

R. Bhattacharya, S. Bandyopadhyay, A review of the causes of Bullwhip effects in a supply chain. Int. J. Adv. Manuf. Technol. **54**(9–12), 1245–1261 (2011)

R.K. Bhote, *Supply Management: How to Make US Suppliers Competitive* (American Management Association Membership Publications Division, New York, NY, 1987)

S. Boonitt, C. Pongpanarat, Measuring service supply chain management processes: The application of the Q-sort technique. Int. J. Innov. Technol. Manag. **2**(3), 217–221 (2011)

S. Borenstein, The evolution of U.S. airline competition. J. Econ. Perspect. **6**(2), 65–77 (1992)

R.N. Boute, S.M. Disney, M.R. Lambrecht, et al., An integrated production and inventory model to dampen upstream demand variability in the supply chain. Eur. J. Oper. Res. **178**(1), 121–142 (2007)

D.N. Burt, D.W. Dobler, S.L. Starling, *World Class Supply Management: The Key to Supply Chain Management*, 7th edn. (McGraw-Hill Irwin, New York, NY, 2003)

S. Chopra, P. Meindl, *Supply Chain Management: Strategy, Planning, and Operations* (Prentice-Hall, Inc., Upper Saddle River, NJ, 2001)

M. Christopher, *Logistics and Supply Chain Management: Strategies for Reducing Cost and Improving Service*, 2nd edn. (Prentice-Hall, Inc., Upper Saddle River, NJ, 1998)

C.L. Chua, H. Kew, J. Yong, Airline code-share alliances and costs: Imposing concavity on translog cost function estimation. Rev. Ind. Organ. **26**(4), 461–487 (2005)

W.G. Cochran, G.M. Cox, *Experimental Designs*, 2nd edn. (John Wiley and Sons, New York, NY, 1992)

D. Cochrane, G.H. Orcutt, Application of least squares regression to relationships containing autocorrelated error terms. J. Am. Stat. Assoc. **44**(245), 32–61 (1949)

R. Croson, K. Donohue, Behavioural causes of the bullwhip effects and the observed value of inventory information. Manag. Sci **52**(3), 323–336 (2005)

Q. Cu, Y. Li, Airline efficiency measures using a dynamic epsilon-based measure model. Transp. Res. A Policy Pract. **100**, 121–134 (2017)

B. David, *Supply Chain Management Best Practices*, 2nd edn. (John Wiley and Sons, Chichester, 2010)

R. Davidson, J.G. MacKinnon, *Estimation and Inference in Econometrics* (Oxford University Press, Oxford, 1993)

A. Davies, K. Lahiri, A new framework for testing rationality and measuring aggregate shocks using panel data. J. Econ. **68**(1), 205–227 (1995)

G.F. Davis, J.A. Cobb, *Resource Dependence Theory: Past and Future* (Stanford's Organization Theory Renaissance, Bingley, 2010)

A.C. Davison, D.V. Hinkley, *Bootstrap Methods and Their Applications* (Cambridge University Press, Cambridge, 1997)

J. Dejonckheere, S.M. Disney, M.R. Lambrecht, et al., Measuring the Bullwhip effects: A control theoretic approach to analyse forecasting induced Bullwhip in order-up-to policies. Eur. J. Oper. Res. **147**(3), 567–590 (2003)

G.G. Dess, G.T. Lumpkin, L. Marilyn, et al., *Strategic Management*, 2nd edn. (McGraw-Hill Irwin, New York, NY, 2005)

R. Doganis, *Flying off Course, Airline Economics and Marketing*, 4th edn. (Routledge, New York, NY, 2010)

C. Dougherty, *Introduction to Econometrics* (Oxford University Press, Oxford, 2007)

P. Drucker, *The Practice of Management* (Harper and Row Publishers Inc., New York, NY, 1954)

L. Ellram, The supplier selection decision in strategic partnerships. J. Purchas. Mater. Manag. **26**(4), 8–14 (1990)

S. Engelberg, *Digital Signal Processing: An Experimental Approach* (Springer, Berlin, 2008)

H. Escaith, N. Lindenberg, S. Miroudot, *International Supply Chains and Trade Elasticity in Times of Global Crisis. ERSD-2010-08* (Economic Research and Statistic Division, World Trade Organization, Geneva, 2010)

L. Fedorco, J. Hospodka, Airline Pricing Strategies in European Airline Market (2013), http://pernerscontacts.upce.cz/30_2013/FeMArco.pdf. Accessed 12 Mar 2014

J.W. Forrester, *Industrial Dynamics* (MIT Press, Cambridge, 1961)

J. Frieden, D. Lake, *International Political Economy: Perspectives on Global Power and Wealth* (Routledge, London, 1995)

R. Ganeshan, T.P. Harrison, *An Introduction to Supply Chain Management* (Penn State University, University Park, 1995)

C. Granger, *Modelling Economic Series: Readings in Stochastic Methodology* (Oxford University Press, Oxford, 1991)

W.H. Greene, *Econometric Analysis*, 7th edn. (Prentice-Hall Inc., Upper Saddle River, NJ, 2012)

A.A. Groppelli, N. Ehsan, *Finance*, 4th edn. (Barron's Educational Series Inc., New York, NY, 2000)

G. Hanson, R. Feenstra, Aggregation bias in the factor content of trade: Evidence from U.S. manufacturing. Am. Econ. Rev. **90**(2), 155–160 (2000)

J.M. Ivancevich, J.H. Donnelly Jr., J.L. Gibson, *Management: Principles and Functions*, 4th edn. (BPI Irwin, Boston, MA, 1989)

K. Jain, L. Nagar, V. Srivastava, Benefit sharing in inter-organizational coordination. Supply Chain Manag. **11**(5), 400–406 (2006)

J.M. Keynes, Characteristics of the airline industry, in *The Airline Industry Challenges in the 21st Century*, ed. by A. Cento, (Springer, Berlin, 2009)

L.R. Kopczak, M.E. Johnson, The supply-chain management effects. MIT Sloan Manag. Rev. **44** (3), 27–34 (2003)

P. Kraljic, Purchasing must become supply management. Harv. Bus. Rev. **61**(5), 109–117 (1983)

P.D. Larson, A. Halldarsson, What is SCM? and where is it? J. Supply Chain Manag. **38**(4), 36–44 (2002)

H.L. Lee, V. Padmanabhan, S. Whang, The Bullwhip effects in supply chains. MIT Sloan Manag. Rev. **38**(3), 93–102 (1997)

H.L. Lee, K.C. So, C.S. Tang, The value of information sharing in a two-level supply chain. Manag. Sci. **46**(5), 626–643 (2000)

K. Matsuyama, A Ricardian model with a continuum of goods under nonhomothetic preferences: Demand complementarities, income distribution, and north-south trade. J. Polit. Econ. **108**(6), 1093–1120 (2000)

P. Milgrom, J. Roberts, *Economics, Organization and Management* (Prentice-Hall, Inc., Upper Saddle River, NJ, 1992)

H. Mintzberg, The design school: Reconsidering the basic premises of strategy formation. Strateg. Manag. J. **11**(3), 171–195 (1990)

J. Pearl, *Causality: Models, Reasoning, and Inference* (Cambridge University Press, Cambridge, 2000)

M.B. Priestley, *Spectral Analysis and Time Series* (Academic Press, Cambridge, 1991)

P.M. Senge, *The Fifth Discipline: The Art and Practice of the Learning Organization* (Currency Doubleday, New York, NY, 1990)

Y. Shiozawa, A new construction of Ricardian trade theory: A many-country, many-commodity with intermediate goods and choice of techniques. Evolut. Inst. Econ. Rev. **3**(2), 141–187 (2007)

D. Simchi-Levi, P. Kaminsky, E. Simchi-Levi, *Designing and Managing the Supply Chain: Concepts, Strategies, and Case Studies* (McGraw-Hill Irwin, New York, NY, 2000)

G.W. Snedecor, W.G. Cochran, *Statistical Methods*, 8th edn. (Iowa State University Press, Ames, 1989)

P. Stoica, R. Moses, *Spectral Analysis of Signals* (Prentice-Hall, Inc., Upper Saddle River, NJ, 2005)

Y.Y. Tesfay, The Bullwhip effect: Applying stochastic models on beer distribution game. Int. J. Appl. Logistics **5**(1), 33–51 (2014)

Y.Y. Tesfay, Modeling the causes of the Bullwhip effect and its implications on the theory of organizational coordination. Supply Chain Forum **16**(2), 30–46 (2015a)

Y.Y. Tesfay, A new econometric model to analyze variations and structural changes in international trade: Applications to the Norwegian import trade across continents and over time. Int. J. Trade Glob. Markets **8**(4), 343–370 (2015b)

Y.Y. Tesfay, Modified panel data regression model and its applications to the airline industry: Modeling the load factor of Europe North and Europe Mid Atlantic flights. J. Traffic Transport. Eng. **3**(4), 283–295 (2016a)

Y.Y. Tesfay, Stochastic evaluation of capacity and demand management of the airline industry: The case of airlines of the AEA for flights of Europe-Africa. Int. J. Math. Eng. Manag. Sci. **3**(2), 1–20 (2016b)

Y.Y. Tesfay, P.B. Solibakke, Spectral density estimation of European airlines load factors for Europe-Middle East and Europe-Far East flights. Eur. Transp. Res. Rev. **7**(2), 14–24 (2015)

Y.Y. Tesfay, P.B. Solibakke, Structure of the Norwegian imports trade concentration: The seemingly unrelated autoregressive regression modelling approach. Glob. Bus. Manag. Res. **8**(2), 19–37 (2016)

T. Uchida, The new development of contract law and general clauses – a Japanese perspective, in *Symposium of Japanese and Dutch Laws Compared, Tokyo, 1993*, (1993)

United Nations (UN), Globalization for Development: The International Trade Perspective (2008), http://unctad.org/en/Docs/ditc20071_en.pdf. Accessed 12 Mar 2014

R.D.H. Warburton, An analytical investigation of the Bullwhip effects. Prod. Oper. Manag. **13**(2), 150–160 (2004)

O.E. Williamson, The economics of organization: the transaction cost approach. Am. J. Sociol. **87** (3), 548–577 (1981)

A. Zellner, T. Ando, A direct Monte Carlo approach for Bayesian analysis of the seemingly unrelated regression model. J. Econ. **159**(1), 33–45 (2010)

Chapter 12
Industrial Applications of Tesfay Dynamic Regression Model

12.1 Introduction

The yield, which measures return per unit of output sold, is an immensely significant airline industry metric. It is the mathematical outcome of two additional fundamental metrics: output sold, and revenue earned. In recent periods, the airline industry's overall yield has declined due to several dynamic factors. The price incentive from the drop accounts for a significant portion of the traffic growth during these conditions, period (Netessine and Shumsky 2002). Very broadly, yields will soften under the following conditions:

- Traffic growth is relatively stable or insufficient to captivate output growth. Low and stable prices are used to sustain higher load factors.
- Intensified competition, lower prices, and yields will toughen when:
  - Load factors are already high, and output is growing no faster than traffic.
  - Traffic growth is outstripping output growth.
  - Lower competition keeps prices unchanged.

The traffic load factor and thus revenue and yield are exaggerated by these modification types, exemplifying how intimately associated the variables are within the context of a vacant output (Talluri and Ryzin 2001).

The primary purpose of this article is that the airline industry's passenger load factor. The load factor quantifies the percentage of an airline's sold output, which, in effect, is a measure of the extent to which supply and demand are balanced at prevalent price points. The industry's load factors hide marked variations between diverse types of airlines, with regional carriers at the lower end of the spectrum and charter airlines at higher load factors than scheduled carriers (Cross 1997). The average load factor for any single airline masks variations between different markets and cabins, with economy/coach achieving a higher load factor. Customers tend to book economy seats in advance and expect lower seat convenience levels than premium cabins. It also conceals prominent daily, weekly, and seasonal variations. Six aspects that mainly drive load factors are as follows.

- The first driver is the industry's production decisions to demand growth. The production growth must be carried into closer configuration with demand growth.
- The second driver is pricing. Contingent upon what decisions are made concerning output, fare discounts usually simulate demand and generate a higher load factor.
- The third driver is the traffic mix. Precisely, the higher the percentage of business travelers carried by airline, the lower the average seat factor. The random element in demand for business travels (unstable) suggests a lower average load factor in business and first-class cabins (McGill and van Ryzin 1999).
- The fourth factor is payment policies. A carrier accepting nonrefundable payment at the time of reservation is likely to have relatively fewer no-shows and a comparatively higher seat factor than selling a more significant portion of tickets on a fully flexible basis.
- The fifth driver is a commercial success. Success in product designs, promotions, marketing communications, distributions, and service conveyance will undoubtedly influence the current load factor (U.S. General Accounting Office 1990).
- The sixth driver is revenue management. The efficiency of a revenue management system (RMS) will influence the load factor. RMS capabilities, specifically, the enhancement of demand forecasting tools, contribute significantly (Marriott Jr. and Cross 2000; Tesfay and Solibakke 2015).

The first driver of the load factor reflects the effectiveness and efficiency of the airline's management efforts. Besides, the other drivers (i.e., two to six) of load factor reflect the airline's demand management efforts' effectiveness and efficiency. The author of this pare evaluates airlines' management capacity by investigating the relationship between the load factor (LF) and available seat kilometer (ASK). Similarly, by analyzing the relationship between the load factor (LF) and revenue passenger kilometer (RPK), this paper evaluates the demand management of airlines (Cynthia et al. 2012; Tesfay and Solibakke 2015).

In the airline industry, yield management describes practices and techniques used to allocate and assign limited resources to various customers to optimize the investment's total revenue. The limited resources are the available seat kilometer (ASK) of an upcoming flight, and the varieties of customers comprise first-class, business, and economy travelers. In short, yield management is concerned with the airline's capacity and demand management (Netessine and Shumsky 2002). Therefore, evaluations of capacity and demand management offer influential contributions to the airline industry's yield management.

Load factor is a metric that measures the success of an airline's capacity and demand management efforts. These efforts are hindered by the fact that, at the same time, demand varies in units of single seat departures in different origins and destination markets and is volatile. Supply can only be produced in units equivalent to the capacity of whichever aircraft type is available to operate the flight legs and routes designed to serve targeted origin and destination markets and is broadly fixed in the short run. Moreover, the requirements to uphold both the high flight completion rate and integrity of network connections and aircraft and crew assignments might impede a scheduled passenger carrier from canceling an impressive number of its lightly loaded flights (Brueckner and Whalen 2000).

Depending on the prevailing market, conditions, load factor, and yield trade-off against each other: except that demand is robust and output growth is under firm control, intensifying the yield is likely related to sliding pressure on load factor. On the other hand, deteriorating profits are often related to higher load factors. Hence, airline carriers naturally want to arrive at a capacity plan with a target load factor that strikes a balance between turning passengers away and costs of meeting all peak demand coming forward and overwhelming the market at other times ("double-edged sword"). It is much easier to manage an airline from an operational viewpoint when load factors are at 64% than at 84% (Cross et al. 2010). A moderate average load factor may be acceptable if the breakeven load factor is sufficiently low, as when, for example, a high yield product is being offered. A high average load factor is not necessarily enough to ensure an acceptable operational performance if the breakeven load factor is high. For example, the unit cost is high, or the yield is low. If the average load factor rises while the yield and unit cost, and therefore the breakeven load factor, remain constant, operating performance will improve (and vice versa).

In the airline line industry, it is crucially important to have quantitative information about the future values of load factor as input in setting airfare decisions. However, the econometric modeling and forecasting of load factors is a challenging task in the industry. Hence, several complicated and dynamic factors directly or indirectly affect the airline industry's load factor. One aspect is that the airline industry is an emblematic example of a cyclical sector, which leads to the load factor to wave (Ľubomír and Jakub 2013). Secondly, airlines apply a dynamic aircraft assignment for any given flight to absorb demand, providing that the airlines capacity management strategy and policy affect the load factor (Bertsimas and Popescu 2003; Li et al. 2007). Third, the industry is dynamic, complex, and significantly subjected to external factors. For example, the dynamics of oil prices directly affect the cost of the airline, and consequently, the airfare, which in turn affects the airline's load factor (Babikian et al. 2002; Borenstein 1992; Chua et al. 2005; Doganis 2010; Flew 2008). Finally, the direct outcome of deregulation in the industry is intense competition among airlines, and the Computer Reservation System (CRS) puts the power in the customer's hands. That leads to airline yield managers' dynamic decisions, which affects the load factor (Doganis 2010; Kahn 1988; John 2009).

Classical stochastic models, such as fixed or random effects panel data regression models, are sufficient in producing an effective and efficient forecast of the airline's load factor. Hence, the load factor is a multifarious variable that must be advanced to modify current modeling techniques. This study extends the existing panel data model by defining dynamic time effects, which are either a linear or a nonlinear function of time (t). Such a model modification helps to capture the essential and hidden variation of an airline's load factor for its geographical flights.

12.2   The Problem

The primary focus of this paper is to apply stochastic models capable of controlling the variability of the load factors for Association European Airlines (AEA) flights in Europe's North Atlantic (NA) and Mid Atlantic (MA). The stochastic model's fit helps forecast the flight load factor in these geographical regions and evaluate the airline's demand and capacity management. Nevertheless, several issues are encountered.

First, appropriate models for the analysis of load factor must be identified. The two most important stochastic quantitative models, which are helpful in forecasting, are regression and time-series analyses. Regression analysis is a type of multivariate analysis that is applied to the cross-sectional (spatial) data to stochastically measure the impact of the exogenous variables (predictors) on the endogenous variables (response variables). Time series analysis studies the endogenous variables' dynamic aspects (Granger 1991; Pearl 2000). However, a time series analysis does not appropriately measure the spatial effect, and regression analysis does not measure the dynamic impact of the exogenous variables on the endogenous variables. Thus, these two fundamental econometric models are complementary. Therefore, in this study's stochastic modeling of the load factor, the author uses a panel data regression model to combine the advantages of both regression and time series analyses (Pearl 2000).

Second, according to the Hausman specification test statistic, the panel data model's time effect can be treated as either fixed or random (Fitzmaurice 2004). However, for this airline's passenger load factor analysis, the Hausman test statistic is insufficient to completely control the time effects' cyclical dynamics on the load factor. Therefore, the major challenge in this stochastic analysis is identifying an autocorrelation configuration for the load factor.

Typically, the serial-correlation diminishes with more distant lags. However, the load factor's autocorrelation structure has various configurations. Hence, it contains both serial and periodic autocorrelations (Tesfay and Solibakke 2015). The classical panel data analysis is extended to express the load factor's time effect as a dynamic linear or nonlinear function of the parameters integrated to geographical NA and MA flights to control such a multifaceted autocorrelation configuration.

The modification helps to prevent the periodic autocorrelation configuration. Moreover, the author applies the Prais–Winsten recursive autoregression estimation (Prais and Winsten 1954) to control the classical serial correlation. The best-fitted dynamic time effects panel data regression model brings new and improved information to AEA European airlines.

12.3   Literature Review

The global airline industry is responsible for effectively providing transportation to passengers, cargo, and postal services to every country in the world. The industry has played an essential role in establishing today's global economy. By itself, the airline industry is a fundamental economic entity, and it supports economic integration worldwide. When one includes the expansion of tourism and recreation, the industry's role has a significant impact in facilitating the world economy (Belobaba et al. 2009).

In general, the global airline industry is complex, dynamic, and profitable (Doganis 2010). It is also subject to immediate changes and innovation, making it both cyclical and subjective to external dynamics. A significant trend is a reliably righteous growth in demand, which is happening at a diminishing rate.

Within the industry, pricing refers to airline products' various service amenities and capabilities, bearing different tariffs in an origin–destination market. Revenue management is the process of determining the number of seats available at each tariff level. Therefore, the airline's revenue management is a function of its tariff strategy and the resulting load factor.

The airline's success is determined by its ability to make unit revenues (Yield) × (LF) that exceed its unit costs (Total Cost/ASK). Consequently, in addition to minimizing the unit cost, the airline manager's central is to take advantage of the yield and load factor quickly. An alternative evaluation of an airline's profitability is its operating ratio. The active percentage signifies an airline's efficiency by comparing its operating expense to its net sales (Brueckner 2004; ICAO 2013; Kellner 2000).

12.3.1   Load Factor Measures the Performance of Airlines

Yield management is a variety of systems, strategies, and tactics that airlines use to manage the demand for their services and products scientifically. Airlines are one of the most prominent users of yield management systems. From its roots in the airline industry, yield management rehearsal has developed to prominence today as an emblematic business toon in other sectors, such as fashion retail, hospitality, energy, and manufacturing (Link 2004).

A bid price is the highest price that a passenger is willing to pay for an air transport service. Consequently, this price is contingent on the type of customer at a time. The bid price reflects the dynamic network models and optimal network solution. Dynamic network models' success involves controlling the optimal revenues in the market (Kaul 2009).

Passenger load factor measures the degree that an airline reaches its passenger-carrying capacity. In other words, the load factor is a measure of an airline's efficiency and performance. The success of a high load is indispensable and a primary indicator of the airline's profitability. Literature regarding the airline industry can be divided into demand structures, fleet, network, revenue modeling, market structures, and operating performance (Dender 2007).

As discussed in Sect. 12.1, this paper's main objective is to focus on the unit revenue, which is inversely proportional to the load factor. That shows the load factor is the degree to which supply and demand are balanced at predominant price points (Distexhe and Perelman 1994). The percentage of seats an airline has in service and must sell at a given yield, or price level, to cover its costs is called the breakeven load factor. To avoid a negative profit, every single airline has a breakeven load factor. The airline's expense is always positive, and thus, the airline's price is negatively correlated with the breakeven load factor (Bamber 2011; Flores-Fillol and Moner-Colonques 2007).

The load factor's magnitude for a given airline directly reflects the competency of that airline. As a result, it is supposed to be infuriating to scrutinize factors that are potentially affecting the airline's load factor. Typically, operational factors play a substantial role in influencing the load factor of airlines. Precisely, an airline's capacity, the journey distance tourist, code-share agreement (aviation business configuration, where more than one airline shares the same flight), and market concentration HHI index (universally recognized measure of market concentration) are the most critical factors. Each has a positive and significant effect on the load factor (Minho et al. 2007).

The GINI index, which measures the price dispersion or price discrimination in the airline of the same flight, is the factor that negatively affects the airline's load factor. Other important factors affecting the load factor include airport features, performance limitations, flight conditions, seasonal demand, traveler schedule, frequency of flights, and dynamic route networks (Minho et al. 2007; Karagiannis and Kovacevic 2000). The data include essential information about available seat–kilometers (ASK), revenue passenger–kilometers (RPK), and load factor (LF). The data are organized to be suitable for the newly modified panel data regression model (see next section).

Knowing and identifying these potential factors will benefit the airline, making more effective and efficient strategic and tactical decisions. These strategic and tactical decisions include the staff training, adapting the airline staff's mindset, determining the optimal number of travel agencies and advertisement, adapting airline management practices, optimizing human resources, and many other related activities (Talluri and Ryzin 2004).

12.4 Data and Methodology

12.4.1 The Data

The dataset downloaded from Research and Statistics (www.aea.be/research/traffic/index.html) is obtained from the Association of European Airlines (AEA) and AEA Traffic and Capacity Data from January 1991 to December 2013.

The AEA is a trustworthy contributor with the following key objectives: raise aviation's roles in Europe's future, increase customer benefit, contribute to move cost-effective regulation, speed up the aviation progress toward a single European Sky, decarbonize aviation to protect the global environment, safeguard circumstances for the fair competition of airlines, and titleholder an international security framework of airlines.

The AEA brings together more than 30 major European airlines and is comprised of Adria Airways (Slovenia), Aegean Airlines (Greece), Air Baltic (Latvia), Air Berlin (Germany), Air France (France), Air Malta (Malta), Air Serbia (Serbia), Alitalia (Italy), Austrian Airlines (Austria), British Airways (United Kingdom), Belgium Brussels Airlines (Belgium), Cargolux (Luxembourg), Croatia Airlines (Croatia), Cyprus Airways (Cyprus), Deutsche Lufthansa (Germany), DHL (Germany), Finnair (Finland), Iberia Airlines (Spain), Icelandair (Iceland), KLM (The Netherlands), LOT Polish Airlines (Poland), Luxair (Luxembourg), Meridiana (Italy), Scandinavian Airlines System (Sweden, Norway, Denmark), Swiss (Switzerland), TAP Portugal (Portugal), Tarom (Romania), TNT Airways (Belgium), Turkish Airlines (Turkey), and Ukraine International Airlines (Ukraine).

Moreover, the North Atlantic–Europe (NA) was defined as any scheduled flights between Europe and North, Central or South America via gateways in the continental USA, including Alaska and Hawaii, and Canada. Mid Atlantic–Europe (MA) is defined as any scheduled flights between Europe and North, Central and South America via gateways in the Caribbean, Central America, or South American mainland north of Brazil (i.e., Bolivia, Colombia including the San Andres Islands, Ecuador, French Guinea, and Guyana).

12.4.2 Methodology

12.4.2.1 Signal Processing

Signal processing is a stochastic process of the time series data formulated as the series of harmonic functions (Hamilton 1994). Specifically, the spectrum decomposes the stochastic process's component into different frequencies, helping to identify periodicities. The signal processing stochastic model for a discrete variable is expressed (Kammler 2000; Priestley 1991; Tesfay 2014).

$$y_t = \mu_t^* + \sum_k [a_k \cos(2\pi \upsilon_k t) + b_k \sin(2\pi \upsilon_k t)] \tag{12.1}$$

where μ_t^* is the expected value of the spectrum at time t, a_k and b_k are coefficients of cosine and sine waves respectively, where the independent zero-mean normal random variables, υ_k is the distinct frequency, and k is the summation index.

The mean and variance of the spectrum are μ_t^* and $E\{(\sum_k [a_k \cos(2\pi \upsilon_k t) + b_k \sin(2\pi \upsilon_k t)])^2\}$, respectively. Moreover, the covariance of the spectrum is expressed as follows:

$$\text{Cov}[y_t, y_{t-\tau}] = E\Bigg\{ \left(\sum_k [a_k \cos(2\pi \upsilon_k t) + b_k \sin(2\pi \upsilon_k t)] \right)$$
$$\times \left(\sum_k [a_k \cos(2\pi \upsilon_k t - \tau) + b_k \sin(2\pi \upsilon_k t - \tau)] \right) \Bigg\} \tag{12.2}$$

To simplify the computation, one assumes time is a continuous variable. Further, one assumes that the spectrum extends infinitely $[T \in (-\infty, \infty)]$ toward time in both directions. Then, the autocorrelation function is expressed as below (Tesfay 2015):

$$\text{Auto}[y_t, y_{t-\tau},] = \frac{\int_{-T}^{T} [a_k \cos(2\pi \upsilon_k t) + b_k \sin(2\pi \upsilon_k t)][a_k \cos(2\pi \upsilon_k t - \tau) + b_k \sin(2\pi \upsilon_k t - \tau)]dt}{\int_{-T}^{T} [a_k \cos(2\pi \upsilon_k t) + b_k \sin(2\pi \upsilon_k t)]^2 dt}$$

$$\tag{12.3}$$

Spectral density analysis illustrates the autocorrelation function's natural configuration of the autocorrelation function on the observed time-series data. That creates an excellent opportunity to observe autocorrelation's exact nature in the Fourier space's time-series data (Boashash 2003). The most crucial advantage of spectral density analysis is its ability to show hidden periodicities of the time series data (Engelberg 2008).

Estimation techniques for spectral density involve parametric or nonparametric approaches based on the time domain or frequency domain analysis. For example, a common parametric technique consists of fitting the observations to an autoregressive model. A common nonparametric approach is a periodogram (Hajivassiliou 2008; Tesfay 2015).

12.4.2.2 Ljung–Box Test

The Ljung–box test simultaneously tests the existence and order of autocorrelation on the time series data.

The null hypothesis (H_0) of the Ljung–Box test procedure is defined as the serial correlation equals zero up to order h versus an alternative hypothesis (H_1). At least one of the serial correlations up to lag h is nonzero (Davidson 2000; Tesfay and Solibakke 2015).

12.4.2.3 Tesfay Dynamic Regression

The Tesfay Dynamic Regression is a dynamic time effects panel data regression model introduced by the author models explicitly the load factor's variations in the airline industry (see Sect. 12.1.1). Using different theoretical and practical quantitative models, the Tesfay Dynamic Regression model is expressed below (Barreto and Howland 2005; Chan et al. 2012; Davidson and Mackinnon 1993; Davies and Lahiri 2000; Fahrmeir et al. 2009; Greene 2002; Hayashi 2000; Hsiao 2003; Luc et al. 2000; Maria and Jim 2014; Schittkowski 2002).

$$y_{it} = \eta_i + \lambda_t + \beta_1 x_{1it} + \beta_2 x_{2it} + \beta_3 x_{3it} + \cdots + \beta_k x_{kit} + \varepsilon_{it} \tag{12.4}$$

$$\lambda_t = f(t; \varphi_i), \varepsilon_{it} = U(\varepsilon_{i,t-1}, \varepsilon_{i,t-2}, \ldots, \varepsilon_{t,t-h}; \rho_{i1}, \rho_{i2}, \ldots, \rho_{ih}) + v_{it}, v_{it} ii\tilde{D}N(0, \sigma_v^2)$$

$$i = 1, 2, 3, \ldots, n, t = 1, 2, 3, \ldots, T, l = 1, 2, 3, \ldots, k$$

where y_{it} is the response from cross-section i at time t, η_i is the ith specific spatial effect, λ_t is the tth specific time effect, x_{1it} represents the exogenous imputes of coefficients $\beta_l, f(\cdot)$ is any real-valued function of time t and a vector of parameter $U(\cdot)$ is a linear function of $\varphi = [(\varphi_{11}, \varphi_{12}, \ldots, \varphi_{1m}), (\varphi_{21}, \varphi_{22}, \ldots, \varphi_{2m}), \ldots, (\varphi_{n1}, \varphi_{n2}, \ldots, \varphi_{nm})]$, $\varepsilon_{i, t-j}, \rho_{ij}$, and $j = 1, 2, 3, \ldots, h$.

The newly introduced specification, that is, $\lambda_t = f(t; \varphi_i)$, causes the existing panel data regression model to have a dynamic time effect. Thus, the modified model is named the Tesfay Dynamic Regression model.

If $\lambda_t = f(t; \varphi_i)$ is a linear function, then, under the complete fulfillment of the Gauss–Markov assumption, results from the least squares dummy variable (LSDV) estimation are the model parameters' best linear unbiased estimators (BLUE) (Baltagi 2008). Otherwise, under the complete fulfillment of the Gauss–Markov– Aitkin assumption, results from the generalized least squares (GLS) estimation are the model parameters' best linear unbiased estimators (BLUE) (Amemiya 1985; Babak et al. 1999; Hsiao 2003; Thomas et al. 2000; Voinov and Nikulin 1993).

If $\lambda_t = f(t; \varphi_i)$ is a nonlinear function, then the following estimation procedure is applied to estimate the model parameters (Billings 2013; Kelley 1999; Meade and Islam 1995; Seber and Wild 1989; Wooldridge 2013).

$$\text{Let } y_{it} = F(t, X; \theta) + \varepsilon_{it} \tag{12.5}$$

where $F(t, X; \theta) = \eta_i + f(t; \phi_i) + \beta_1 x_{1it} + \beta_2 x_{2it} + \beta_3 x_{3it} + \cdots + \beta_k x_{kit}$, $\theta = [\theta_1, \theta_2, \theta_3, \ldots, \theta_{n(m+1)+k}] = [\eta_i, \phi, \beta]$ is the vector of model parameters.

Now minimizing the total sum of the errors, one achieves

$$\min\left[\sum_{i=1}^{n}\sum_{t=1}^{T}[y_{it} - F(t, X; \theta)]^2\right] \tag{12.6}$$

$$\Rightarrow \frac{\partial}{\partial \theta_s} \left[\sum_{i=1}^{n} \sum_{t=1}^{T} [y_{it} - F(t, X; \theta)]^2 \right] = 0, \quad s = 1, 2, 3, \cdots, n(m+1) + k$$

$$\Rightarrow \sum_{i=1}^{n} \sum_{t=1}^{T} [y_{it} - F(t, X; \theta)] \left[\frac{\partial F(t, X; \theta)}{\partial \theta_s} \right] = 0$$

$$\Leftrightarrow \sum_{i=1}^{n} \sum_{t=1}^{T} \left[y_{it} \left[\frac{\partial F(t, X; \theta)}{\partial \theta_s} \right] - F(t, X; \theta) \left[\frac{\partial F(t, X; \theta)}{\partial \theta_s} \right] \right] = 0 \qquad (12.7)$$

To solve Eq. (12.7), the Newton–Raphson numerical optimization recursive algorithm is applied by defining a new function as:

$$G_s(t, X; \theta) = \sum_{i=1}^{n} \sum_{t=1}^{T} \left[y_{it} \left[\frac{\partial F(t, X; \theta)}{\partial \theta_s} \right] - F(t, X; \theta) \left[\frac{\partial F(t, X; \theta)}{\partial \theta_s} \right] \right] \qquad (12.8)$$

$$G = [G_s(t, X; \theta)]_{n(m+1)+k}$$

Afterward derive the Jacobean matrix (Hazewinkel 2001) from Eq.(12.8)

$$J_G = \begin{bmatrix} \dfrac{\partial G_1(t; X; \theta_1)}{\partial \widehat{\theta}_1}, & \dfrac{\partial G_1(t; X; \theta_1)}{\partial \widehat{\theta}_2}, & \cdots, & \dfrac{\partial G_1(t; X; \theta_1)}{\partial \widehat{\theta}_{n(m+1)+k}} \\ \dfrac{\partial G_2(t; X; \theta_2)}{\partial \widehat{\theta}_1}, & \dfrac{\partial G_2(t; X; \theta_2)}{\partial \widehat{\theta}_2}, & \cdots, & \dfrac{\partial G_2(t; X; \theta_2)}{\partial \widehat{\theta}_{n(m+1)+k}} \\ \vdots & \vdots & \cdots & \vdots \\ \dfrac{\partial G_{n(m+1)+k}(t; X; \theta_{n(m+1)+k})}{\partial \widehat{\theta}_1}, & \cdots, & & \dfrac{\partial G_{n(m+1)+k}(t; X; \theta_{n(m+1)+k})}{\partial \widehat{\theta}_{n(m+1)+k}} \end{bmatrix}_{[n(m+1)+k] \times [n(m+1)+k]}$$

$$(12.9)$$

To get the numerical solution of the model parameters', apply the Newton–Raphson recursive algorithm (Bonnans et al. 2006; Ortega and Rheinboldt 2000).

$$\left[\widehat{\theta}_s \right]_{r+1} = \left[\widehat{\theta}_s \right]_r - [J_G^{-1}][G], \quad r = 1, 2, 3, \ldots \qquad (12.10)$$

where J_G^{-1} is the inverse of the Jacobian matrix.

12.4.3   Model Adequacy and Diagnostics

This section involves removing the serial correlation from the time-series data to obtain efficient model parameters. To remove the serial correlation, the following algorithm is used.

Step 1: Compute residuals as below (Weisberg 1985; Cook et al. 1982).

$$\widehat{\varepsilon}_{it} = y_{it} - F\left(t, X; \widehat{\theta}\right) \tag{12.11}$$

Step 2: Test the presence of a serial correlation on the estimated residuals. Here, one applies the Ljung–box test of serial autocorrelation.

Step 3: If autocorrelation is identified, then the Prais–Winsten methodology is applied to remove the serial correlation (Amemiya 1985; Davies and Lahiri 1995; Frees 2004; Prais and Winsten 1954; Verbeek 2004; Wooldridge 2013).

Step 4: Repeat steps 1–3 unless the Ljung–Box test of autocorrelation approves that there is no serial correlation on the estimated residuals.

12.5   Results and Discussion

12.5.1   *Evaluation of the Load Factor's Regional Characteristics*

Before fitting the panel data model, it is vital to analyze the relationship between the load factor (LF) of both the Europe–North Atlantic (NA) and Europe–Mid Atlantic (MA) airlines with reverence to the available seat kilometers (ASK) and revenue passenger–kilometers (RPK).

Bootstrap result estimates of the RPK and ASK of NA and MA flights are given in Table 12.1, which shows that the mean RPK estimates (millions) of NA and MA flights are 13,514 (with a bias of −18.82) and 3265.5 (with a bias of −1.11), respectively. The mean ASK estimates (millions) of NA and MA are 17,018.9 (with a bias of +0.92) and 4090.49 (with a bias of +4.68), respectively. These results confirm that both the RPK and ASK of NA flights are higher than MA flights. Furthermore, an average of 3504.9 ASK and 824.99 ASK (millions) is out of use every month for NA and MA flights, respectively.

Referring to the results in Table 12.2, average NA flights have 10,248.52 RPK (millions) and 12,928.45 ASK (millions) more than MA flights. Moreover, the results confirm that the load factor of NA and MA flights are statistically equal.

The estimation results from Tables 12.3 and 12.4, along with Figs. 12.1 and 12.2, show a strong and positive linear relationship between RPK and ASK, with determination coefficients of 90.7% and 97.5%, respectively, for both NA and MA flights. Results also show that, generally, the airlines' operational decisions to balance these geographical flights' supply and demand are good. The linear regression fit using the Prais–Winsten recursive parameter estimation indicates that from one million increases in ASK, the RPK of the NA and MA flights is increased by 1.067 and 0.907 million, respectively.

To evaluate the airlines' managerial performance, one must analyze the results reported in Figs. 12.3 and 12.4. The estimation results from Table 12.5, together with

Table 12.1 Mean estimates of RPK (in million) and ASK (in million) of the NA and MA flights

Flight	Statistic	Estimates of RPK (millions)					Estimates of ASK (millions)				
		Est.	Bias	Std. error	95% Confidence interval		Est.	Bias	Std. error	95% Confidence interval	
					Lower	Upper				Lower	Upper
NA	Mean	13,514	−18.82	246.31	13,003.8	14,003	17,018.9	0.92	239.33	16,560.50	17,469.18
	Std. dev	4115.4	−7.89	143.22	3819.9	4384.3	3940.39	−8.91	132.62	3674.34	4203.10
MA	Mean	3265.5	−1.11	72.47	3128.61	3404.1	4090.49	4.68	80.90	3922.51	4259.37
	Std. dev	1214.3	−1.81	38.46	1138.7	1286.9	1336.16	−3.33	47.25	1236.82	1419.24
Estimation method		Bootstrap results are based on 1000 bootstrap samples									

Table 12.2 Comparison of ASK, RPK, and load factor of NA and MA flights

Variable	Comparison of flights	Mean difference	Std. error	t-cal	Sig. (two-tailed)	95% Confidence interval	
						Lower	Upper
ASK (millions)	NA vs. MA flights	12,928.450	254.243,72	50.85	0.0010	12,444.25	13,440.09
RPK (millions)	NA vs. MA flights	10,248.520	267.221,49	38.49	0.0010	9697.97	10,753.41
LF (%)	NA vs. MA flights	−0.281	0.575,51	−0.49	0.642	−1.42	0.835,45
Estimation method	Bootstrap results are based on 1000 bootstrap samples						

Table 12.3 Prais–Winsten recursive parameter estimation of an RPK (millions) linear regression in response to the ASK (millions) of NA flights

	B	Std. error	T	App. sig	R^2
ASK-NA	1.067	0.021	51.451	0.000	0.907
(Constant)	−4634.960	368.984	−12.561	0.000	

The Prais–Winsten autoregression estimation method is used
Dependent variable: RPK

Table 12.4 Prais–Winsten recursive parameter estimation of an RPK (millions) linear regression in response to the ASK (millions) of MA flights

	B	Std. error	T	App. sig	R^2
ASK-MA	0.907	0.009	102.809	0.000	0.975
(Constant)	−446.588	38.132	−11.712	0.000	

The Prais–Winsten estimation method is used
Dependent variable: RPK

Fig. 12.1 Time series plot of ASK and RPK (millions) of NA flights

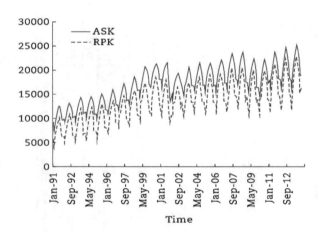

Fig. 12.2 Time series plot of ASK and RPK (millions) of MA flights

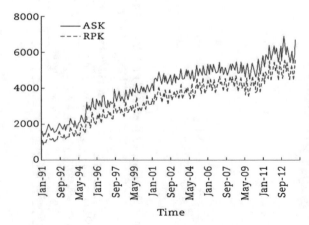

Fig. 12.3 Scatter plot of the load factor versus RPK

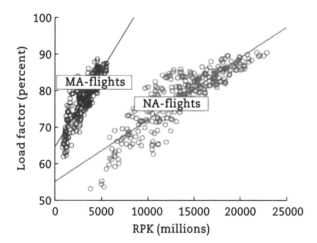

Fig. 12.4 Scatter plot of the load factor versus ASK

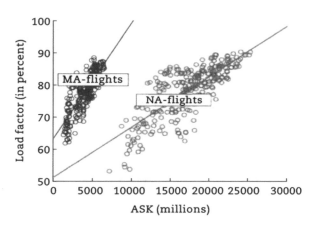

Table 12.5 Prais–Winsten recursive parameter estimation of a linear regression of the load factor (percent) in response to the RPK (millions)

Flight	Predictor	Estimator	Std. error	t-cal	Sig.	R^2	Model std. error
NA	RPK-NA	0.002	0.0001	23.738	0.000	0.674	2.890
	Constant	45.975	1.600	28.729	0.000		
MA	RPK-MA	0.004	0.0002	20.260	0.000	0.601	2.678
	Constant	63.571	0.780	81.527	0.000		

Fig. 12.3, show that there exists there is a moderate and positive linear relationship between LF and PRK for both NA and MA flights with determination coefficients of 67.4% and 60.1%, respectively. The linear regression fit using the Prais–Winsten recursive parameter estimation indicates that, for a one million increase in RPK, LF is improved by 0.002% and 0.003% for NA and MA flights, respectively. Moreover, Table 12.6 estimation results, together with Fig. 12.4, show a significant and positive linear relationship between LF and ASK for both NA and MA flights with

Table 12.6 Prais–Winsten recursive parameter estimation of a linear regression of the load factor (percent) in response to the ASK (millions)

Flight	Predictor	Estimator	Std. error	t-cal	Sig.	R^2	Model std. error
NA	ASK-NA	0.002	0.0001	13.902	0.000	0.415	3.967
	Constant	47.067	2.314	13.902	0.000		
MA	ASK-MA	0.004	0.0002	16.161	0.000	0.489	3.126
	Constant	63.507	0.964	65.892	0.000		

determination coefficients of 41.5% and 48.9%, respectively. The linear regression fit using the Prais–Winsten recursive parameter estimation indicates that, for a one million increase in ASK, LF is improved by 0.002% and 0.004% for NA and MA flights, respectively. The overall analysis of Tables 12.4 and 12.5 confirms that airlines have enhanced NA and MA flights' demand and capacity management.

12.5.2   Structure of the Load Factor's Autocorrelation

The time series econometric analysis is necessary to obtain the time series' exact autocorrelation configuration. The spectral density analysis is a powerful method of ascertaining the alignment of the autocorrelation function. The nonparametric spectral density and analysis provide graphical information about how the autocorrelation function behaves in Fourier space.

One of the graphical methods is the periodogram of the autocorrelation function of the time series observation frequency. This method is tremendously sensitive to the series' optimum autocorrelation. Another way is the response of density of the autocorrelation function of the time series observation frequency. This technique is sensitive to the weighted autocorrelation of the series. Thus, both plots provide imperative information about the autocorrelation structure of the load factor. The spectral density estimation and the Ljung–Box test for NA and MA flights' load factor are given in Table 12.7.

12.5.2.1   The Load Factor's Autocorrelation Configuration of North Atlantic–Europe Flights

The nonparametric plot of the periodogram and spectral density of the LF of the NA flights indicate a strong periodic autocorrelation, which is observed after jumping a certain period of months. The LF recurrent yearly plot of NA flights over several months shows that, from year to year, there is a robust periodic pattern with small variance.

In the recurrent yearly plot of LF over several months, one noticed noticeable patterns. The smallest LF is observed in January, and then it underway grows until July before deteriorating until December.

Table 12.7 The structure of autocorrelation of load factor of the NA and MA flights

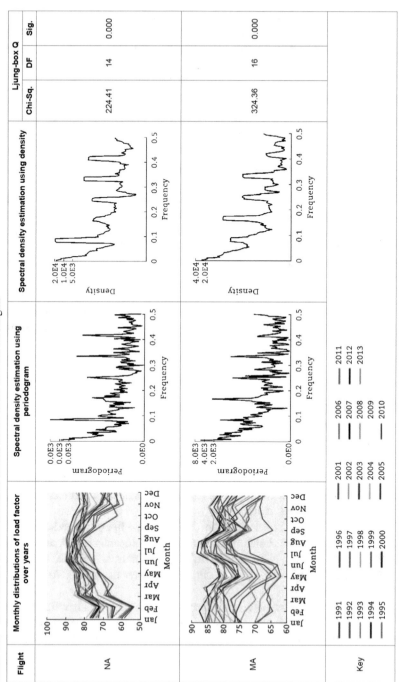

The periodogram plot and spectral density suggest that NA flights' LF distribution is serially correlated up to a certain month lag. Moreover, the Ljung–Box parametric test proposes a significant serial correlation of 17 months, which is unrestrained after the 18th month.

12.5.2.2   The Load Factor's Autocorrelation Configuration of Mid Atlantic–Europe Flights

The nonparametric plot of the periodogram and spectral density of the LF of flights of MA indicates that there exists a strong periodic autocorrelation, which is observed after jumping a certain period of months. The repeated yearly LF plot of MA flights over several months shows a strong periodic pattern.

In the recurrent yearly LF plot over several months, one observes a significant pattern: The smallest LF is observed in November, December, and January. Then LF starts to grow during July, August, and September before deteriorating until November.

The periodogram plot and spectral density suggest that MA flights' LF distribution is serially correlated up to a certain month lag. Furthermore, the Ljung–Box parametric test proposes a significant serial correlation of order 15 months, dissipated after the 16th month.

12.5.3   Fitting the Load Factor's Panel Data Regression Model

Analysis from Sect. 12.4.1 identifies that both the RPK and ASK of NA flights are higher than MA flights. Moreover, the average LF of MA flights is higher than the average of NA flights. That confirms that the panel data model's spatial effects exert control for such significant variability.

The analysis of Sect. 12.4.1 describes how LF that the characteristics behave differently for MA and NA flights. The load factor significantly correlates with both the RPK and ASK for the NA and MA flights in different magnitude. Using these variables (i.e., both RPK and ASK) as common exogenous imputes to predict the load factor is inadequate.

The autocorrelation analysis's configuration in Sect. 12.4.2 identifies that both periodic and serial correlations exist on the LF. The autocorrelation structure is different for both MA and NA flights. That shows that the time effect on LF is not merely fixed or random. Instead, it is dynamically and uniquely associated with regional flights.

Therefore, the appropriate model fit to analyze the LF of airline flights is the dynamic time effects two-way panel data regression model. The dynamic time effect two-way panel data regression model's fit for the load factor of MA and NA flights is given in Table 12.8 and Fig. 12.5.

Table 12.8 The fit of T-panel data regression model of load factor of the NA and MA flights

Parameter estimates	Spatially integrated dynamic time effects	Estimates	Std. error	t-cal	Approx. sig.	Model S.E.	Month	Forecasting of Load factor of 2014 (in %)		
									95% Prediction interval	
								Expected	LB	UP
Rho (AR1)		0.25622	0.05931	4.31968	0.00002	3.53325	Jan	76.79741	69.84714	83.74768
Time function coefficients of the NA flights	t	0.03602	0.00748	4.81731	0.00000		Feb	76.97245	70.02218	83.92272
	$\ln(t)$	1.83815	0.61281	2.99952	0.00296		Mar	80.95242	74.00215	87.90269
	$\sin(\omega_2 t)$	−0.69541	0.29154	−2.38533	0.01776		Apr	84.84587	77.89560	91.79614
	$\sin(\omega_6 t)$	−4.05655	0.38286	−10.59536	0.00000		May	87.16869	80.21842	94.11896
	$\cos(\omega_2 t)$	0.96537	0.29145	3.31234	0.00105		Jun	90.17115	83.22088	97.12142
	$\cos(\omega_6 t)$	−6.34351	0.38093	−16.65287	0.00000		Jul	93.48762	86.53735	100.00000
	$\cos(\omega_5 t)$	−0.70343	0.37299	−1.88594	0.06039		Aug	93.31471	86.36444	100.00000
Spatial effect of NA flights		64.48071	2.01055	32.07111	0.00000		Sep	89.20248	82.25221	96.15275
							Oct	85.12541	78.17514	92.07568
							Nov	82.67023	75.71996	89.6205
							Dec	79.66971	72.71944	86.61998
Rho (AR1)		0.34176	0.05685	6.0117	0.0000	2.5587	Jan	84.58222	79.54898	89.61545
Time function coefficients of the MA flights	t	0.02931	0.00605	4.84263	0.0000		Feb	87.05071	82.01748	92.08394
	$\ln(t)$	2.59844	0.49259	5.2751	0.0000		Mar	87.95908	82.92585	92.99231
	$\sin(\omega_2 t)$	−1.62055	0.20627	−7.85658	0.0000		Apr	84.27398	79.24074	89.30721
	$\sin(\omega_6 t)$	2.27482	0.2477	9.18377	0.0000		May	80.79658	75.76335	85.82982
	$\cos(\omega_2 t)$	0.54307	0.2062	2.63374	0.00894		Jun	83.20649	78.17326	88.23972
	$\cos(\omega_6 t)$	−1.35234	0.2472	−5.4706	0.0000		Jul	88.05483	83.0216	93.08807
Spatial effect of MA flights		62.17773	1.61428	38.51721	0.0000		Aug	88.36816	83.33493	93.40139

(continued)

Table 12.8 (continued)

Parameter estimates	Spatially integrated dynamic time effects						Forecasting of Load factor of 2014 (in %)			
		Estimates	Std. error	t-cal	Approx. sig.	Model S.E.	Month	Expected	95% Prediction interval	
									LB	UP
							Sep	84.94909	79.91586	89.98233
							Oct	83.41876	78.38553	88.45199
							Nov	84.26842	79.23519	89.30165
							Dec	84.52316	79.48993	89.55639

Where $t = 12$ (*Current year* − *1991*) + *Current month* and $\omega_i = \frac{\pi}{i}, i = 1, 2, 3, \ldots$ are the periods

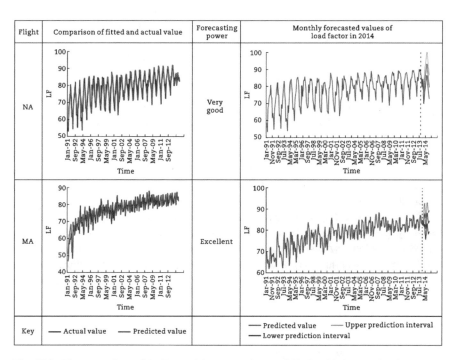

Fig. 12.5 The comparison of the T-panel data regression model's fit with the actual values

The significance of the time function harmonic component suggests that the LF is seasonal by its nature. Generally, the model's fit indicates that the load factor improves (i.e., growing) with time for both MA and NA flights. Specifically, the load factor's spatial effect on MA flights (62.18%) is smaller than those of the NA flights (64.48%).

12.6 Conclusions and Recommendations

12.6.1 Conclusions

This paper applies advanced and modified stochastic analysis to the load factor (LF) of flights AEA airline's MA and NA. The research provides the following conclusions.

The bootstrap estimation results suggest that the mean RPK (estimates millions) of NA and MA flights are 13,514.0 (with a bias of −18.82) and 3265.5 (with a bias of −1.11), respectively. The mean ASK estimates (millions) of NA and MA flights are 17,018.9 (with a bias of +0.92) and 4090.49 (with a bias of +4.68), respectively. The results confirm that both the RPK and ASK of NA flights are higher than MA flights. Furthermore, an average of 3504.9 and 824.99 ASK (millions) is out of use for every month in the NA and MA flights, respectively.

LF for both NA and MA flights are significantly and positively correlated with RPK and ASK. The results suggest that airlines have a better demand and capacity management for both NA and MA flights.

LF of both MA and NA flights have periodic (i.e., season to season) correlations. The smallest LF is observed in November, December, and January then starts to grow during July, August, and September before declining until November. The smallest LF of NA flights is observed in January, then rises until July, and then falls until December. Furthermore, the LF of both MA and NA flights has serial (i.e., month to month) correlations. The LF of MA and NA flights have an order of 15 and 17 months, respectively.

The overall autocorrelation structure suggests that the dynamic time effect and two-way panel data regression model is an appropriate and realistic forecasting model for MA and NA flights' load factor. The fitted model forecast monthly values for LF with upper and lower 95% prediction intervals for 2014. Furthermore, Fig. 12.5 provides the LF's monthly predicted values (with upper and lower 95% prediction intervals) for 2014.

12.6.2   Recommendations and Policy Implications

This paper applies a stochastic model for the LF AEA airliners of the MA and NA flights. Results from the study have important implications, helping to form the following policy recommendations.

The forecasting power of the T-panel data regression model of the airline industry's load factor is high. Consequently, the T-panel data regression model is vital for solving complex problems in the airline industry. Therefore, the author strongly recommends that engineers and researchers employ the model to analyze the airline industry's efficiency.

Using the Tesfay Dynamic Regression model, the airline's load factor is more robust and realistic. Therefore, AEA may use the model to predict the LF to distribute MA and NA flights. In this regard, airlines should apply the model to their regional flights across the globe to forecast the load factor more precisely.

In addition to reducing costs in the airline industry, any given airline's profitability relies on yield and LF's joint maximization. To push up the LF and yield simultaneously and produce strategic decisions regarding profitability, AEA may extend the LF analysis by considering individual airlines (i.e., spatial effects) into the T-panel data regression model. Such research can give rigorous information about the LF of each airline. Consequently, the AEA will have improved quantitative input regarding restructuring yield management, network design, etc., concerning their specific flights.

The stochastic analysis indicates that the airlines' response adopts the demand is found generally good for both the MA and NA flights. Furthermore, the paper found that airlines have a good demand and capacity management of both regional flights. It is recommended that one keep up with the existing demand and capacity management strategy.

Finally, as suggested by many scholars, the airline industry is seasonal. This paper found that the LF of MA and NA flights are both seasonal and different. That implying that LF is far from stable. In aggregate, the airlines stabilizing policies have so far failed. AEA may, therefore, continuously focus on LF stabilization and improvement of the LF.

12.6.3 Further Studies

Since it is crucial to modify the existing panel data regression model to analyze the airline industry's load factor, it is essential to recognize that the revised model specifications have certain limitations. This stochastic analysis's main limitations are detailed first, this study does not include all essential variables of the airline industry with a potential impact on the load factor. The author focused the analysis on all member airlines' aggregated load factor of the Association of European Airlines (AEA). Therefore, it is recommended that other important variables are included when engineers or researchers make a similar analysis of airlines. Secondly, the introduction of dynamic time effects in the existing panel data regression model demands specific formulation. Therefore, engineers or researchers must recognize and model the airline's load factors with Tesfay Dynamic regression model requires advanced acquaintance of curve estimation and numerical mathematical methods.

Bibliography

T. Amemiya, *Advanced Stochastics* (Harvard University Press, Cambridge, 1985)

R. Babikian, S.P. Lukachko, I.A. Waitz, The historical fuel efficiency characteristics of regional aircraft from technological operational and cost perspectives. J. Air Transp. Manag. **8**(6), 389–400 (2002)

B.H. Baltagi, *Stochastic Analysis of Panel Data*, 4th edn. (John Wiley and Sons, New York, NY, 2008)

G.J. Bamber, *Up in the Air: How Airlines Can Improve Performance by Engaging their Employees* (ILR Press, Ithaca, 2011)

C. Barnhart, D. Fearing, A.R. Odoni, et al., Demand and capacity management in air transportation. EURO J. Transp. Logistics **1**(1/2), 135–155 (2012)

H. Barreto, F. Howland, *Dummy Dependent Variable Models. Introductory Stochastics: Using Monte Carlo Simulation with Microsoft Excel* (Cambridge University Press, Cambridge, 2005)

P. Belobaba, A. Odoni, C. Barnhart, *The Global Airline Industry* (John Wiley and Sons, New York, NY, 2009)

D. Bertsimas, I. Popescu, Revenue management in a dynamic network environment. Transp. Sci. **37**(3), 257–277 (2003)

S.A. Billings, *Nonlinear System Identification: NARMAX Methods in the Time, Frequency, and Spatio-Temporal Domains* (John Wiley and Sons, New York, NY, 2013)

B. Boashash, *Time-Frequency Signal Analysis and Processing: A Comprehensive Reference* (Elsevier, Oxford, 2003)

J.F. Bonnans, J.C. Gilbert, C. Lemaréchal, et al., *Numerical Optimization: Theoretical and Practical Aspects*, 2nd edn. (Springer, Berlin, 2006)

S. Borenstein, The evolution of U.S. airline competition. J. Econ. Perspect. **6**(2), 45–73 (1992)

J.K. Brueckner, Network structure and airline scheduling. J. Ind. Econ. **52**(2), 291–312 (2004)

J.K. Brueckner, W.T. Whalen, The price effects of international airline alliances. J. Law Econ. **43**(2), 503–545 (2000)

A. Cento, *The Airline Industry Challenges in the 21st Century. Characteristics of the Airline Industry* (Springer, Heidelberg, 2009)

J.C.C. Chan, G. Koop, R. Leon-Gonzales, et al., Time varying dimension models. J. Bus. Econ. Stat. **30**(3), 358–367 (2012)

M. Cho, M. Fan, Y.P. Zhou, *An Empirical Study of Revenue Management Practices in the Airline Industry* (University of Washington, Washington, DC, 2007)

C.L. Chua, H. Kew, J. Yong, Airline code-share alliances and costs: Imposing concavity on translog cost function estimation. Rev. Ind. Organ. **26**(4), 461–487 (2005)

R.D. Cook, S. Weisberg, Residuals and influence in regression. Biom. J. **27**(1), 80 (1985)

R.G. Cross, *Revenue Management: Hard-Core Tactics for Market Domination* (Broadway Books, New York, NY, 1997)

R.G. Cross, J.A. Higbie, Z.N. Cross, Milestones in the application of analytical pricing and revenue management. J. Revenue Pricing Manag. **10**(1), 8–18 (2010)

J. Davidson, *Stochastic Theory* (Blackwell Publishing, Oxford, 2000)

R. Davidson, J.G. Mackinnon, *Estimation and Inference in Stochastics* (Oxford University Press, Oxford, 1993)

A. Davies, K. Lahiri, A new framework for testing rationality and measuring aggregate shocks using panel data. J. Stochastics **68**(1), 205–227 (1995)

A. Davies, K. Lahiri, *Re-Examining the Rational Expectations Hypothesis Using Panel Data on Multi-Period Forecasts. Analysis of Panels and Limited Dependent Variable Models* (Cambridge University Press, Cambridge, 2000)

K.V. Dender, Determinants of fares and operating revenues at US airports. J. Urban Econ. **62**(2), 317–336 (2007)

V. Distexhe, S. Perelman, Technical efficiency and productivity growth in an era of deregulation: The case of airlines. Revue Suisse d'Economie Politique **130**(4), 669–689 (1994)

R. Doganis, *Flying Off Course: Airline Economics and Marketing*, 4th edn. (Routledge, London, 2010)

S. Engelberg, *Digital Signal Processing: An Experimental Approach* (Springer, Heidelberg, 2008)

L. Fahrmeir, T. Kneib, S. Lang, *Regression Model and Method*, 2nd edn. (Springer, Heidelberg, 2009)

G.M. Fitzmaurice, *Applied Longitudinal Analysis* (John Wiley and Sons, New York, NY, 2004)

T. Flew, *New Media: An Introduction* (Oxford University Press, Oxford, 2008)

R. Flores-Fillol, R. Moner-Colonques, Strategic formation of airline alliances. J. Transport Econ. Policy **41**(3), 427–449 (2007)

E. Frees, *Longitudinal and Panel Data: Analysis and Applications in the Social Sciences* (Cambridge University Press, Cambridge, 2004)

C. Granger, *Modelling Economic Series: Readings in Stochastic Methodology* (Oxford University Press, Oxford, 1991)

W.H. Greene, *Stochastic Analysis*, 5th edn. (Prentice Hall, Upper Saddle River, NJ, 2002)

V.A. Hajivassiliou, *Computational Methods in Stochastics. The New Palgrave Dictionary of Economics*, 2nd edn. (Palgrave Macmillan, Heidelberg, 2008)

J.D. Hamilton, *Time Series Analysis* (Princeton University Press, Princeton, 1994)

B. Hassibi, A.H. Sayed, T. Kailath, *Indefinite Quadratic Estimation and Control: A Unified Approach to H2 and H-Infinity Theories* (Society for Industrial and Applied Mathematics (SIAM), Philadelphia, 1999)

F. Hayashi, *Stochastics* (Princeton University Press, Princeton, 2000)

M. Hazewinkel, *Jacobian-Encyclopedia of Mathematics* (Springer, Heidelberg, 2001)

C. Hsiao, *Analysis of Panel Data*, 2nd edn. (Cambridge University Press, Cambridge, 2003)

ICAO, Airport Economics Manual (2013), www.icao.int/sustainability/MAcuments/MAc9562_en.pdf. Accessed 21 Apr 2014

A.E. Kahn, Surprises of airline deregulation. Am. Econ. Rev. **78**(2), 316–322 (1988)

T. Kailath, A.H. Sayed, B. Hassibi, *Linear Estimation* (Prentice-Hall, London, 2000)

M. Kalli, J.E. Griffin, Time-varying sparsity in dynamic regression models. J. Econ. **178**(2), 779–793 (2014)

D. Kammler, *A First Course in Fourier Analysis* (Prentice Hall, London, 2000)

E. Karagiannis, M. Kovacevic, A method to calculate the jackknife variance estimator for the gini coefficient. Oxf. Bull. Econ. Stat. **62**(1), 119–122 (2000)

S. Kaul, Yield management: Getting more out of what you already have. Ericsson Bus. Rev. **2**, 17–19 (2009)

C.T. Kelley, Iterative methods for optimization. Front. Appl. Math. **41**(9), 878 (1999)

L. Kellner, Building a global airline brand, in *2000 Transport Conference, London*, (2000)

M.Z.F. Li, T.H. Oum, C.K. Anderson, An airline seat allocation game. J. Revenue Pricing Manag. **6**(4), 321–330 (2007)

H. Link, PEP-a yield-management scheme for rail passenger fares in Germany. Japan Railway Transport Rev. **38**, 54 (2004)

F. Ľubomír, H. Jakub, Airline pricing strategies in European airline market (2013), pernerscontacts.upce.cz/30_2013/FeMArco.pdf. Accessed 12 Mar 2014

B. Luc, L. Michel, R. Jean-François, Bayesian Inference in Dynamic Stochastic Models (2000), Oxford Scholarship Online. Accessed 1 Oct 2011

J. Marriott Jr., R. Cross, Room at the revenue inn, in *Book of Management Wisdom: Classic Writings by Legendary Managers*, ed. by P. Krass, (Wiley, New York, NY, 2000)

J. McGill, G. van Ryzin, Revenue management: Research overview and prospects. Transp. Sci. **33**(2), 233–256 (1999)

N. Meade, T. Islam, Prediction intervals for growth curve forecasts. J. Forecast. **14**(5), 413–430 (1995)

S. Netessine, R. Shumsky, Introduction to the theory and practice of yield management. INFORMS Trans. Educ. **3**(1), 34–44 (2002)

J.M. Ortega, W.C. Rheinboldt, *Iterative Solution of Nonlinear Equations in Several Variables* (SIAM, Philadelphia, 2000)

T. Oum, C. Yu, Cost competitiveness of the world's major airlines: An international comparison. Transp. Res. A Policy Pract. **32**(6), 407–422 (1998)

J. Pearl, *Causality: Models, Reasoning, and Inference* (Cambridge University Press, New York, NY, 2000)

S.J. Prais, C.B. Winsten, *Trend Estimators and Serial Correlation* (Cowles Commission, Chicago, IL, 1954)

M.B. Priestley, *Spectral Analysis and Time Series* (Academic Press, Amsterdam, 1991)

K. Schittkowski, *Numerical Data Fitting in Dynamical Systems* (Kluwer, Boston, MA, 2002)

G.A.F. Seber, C.J. Wild, *Nonlinear Regression* (John Wiley and Sons, New York, NY, 1989)

K. Talluri, G.V. Ryzin, Revenue management under a general discrete choice model of consumer behavior. Manag. Sci. **50**(1), 15–33 (2001)

K. Talluri, G.V. Ryzin, *The Theory and Practice of Revenue Management* (Kluwer Academic Puslishers, Amsterdam, 2004)

Y.Y. Tesfay, The Bullwhip effect: Applying stochastic models on Beer distribution game. J. Appl. Logistics **5**(1), 19 (2014)

Y.Y. Tesfay, Modeling the causes of the bullwhip effect and its implications on the theory of organizational coordination. Supply Chain Forum **16**(2), 30–46 (2015)

Y.Y. Tesfay, P.B. Solibakke, Spectral density estimation of European airlines load factors for Europe-Middle East and Europe-Far East flights. Eur. Transp. Res. Rev. **7**(2), 1–11 (2015)

U.S. General Accounting Office, *Airline Competition: Higher Fares and Reduced Competition at Concentrated Airports* (U.S. General Accounting Office, New York, NY, 1990)

M. Verbeek, *A Guide to Modern Stochastics*, 2nd edn. (John Wiley and Sons, New York, NY, 2004)

V.G. Voinov, M.S. Nikulin, *Unbiased Estimators and Their Applications*, Univariate case, vol 1 (Kluwer Academic Publishers, Amsterdam, 1993)

S. Weisberg, *Applied Linear Regression*, 2nd edn. (John Wiley and Sons, New York, NY, 1985)

J.M. Wooldridge, *Introductory Stochastics: A Modern Approach*, 5th international edn. (South Western, Australia, 2013)

Index

Printed in the United States
by Baker & Taylor Publisher Services